Precision Farming in Horticulture

NIPA® GENX ELECTRONIC RESOURCES & SOLUTIONS P. LTD.
New Delhi-110 034

Precision Farming in Horticulture

Editors :

Jitendra Singh
Associate Professor, Fruit Science

S. K. Jain
Associate Professor, PHT

L. K. Dashora
Dean

College of Horticulture & Forestry
Jhalarapatan, Jhalawar - 326 023, Rajasthan

B.S. Chundawat
C-45, Pratapnagar, Udaipur - 313 001, Rajasthan

NIPA® GENX ELECTRONIC RESOURCES & SOLUTIONS P. LTD.
New Delhi-110 034

NIPA® GENX ELECTRONIC
RESOURCES & SOLUTIONS P. LTD.

101,103, Vikas Surya Plaza, CU Block
L.S.C. Market, Pitam Pura, New Delhi-110 034
Ph : +91 11 27341616, 27341717, 27341718
E-mail: newindiapublishingagency@gmail.com
web: www.nipabooks.com

For customer assistance, please contact
Phone: + 91-11-27 34 17 17
Fax: + 91-11- 27 34 16 16
E-Mail: feedbacks@nipabooks.com

ISBN: 978-81-19254-08-8

Composed and Designed by NIPA®.

डॉ. गोरख सिंह
DR. GORAKH SINGH

बागवानी आयुक्त
भारत सरकार, कृषि मंत्रालय
(कृषि एवं सहकारिता विभाग)
कृषि भवन, नई दिल्ली - 110001
Horticulture Commissioner
Government of India,
Ministry of Agriculture
(Department of Agriculture & Cooperation)
Krishi Bhawan, New Delhi-110001
फोन /Phone : 011-23381012
फैक्स/Fax : 011-23383712
ईमेल /email : gorakh.singh@nic.in
singh_gorakh@yahoo.co.in

Dated ________________

FOREWORD

India is the second largest producer of fruits, vegetables and flowers after China. Still, there is a wide gap in yield potential being really harvested and the potential realized at experimental level. The yield of fruits in cultivation in India is worth transformation 216 per cent higher than what is being realized at field level at present. In case of apple, it may go up to 561 per cent higher, banana 146 per cent, mango 280 per cent, papaya 196 per cent and in pineapple 382 per cent by resorting precision management practices. Yield of such a high enormity is being harvested in Indonesia in banana, papaya and pineapple, in Italy in apple, and in mango in Brazil.

Likewise, in case of vegetable, the higher yield transformation of the tune of 214 per cent can be achieved. It may 197 per cent higher in brinjal, 303 per cent in cabbage, 118 per cent in cauliflower, 149 per cent in okra, 330 per cent in onion, 136 per cent in pea, 232 per cent in potato and 432 per cent higher in tomato when compared to yield realized in advanced countries. Countries like Spain, Japan, Korea, Italy, Egypt, USA, Germany and France are harvesting quite high yield than what is harvested in our country.

The figures distinctly quantify the need of revamping the existing scenario prevalent under diversified horticultural crops. As per census report, there are 120.82 million operational holdings across 641000 villages in the country. Hardly variable rate application technology emphasizing the use of input and resources focussed highly on analytical results is followed. This is the dauntic task in breaking the barrier of realizing really vested potential of any crop flora. In needs hindsight. In the present book, compilation concentred on precision farming has been made. I compliment the editors: Dr. Jitendra Singh, Dr. S.K. Jain, Dr. L.K. Dashora faculties from College of Horticulture and Forestry (MPUAT, Udaipur), Jhalawar (Raj.) and Dr. B. S. Chundawat, Ex - Vice Chancellor, SDAU, Banaskantha, Gujarat for making the information in line to precision farming in horticulture disseminable. I hope, the accomodated material will be found useful for categories of readers in the field of commerical horticulture.

New Delhi

September 14, 2012

(Gorakh Singh)

PREFACE

India is the second largest producer of fruits, vegetables and flowers after China. It is the largest producer of cashewnut and arecanut in the world. The country is the largest producer, consumer and exporter of spices. It occupies first place in production of mango, banana, litchi, papaya, pomegranate, sapota and *aonla*, second in limes and lemons and fifth place in pineapple production. The grapes productivity of the country is highest in the world. India has the higher national average productivity in banana and sapota as compared to world average productivity while in citrus, mango, apple, guava, pineapple and papaya, it has substantially low productivity in comparison to the world. In respect to okra, the country is the largest producer; the second largest producer of brinjal, cabbage, cauliflower, pea, onion, and tomato, and it is the third largest producer of potato in the world. Except productivity of tomato, we fall behind in productivity of other vegetables.

Demographic trends indicate continuous rise in population pressure of the country. Our population is touching 1.2 billion. By 2011-12, we need to double horticultural production to the level of 300 million tonnes as against 214.7 million tonnes as in 2008-09. To gear up production attuned with requirements, National Horticulture Mission (NHM) commissioned in the country during 2004, envisages bringing 33 lac hectare area under new horticultural plantation and by 2009-12 rejuvenation of 16 lac hectare senile orchards under mission mode approach. There are underlying challenges which mar the strategy of hastening production. Environment protection, mixed/multiple cropping, micro-irrigation, INM, IPM, IWM, stress management, organic farming, mechanized farming etc are need of the hour.

Climatic variations often tend to have adverse effect on the yield and production of crops. Efforts have, therefore, been on for harnessing this natural resource through artifical means for increasing crop productivity. One such technology is protected cultivation. This technique is well adopted in Europe and USA and now China and Japan are leading in controlled sphere production of horticultural crops. In India, the technology is making breakthrough in Karnataka and Maharashtra in protected cultivation of pepper, tomato, cucumber, muskmelon, baby corn etc.

With boom in retail sector, there is growing demand of high quality speciality produce. Change in food habit has further hastened the requirements of horticultural produce. With veering vigilance towards health, paradigm shift in dietary consumption pattern is witnessed. The consumption of cereal which was 192 kg/per capita has reduced to 152 kg/per capita in rural India and from 147 to 125 kg/per capita in urban India. Fruit consumption increased by 533 per cent and those of vegetables by 167 per cent. All these have widened the gap of demand and supply and drive towards more production is on incessantly without paying much care for mother earth.

Our earth planet on an average is capable to bear load of 5.0 billion global population only. Let us have a quick recap of productivity of soil. In the wake of making the country self reliant in food production, excessive use of fertilizers and pesticides proved degradative to the innate productivity of the soil. The fertilizer consumption which was barely 70 thousand tonnes in 1950-51 reached to 231.5 lakh tonnes (300% increase) during 2008-09. It though hastened the production, but turmoiled the health of soil. As per FAO report, half of the cultivable land is getting exposed to degradation owing to lack of specific variable management practices. How to preserve and maintain space-time continuum to production system is a big question. Water, the elixir of life, may it be animal or plant, is being exploited beyond recuperation level. The flood irrigation is still in practice when underground aquifer is deepening and the area under dark zone is increasing day by day. The water use efficiency in flood irrigation system is 25-40 per cent only. Our irrigation is restricted to 25 per cent of cultivable land. Insect-pests and weeds have further emerged as major bottlenecks in crop production.

Production sphere has to be managed precisely. This is precision farming. It is defined as the cultivation by adopting technologies which give maximum precision in production of a superior crop with a desired yield levels and quality at competitive production. These include use of genetically modified crop varieties, micropropagation, integrated nutrient, water and pest managements, protected cultivation, organic farming, hi-tech horticulture, and post harvest technology.

Post-harvest sector needs lot of precision. Peels, rags, pomace etc go waste. Many times, peels being rich in polyphenols, colouring pigment, nutrients etc are richer in antitoxidant than what we actually eat. Here, we need precision. Precision in management, precision in product diversification, precision in value addition etc are much sough after aspect.

With all such considerations, to have an stock of information, it was attempted to organize national seminar on precision farming in horticulture during Dec. 28-29, 2010. The book has been derived out of lead lectures received for the seminar. In present stab, information has been synthesized putting them in order in form of book. Editors express their deep sense of gratitude to Dr. Gorakh Singh, Horticulture Commissioner, Govt. of India, Ministry of Agriculture, New Delhi, who contributed 'Foreword' which constitutes the premier formative part of the book. This has been portrayed due to strenuous effort put in by authors of respective chapters who really spared themselves to contribute presentable materials as regard to precision horticulture. We convey our deep sense of indebtedness to all worthy authors whose chapters have formed content of the book. Due gratefulness is also extended to one and all concerned in materialization of the present text. It is expected that readers will find the book worth in conformity to their eagerness in the field of precision farming in horticulture.

Editors

LIST OF CONTRIBUTORS

A. K. Srivastava

National Research Centre for Citrus, Nagpur, 440 010 (Maharashtra)

Atul Chandra

Krishi Vigyan Kendra, SKRAU, Beechwal, Bikaner - 334 006 (Rajasthan)

A. Mishra

College of Horticulture and Forestry (MPUAT), Jhalarapatan,
Jhalawar - 326 023 (Rajasthan)

A. V. Barad

Department of Horticulture, College of Agriculture, Junagadh Agricultural University,
Junagadh - 326 001 (Gujarat)

A. K. Singh

Central Horticultural Experiment Station, Panchmahals, Vejalpur,
Godhra - 389 340 (Gujarat)

A. R. Gaikwad

ASPEE College of Horticulture and Forestry, Navsari Agricultural University,
Navsari - 396 450 (Gujarat)

Alka Singh

ASPEE College of Horticulture and Forestry, Navsari Agricultural University,
Navsari - 396 450 (Gujarat)

Amrendra Kumar

National Research Centre for Litchi, Muzaffarpur - 842 002 (Bihar)

Babita Singh

Institute of Horticulture Research, Amity University, Noida (Uttar Pradesh)

B. Narsimha Rao,

Directorate of Oil Palm Research, Pedavegi - 534 450 (Andhra Pradesh)

B. N. Patel

Regional Horticultural Research Station, ASPEE College of Horticulture and Forestry,
Navsari Agricultural University, Navsari - 396 450 (Gujarat)

B. D. Sharma

Central Institute for Arid Horticulture, Beechwal, Bikaner - 334 006 (Rajasthan)

Deepa H. Dwivedi

Department of Applied Plant Science, BBA University, Vidya Vihar, Lucknow - 226 025 (Uttar Pradesh)

Dinesh Kumar

National Research Centre for Citrus, Nagpur - 440 010 (Maharashtra)

D.S. Khurdiya.

DDA, SFS Flat - 149, Pocket - 10, Sector - 11, Golden Jubilee Apartment, Rohini, New Delhi - 110 085

G. D. Patel

Regional Horticultural Research Station, ASPEE College of Horticulture and Forestry, Navsari Agricultural University, Navsari - 396 450 (Gujarat)

Hari Prakash

Quality Council of India, 2nd Floor, Institution of Engineers Building, 2, Bahadur Shah Jafar Marg, New Delhi - 110 002

I. B. Maurya

College of Horticulture and Forestry (MPUAT), Jhalarapatan, Jhalawar - 326 023 (Rajasthan)

I.S. Singh

ICAR Research Centre for Makhana, Basudeopur Farm, Darbhanga - 846 005 (Bihar)

Jitendra Singh

College of Horticulture and Forestry (MPUAT), Jhalarapatan, Jhalawar - 326 023 (Rajasthan)

J. Shankara Swamy

Department of Horticulture, College of Agriculture, Junagadh Agricultural University, Junagadh - 326 001 (Gujarat)

Kuldeep Singh

School of Agricultural Biotechnology, Punjab Agricultural University, Ludhiana - 141 004 (Punjab)

Kanak Lata

Krishi Vigyan Kendra, Panchmahals, Vejalpur, Godhra - 389 340 (Gujarat)

K. Suresh

Directorate of Oil Palm Research, Pedavegi - 534 450 (Andhra Pradesh)

K. P. Singh

Indian Institute of Vegetable Research, P.O. Jakhini - Shahanshahpur, Varanasi - 221 305 (Uttar Pradesh)

Krishan Kumar

School of Agricultural Biotechnology, Punjab Agricultural University, Ludhiana - 141 004 (Punjab)

L. K. Dashora

College of Horticulture and Forestry (MPUAT), Jhalarapatan, Jhalawar - 326 023 (Rajasthan)

Lallan Ram

National Research Centre for Citrus, Nagpur - 440 010 (Maharashtra)

M. C. Jain

College of Horticulture and Forestry (MPUAT), Jhalarapatan, Jhalawar - 326 023 (Rajasthan)

M. K. Kaul

S. K. Rajasthan Agricultural University, Agricultural Research Station, Sri Karni Marg, Sriganganagar (Rajasthan)

M. K. Verma

Division of Fruits and Horticultural Technology, Indian Agricultural Research Institute, Pusa, New Delhi - 110 012

Mam C. Singh

Division of Floriculture, Indian Agricultural Research Institute, PUSA, New Delhi-110 012

M. I. S. Gill

Department of Horticulture, Punjab Agricultural University, Ludhiana - 141 004 (Punjab)

O.P. Awasthi

Division of Fruits and Horticultural Technology, Indian Agricultural Research Institute, PUSA, New Delhi - 110 012

P. L. Saroj

Director, Directorate of Cashew Research, Puttur - 574 202 (Karnataka)

Pravisha Lahoty

College of Horticulture and Forestry (MPUAT), Jhalarapatan, Jhalawar - 326 023 (Rajasthan)

P. S. Sisodia

Central Horticultural Experiment Station, Panchmahals, Vejalpur, Godhra - 389 340 (Gujarat)

Ram Asrey

Division of Post Harvest Technology, Indian Agricultural Research Institute, New Delhi - 110 012

R.K. Pathak

11, Ram Kunj, Faridi Nagar, P. O. CIPAP, Lucknow - 226 015 (UP)

Rita Markam

Department of Horticulture, College of Agriculture, Junagadh Agricultural University, Junagadh - 326 001 (Gujarat)

R. K. Giri

Indian Meteorological Department, Mausam Bhawan, New Delhi

Raj Kumar

Central Horticultural Experiment Station, Panchmahals, Vejalpur, Godhra - 389 340 (Gujarat)

S. K. Sharma

Central Institute for Arid Horticulture, Beechwal, Bikaner - 334 006 (Rajasthan)

S. L. Chawla

ASPEE, College of Horticulture and Forestry, Navsari Agricultural University, Navsari - 396 450 (Gujarat)

S. K. Malhotra

Division of Horticulture, Indian Council of Agricultural Research, Pusa, New Delhi - 110 012

Sanjay K. Dwivedi

DRDO, CEPTAM, Metcalfe House, New Delhi - 110 054

S. Sunitha

Directorate of Oil Palm Research, Pedavegi - 534 450 (Andhra Pradesh)

S. K. Jain

College of Horticulture and Forestry (MPUAT), Jhalarapatan, Jhalawar - 326 023 (Rajasthan)

Sanjay Singh

Central Horticultural Experiment Station, Panchmahals, Godhra - 389 340 (Gujarat)

Virendra Singh

Department of Horticulture, College of Agriculture, Junagadh Agricultural University, Junagadh - 326 001 (Gujarat)

Vishal Nath

National Research Centre for Litchi, Muzaffarpur - 842 002 (Bihar)

CONTENTS

Chapter – 1

Precision Farming : A Promising Technology

J. Singh; S.K. Jain; I.B. Maurya; A. Mishra; L.K. Dashora; M.C. Jain and P. Lahoty

1.1 INTRODUCTION

The share of agriculture in GDP is sharp declining. Its share which was 55.1 per cent in 1950-51 came down to 21.6 per cent in 2003-04 and it further reduced down standing barely at 14.6 per cent in 2009-10. Horticulture acconts for 29 per cent share in gross GDP obtained from agriculture sector from about 13.08 per cent area put forth under horticultural plantations (Anon., 2010). To achieve the targetted growth rate of 4 per cent in agriculture, horticulture has to grow at the accelerated growth rate above 7 per cent. As a result of R&D initiatives made, there has been considerable increase in horticultural produce which is acclaimed as golden revoution. Horticulture sector has emerged as the frontier area in dealing with the situation of livelihood security and the employment generation. However, looking to the pressure arising from rising population and erratic climatic variation, more attention is required towards the development of technology driven horticulture. Precision farming is being viewed as a promise in this regard.

Precision farming is the process by which exact or accurate results of farming can be obtained. It is a management strategy that relies upon detailed, site specific information to precisely manage production inputs.

Maximisation of quality yield by managing site variable specific management practices is the basic of precision farming. More emphasis needs to be given to make the farming sustainable. In order to make precision farming a reality the following area needs due consideration.

- **Water management**

  Micro/Drip irrigation

  Sprinkler irrigation

Lining of water storage and conveyance system with plastics

PVC water distribution network

Sub –surface drainage

- **Surface covered cultivation**

  Mulching

  Soil solarization

- **Controlled environment agriculture**

  Green House.

  Tunnels/Low tunnels

- Organic farming
- Precision farming practices
- Environmental issues
- Post- harvest management
- Packaging and Storage
- Remote Sensing
- Global information systems (GIS), Global positioning system, Information Technology (IT) applications.
- Technology Dissemination and allied subjects

The concept of precision farming is of recent origin and trial on it was started in western countries, where land holdings were very large, farmers could afford farm machineries, land was better levelled and shaped, water use efficiency was better and had better information on their farm resources.

Precision farming uses modem technology like remote sensing and in country like India where majority of the farmers are having land holding average of 1-2 acres only, it was felt difficult to use aerial remote sensing to get detailed information of natural resources, particularly soil and other alike details. Remote sensing data with increased spatial solution to the possibility of applying precision farming method to smaller areas have also been made applicable (Narayan 2005).

Precision farming is supposed to be highly site-specific or plant specific with all information on the type of soil, water and cropping system.

There is new challenges and opportunities when this kind of modern precision farming would have to be mooted right now with proper region-specific crop planning.

Weather forecasting facilitates precision farming. Reliable wheather forecasting not only means better quality of life to the formers but also translates into economic gains by farming sector.

Farmers of the country need location specific wheather forecast based on crop-specific advice for them in real time. It needs to be established.

Right from selecting the crop cultivars, decision, on their crop planning throughout the year, planting time of suitable crops to their harvesting time, application of costly farm input like fertilizers, plant protection measures with pesticides, irrigation and intercultural operation to a practicular site is favourable for precision farming.

1.2 RELEVANCE

Precision agriculture is not necessarily applicable to every farm, or even every paddock on a farm. There are three important questions that determine if precision agriculture is an appropriate management technique for a given situation:

- Is the yield variation across the paddock large enough to justify different management decisions in different parts of the paddock in an effort to improve productivity?
- Can the factors causing the variation actually be managed with variable management?
- Does the pattern of the variability allow practical variable management of different areas?

If the answer to any of these questions is "no", then the application of precision agriculture techniques is unlikely to be appropriate.

It would be reasonable to add another question to the list - "Is it possible to manage or minimise the variability in some other way which may be cheaper, easier or more effective?"

This last question is particularly important, as it has been found that the adoption of controlled traffic farming systems can have a significant effect on reducing yield variability in paddocks, even in situations of different soil types and drainage characteristics. Many people who have adopted controlled traffic believe that there is no point even considering precision agriculture techniques unless controlled traffic has been adopted first.

1.3 SCOPE

Although the broad acre grain industries have been the major focus for precision agriculture development, there is a lot of potential for application of the techniques in intensive horticulture. Compared to grain production, the yields, costs and returns per hectare are high, giving large scope for economic gain through reducing production variability and tailoring inputs on the basis of paddock variation.

Improving irrigation application uniformity may be one of the most important management improvements to come from precision agriculture. The introduction of yield mapping may highlight subtle yield variations arising from non-uniform water application.

It is likely that precision agriculture will initially be adopted by a vegetable grower with a large area under crop, or a grower/contractor. This is purely related to the opportunity to spread the capital cost of equipment over a greater area and volume of production. Precision agriculture may also allow contractors to provide a value-added service to clients.

The adoption of precision agriculture will depend on the economic benefits offered, which will, in turn, depend on the degree of variability present in the field, and the opportunity to manage that variability. However, the adoption of precision agriculture techniques and technologies will have limited benefit unless a controlled traffic farming system has been adopted first.

1.4 HIGH DENSITY PLANTING: PRECISE SPACE UTILIZATION

It consists of planting more number of plants per unit area than the normal planting. It is of different types:

1.4.1 Low density planting

It is a non-intensive system of cultivating fruit crops. It is an age old planting system in which trees are planted at wide spacing accommodating about 100-250 trees on a hectare. Pruning is resorted at a minimal level and orchards are nurtured so as to favour maximum development of tree. Dwarfing rootstocks don't find place in planting. Trees acquire commercial production potential after 10-15 years of planting. The output from the orchard during early 10-15 years is less. The system being less input and care intensive holds popularity among growers.

1.4.2 Medium density planting

Trees are planted at highly minimized distance covering 250-500 plants/ha. Proper pruning is undertaken to manage the tree in desirable shape. The system being more care-intensive the labour requirement is more and so also the yield obtained is more. This system is taking lead in output conscientious growers to produce amenable fruits crops like pomegranate, citrus, guava, papaya, banana, pineapple etc with assured production management system.

1.4.3 High density planting

This system is very condensing in planting accommodating 500-100000 plants on a hectare basis. Different types of high density planting are as under (Mohammed and Wilson, 1984):

- Medium high density : 500-1500 plants/ha.
- Optimum high density : 1500-10000 plants/ha.
- Ultra-high density : 100000 plants/ha.

This system relies heavily on rigorous training and pruning. Maintenance pruning is very heavy in this system. Use of dwarfing rootstocks and chemicals find place in the high-density planting. Passo *et al.* (1979) reported inoculation of some viruses and viroids in controlling the size of plant. The expenses per unit are high and so is the yield.

1.4.4 Meadow orcharding

The term meadow literally means grassland. Analogous look of grassland covering every bit of land forms the basis for naming. This is ultra-high density planting system in which 20000-100000 plants/ha is accommodated. In order to maintain tree form, severe top pruning is practised similar to mowing of grassland. The plant intended to produce yield after two years age, heavy use of growth regulator is made in meadow orchard.

In case of apple, during first year attempt is made to induce shoots to reproductive stage and during second year, plants flower and produce fruits. When fruiting is over, they are cut back severely in stump form and again new shoots appear and repeat the biennial cycle of fruiting. Mechanized harvesting is followed using combine which separates the fruit from the shoot. However, the system being capital intensive, it may not suit the small growers. Suiting to the requirement, two systems of meadow orcharding; mechanized system for a large farm and intensive system for the small farm are followed. In the mechanized system only one stem is allowed to bear fruit. When the fruiting is over, this is cut from ground level and over new shoot, hormonal spray is made to accomplish fruiting in two years cycle. In intensive system the tree is developed to two shoots. Out of these two shoots, one is allowed to produce fruits and another is pruned to a stump. The pruned shoot regenerates to produce new growth and flower bud during the growing season. Thus, every shoots produce fruits every second year. The tree managed by intensive system yields more in comparison to mechanized system due to increased flower bud, fruit set and yield per tree. Fruit plants which are very easy to propagate by cutting or air layering are especially suitable for HDP. The species which are propagated by deploying rootstock, removal of sprouting appearing out of it, makes meadow orcharding difficult.

1.5 PLANT COMPETITION IN HDP

Increasing number of plants per unit area produces more yields in proportion to the increased number of plants until the competition between plants sets in. When the plant population per unit area increases and the competition sets in for recourses, there is reduction in yield in proportion to the degree of competition between plants. With increase in population beyond the optimal, the plants respond to population pressure in form of decreased number, reduced size, reduced colour of fruits etc. Even though with reduced fruiting surface, there is progressive decline in per plant yield, the yield per unit area is higher in HDP owing to higher plant population.

1.6 COMPONENT OF HDP

HDP is resorted by adopting - a) dwarf scion varieties; b) dwarf rootstock; c) training and pruning; d) suitable crop management practices; e) use of bio-regulators (Singh, 2011).

1.6.1 Dwarf scion varieties

Apple	:	Red Spur, Star Crimson Spur, Gold Spur, Well Spur, Oregon Spur, Silver Spur, Red Chief, and Hardi Spur.
Peach	:	Red Heaven, Candor
Cherry	:	Compact Lambert, Compact Stella
Pecan	:	Western, Desirable, Cheyenne,
Walnuts	:	Hartely, Chico, Vina, Howard
Sour cherry	:	English, Morello
Papaya	:	Pusa Nanha
Banana	:	Dwarf Cavendish
Mango	:	Amrapali
Sapota	:	PKM-1, PKM-2

1.6.2 Dwarf rootstock varieties

HDP is possible using dwarf growing rootstocks. These rootstocks for different fruits are as under:

Apple	:	M9, M26, M4, M7, M9, MM106, M 27 (ultra-dwarfing)
Pear	:	Quince
Cherry	:	Mahaleb
Mango	:	Vellaikolumban (for Alphonso),Olour (for Himsagar and Langra)
Guava	:	Pusa Srijan, *Psidium friedrichsthalianum*, Aneuploid No.82
Citrus	:	Trifoliate orange, Sour orange,Citranges
Avocado	:	Mexican rootstocks
Loquat	:	Quince
Ber	:	*Zizyphus nummularia*

1.6.2.1 Physiology of dwarfing

The dwarfing effect of rootstock is impacted by the translocation of auxin. The translocation of auxin from shoots and young leaves to roots is governed by the amount of IAA oxidase, peroxidase and phenols and other compounds present in the phloem and cambial cells. The levels of these compounds are genetically controlled and may vary from one species to another. Such variation is responsible for varying growth performance. The rootstocks with thicker bark and high per cent of starch indicate low levels of auxins. Dwarfing rootstock or interstock may have dwarfing effect by controlling flow of auxin through bark of rootstock or interstock (Lockard and Schneider 1981). The bark of different genetic compositions allows different amounts of auxin to pass through. It affects root growth. This ultimately affects the cytokinin synthesis. The level of cytokinin synthesis affects shoot growth leading to small or large tree.

1.6.3 Training and pruning

Training and pruning are effective tools in HDP by virtue of their impact on shape and size control of the tree. HDP amenable training and pruning methods are described as under:

1.6.3.1 Bush system

In this system, the height of the plant is kept to 2.0 metres. During first year, the plant is cut at a height of 70 cm. No shoot is allowed to grow up to a height of 25 to 30 cm. Above this height, 3 to 4 branches are allowed to grow over which number of branches emerge out. The plants acquire the shape of bush. The centre of the plant is kept open. This system is suitable for apple.

1.6.3.2 Pyramid system

In this system the plants are trained in a fashion so that lower branches may remain longer and higher branches gradually smaller. The alternative tiers of horizontal branches radiating from main stem scattered all around, gives the plant an appearance of pyramid. The branches are allowed to grow on main stem at 20 cm height from ground level. The plants are pruned from the tip of main stem and branches to maintain pyramid shape.

1.6.3.3 Espalier system

The word espalier is French in origin meaning a fence, a fruit wall or paling. It refers to the support used for training trees especially apples and pears. The tree trained through this system consists of three to six tiers of horizontal branches trained to grow one foot apart from one another at right angles to the main stem. Thus, the branches grow parallel to the ground. In this system using poles, three to six rows of wires are stretched one above the other. The first row of wire is stretched at a height of 60 to 70 cm, the second row at 130 to 140 cm height and the third row is stretched at a height of 200 cm from ground level. Over these wires, the branches of the trees are trained in both the directions parallel to the ground. In this system, the line to line distance of the plant is kept less as the plants are grown only in two directions along with wire.

1.6.3.4 Cordon system

Cordon refers to closely spurred single stemmed tree tied to a support e.g. wires or bamboo canes, either in vertical, oblique or horizontal position. This system usually finds favour in apples and pears. The trained plants bear early crop as compared to dwarf pyramid and bush system. The plants are planted at a distance of 1 to 1.5 m. The stem of the plant is tied with wire. The wires of 12 to 13 gauges are fixed to the ground using cement and concrete at 4.5 to 6.0 metres interval. The plants are maintained single stemmed by practising severe pruning of emerged branches during winter and summer. Depending upon the number of main stem trained along with wire, the system is known as single cordon and triple cordon.

1.6.3.5 Tatura trellis

This is system of training fruit trees along with trellis of wire to harvest early and high yield without use of dwarfing rootstock. The alignment of tree lies in the centre of trellis and its branches are trained along the wires of the trellis. The system was developed by David Chalmers, Ban Van den Ende and Leo van Heek during the year 1973 at Irrigation Research Institute, Tatura, Victoria, Australia. The trellis of wire is erected using iron pole of 10.5 ft height over which $12^1/_2$ guage high tensile steel wire is stretched. In between two trellises, gap of 7 ft is maintained. The orientation of trellises is kept to north-south directions. Common orchard cultural practices are followed. Canopy of tree is maintained under desired frame using mechanical hedger. Harvesting is done mechanically. However, the system is highly labour intensive, as it requires hand training and hand thinning also, to get more yields of large, high quality fruit. The system favours high yield owing to optimum light interception and close planting. This is suitable for training and maintaining orchard of apple, pear, peach, plum, apricot, nectarine, sweet cherry, kiwifruit, grape etc. which sustain the rigour of pruning.

1.6.3.6. Hedge row system

As the name indicates, the system accommodates plants similar to as in hedge planting. The distance between trees is set 50 cm-2 m and that between rows 3-4 metres. The close planting of tree develop in form of compact wall looking like hedge. The orientation of row should be North-South so as to have ensured availability of light in both sides of the hedge walls of the tree. The tree can be arranged in single, double or multiple rows system. However, arrangement in single row is favorable in view of maximum availability of solar radiation to the trees. Reduction in spacing within row between plants reduces yield and also the size of fruits. Multi-row system though favours light interception and so also the yield but the fruit colour is less desirable. Added to this, the canopy being too compact, it becomes difficult to follow intercultural operations and the chance of invasion of pests/diseases increases.

1.6.4 Use of bio- regulators

Different growth retardants are used to control shape and size of trees besides favouring reproductive growth at the expense of vegetative growth. They find uses in manipulating different physiological activities as under:

1.6.4.1 Prolonging dormancy

Tender growth of plant is susceptible to winter injury if appears during severe winter. Growth retardants like Abscisic acid (ABA), SADH, (Succinic acid-2, 2-dimethylhydrazine, also known as B-9, B-995 or Alar), CCC (Chlormequat chloride, also known as cyocel, chlorocholine or 2-chloroethyl ammonium chloride) when applied to foliage, delays bud break. They are of use in citrus and grape.

1.6.4.2 Reducing vegetative growth

In order to facilitate better management practice in orchard, reduction in vegetative growth is envisaged. AMO 1618 (ammonium, 5-hydroxycarvacryl) trimethly chloride piperidine carboxylate, CCC, Paclobutrazol, B-9, etc. have been found effective in reducing the growth of pear, peach, lemon, litchi, mango, apple, plum and apricot.

1.6.4.3 Flowering

Many bio-regulators play key role in induction of flowering. Ethylene 200-400 ppm induces flowering in pineapple. SADH promotes flowering in apple, pear, peach, blue berry and lemon. Paclobutrazol is effective in promoting flowering in lower nodes of papaya.

1.6.4.4 Reducing fruit drop

SADH when sprayed at full bloom reduces fruit drop. The chemical also minimizes development of water core and cold strong scales in fruits.

SADH at 1000-2000 ppm, ccc at 2000-3000 ppm, paclobutrazol (Cultar or Bonzi or PP333) 1000-2000 ppm, MH at 1000-1500 ppm have produced desired results in HDP. The bio-regulators alter assimilate partitioning more in favour of reproductive shoot. Less shoot growth favours reducing cost incurred in pruning. In fruit trees in which neither dwarfing rootstock nor compact scion cultivars are available, bio-regulators are especially useful.

1.6.5 Adoption of suitable crop management practices

HDP being productive, it always calls adoption of proper crop management practices. Mulching, fertigation, organic farming, INM, IPM etc. like hi-tech strategy prove boonful in harvesting proper yield and potential of a particular fruit crop.

1.7 COMPARISON BETWEEN TRADITIONAL AND HDP ORCHARDING

HDP orcharding being labour and cost intensive though yield obtained is more (Table1), it has not yet trickled to majority of farm families. The detail of various features of HDP and Traditional system of planting is produced below:

S. No.	Parameters	Traditional	HDP
1.	Plant population	Less (150-200/ha.)	More (500-100000/ha.)
2.	Production	Low (15-25 t/ha)	High (30-50 t/ha)
3.	Management	Large tree size difficult to control	Small tree size convenient to manage
4.	Labour requirement	More	Less
5.	Harvesting	Difficult, manual	Easy by machine
6.	Quality	Large canopy, poor sunlight penetration, poor quality	Small canopy, better sunlight
7.	Establishment	Cost less	Cost high
8.	Machinery	Doesn't require expensive machine	Requires expensive machine
9.	Bioregulator	Use is not essential	Use is essential

Source : Ray, P. K. (1999)

Table 1 : Fruit productivity in traditional and high- density orchards

Fruit crop	Plant distance (m)	Trees (/ha)	Tree age (year)	Cultivar	Yield (t/ha)*
Almond	5.4 × 5.4 – 7.5 × 7.5 5.0 × 2.5	177-342 800	5	Tuono	5.0- 10.0 Higher
Apple	5.0 × 5.0 – 8.0 × 8.03 8.03. × 0.75	227-400 2222		Jonathan	30-50 68.75
Apricot	6.0 × 6.0 4.5 × 1.5	277 1481	10-13	Tyrinthos	13-22 64.40
Avocado	7.0 × 7.05 5.0 × 2.5	204 800		Hass	10-15 34.4
Banana	2.0 × 2.0 1.2 × 1.2	2500 6944		Robusta	35-36 174.39
Ber	6.0 × 6.0 – 8.0 × 8.0 2.0 × 2.0	227-227 2500	10-20	Jinsixlaozao	25 - 40 250 - 400
Cherry	8.0 × 5.04.0 × 2.0	250 1250	3	Van	6.3-9.0 16.0
Grape	5.0 × 3.0 – 6.0 × 3.0 1.8 × 0.8	555-667 7000	9-14	Grenache Noir	15 - 20 Higher
Grape fruit	6.0 × 8.0 – 8.0 × 8.0 3.3 × 1.5	156-227 2020	7	Redblush	10-15 58.25
Guava	6.0 × 6.0 2.0 × 2.0	277-2500	5	Sardar	15.70 58.30
Kiwi fruit	5.0 × 5.0 – 6.0 × 6.0 4.5 × 1.5	277-400 1481	1	Hayword	25 37

Contd...

Lemon	6.0 × 6.0 – 8.0 × 8.0 3.47 × 3.47	156-227 832	4		10-15 68.2
Litchi	10.0 × 10.0 5.0 × 5.0	100 400	10	Mauritius	9-10 12
Mandarin	6.0 × 6.0 – 8.0 × 8.0 3.0 × 2.2 – 2.5 × 2.0	156-227 1500-2000	5		10-15 55
Mango	7.5 × 7.5 – 12.5 × 12.5 3.0 × 2.5	64-177 1333	14 24	Dashehari Dashehari	4.25-9.0 34.74
Orange	6.0 × 6.0 – 8.0 × 8.0 3.3 × 1.5	156-227 2020	7	Valencia	10-15 23-75
Papaya	2.0 × 2.0 1.2 × 1.2	2500 6944			40-50 75
Pomegranate	5.0 – 6.0 × 5.0 – 6.0 5.0 × 2.0	277-400 1000			7-12 30
Sapota	10.0 × 10.0 5.0 × 5.0	100 400	6-12	Kalipatti	15-20 13.61
Strawberry	45 × 90 (cm) 25 × 60 (cm)	24691 66666			17.5 30
Tanger	6.0 – 8.0 × 6.0 – 80 11 × 5 feet	156-227 800	7		10-15 52
Walnut	10.0 × 10.0 6.1 × 3.0	100 519	5	Tulare	7.8

Source: Adapted from Singh (2006)

**The upper value in each column connotes to yield under traditional planting while the lower under HDP.*

1.8 SUCCESS IN PRECISION FARMING: A CASE STUDY

1.8.1 Tomato

Scientists at the Tamil Nadu Agricultural University (TNAU), Coimbatore, have developed a technology called precision farming, said to be a first of its kind in India. The technology is being presently implemented in several major areas of Tamil Nadu. At present precision technology is implemented in about 200 hectares in Dharmapuri district in Tamil Nadu. More than 100 farmers have been benefited by this project. Hybrid chilli in about 30 hectares, hybrid tomato in about 20 hectares and Cavendish (green) banana variety in about 10 hectares were raised successfully under precision farming system.

1.8.2 Technical guidelines

Under this farming concept, the University offers technical advice to farmers on soil and water conditions, weather mapping, forewarning of pest attacks and also supplies water soluble fertilizers to be applied along with drip irrigation.

Mr. P.M. Chinnasamy, is one such beneficiary farmer in Dharmapuri, who has harvested about 135 tonnes of tomato from his one hectare field in nine months using this technology.

1.8.3 Higher yield

The yield is almost triple than that of the farmers growing tomato under the conventional system of planting. Under the conventional system the crop is grown under channel irrigation and only about 25-30 tonnes of fruits can be expected, according to Mr. Chinnasamy.

The life of the tomato crops grown under normal cultivation (channel or furrow irrigation) is only 3-4 months, whereas the life span of the same crop grown under precision technology can be extended to 8-9 months.

The seeds were first sown in small plastic pots, and kept under a shade net in a nursery. Cocopeat was used as a growing medium. The cocopeat was completely sterilised with chemical or under steam treatment before being used.

The seedling growth was uniform and without any pest and disease infestations. The main field was installed with an in-line drip irrigation system at a spacing of 1.5m lateral intervals. Drippers at 60 cm spacing were used. About one hour of irrigation was found sufficient to cover one hectare of land.The field was ploughed well by mixing about 25 tonnes of farmyard manure, 2 kg of azospyrillum and phospobacteria each, about 2.5 kg of pseudomonas and 1,172 kg of super phosphate.

Raised beds of about one foot height were formed and the laterals were placed at the centre of the beds. Planting was done on either side of the laterals at a spacing of 90 x 60 x 60 cm. About 20,000 plants were planted in one hectare. Drip fertigation (application of water soluble fertilizer along with water) was done through a 60-litre capacity fertilizer tank and once in three days starting from the day of planting.

1.8.4 Flower drop

Weeding was done once after 30 days and again one month later. The plants were tied to poles for support on the 30th day after planting. About 1.25ml of triacontanol and 0.25ml planofix hormone diluted in one litre of water were sprayed on the 30th, 60th and 75th day to control flower dropping.

1.8.5 Infestations control

The crop was found susceptible to infestations from sucking pest, leaf minor and cutworm.

Spraying 2ml of triozophos diluted in one litre of water was found effective in managing sucking pest, leaf minor and cutworms infestations, while fruit borer infestation was controlled by spraying 2ml of chloripyriphos diluted in one litre of water.

The first harvest was done on the 65th day after planting. The fruits were uniform in size and the skin colour was deep red.

Net income of about Rs. 5 lakhs from tomato cultivation in nine months.

1.9 REFERENCES

Anonymous (2010) *Souvenir*. 4th Indian Horticulture Congress, New Delhi, p.xxviii Erez, A. (1982) *Hort. Science*, 17 : 138-142

Jana, B. L. (2008) *Precision farming*. Publd. from Agrotech Publishjing Academy, Udaipur, 1040 pp.

Lockard R.G. and Schneider G.W. (1981) Stock and scion growth relationship and the dwarfing mechanism of apple. *Horticulture reviews,* Jules Janick (Ed.) Wiley, 3 : 315-75.

Mohammed, S and Wilson, S.A. (1984) *Tropical Agric*., 61:137-42.

Narayan, L.R.A. (2005) Remote sensing-providing important inputs. *The Hindu Survey of Indian Agriculture*, pp.168-170.

Passos, O.S.. Almas, C.des and Boswell, S.B. (1979) *Citrograph*, July, pp. 211-218.

Ray, P.K. (1999) Orchard establishment. In *Tropical Horticulture,* Vol-1, Naya Prokash, Kolkatta, 22 pp.

Singh, C. P. (2006) Establishment and management of high density orchards. In *integrated crop management in subtropical fruits* (Eds. J. P. Tiwari and R. Srivastava), published from Agrotech publishing academy,Udaipur, pp., 53 – 61.

Singh, Jitendra (2011) Basic Horticulture, Kalyani Publishers, Ludhiana, 14p.

Chapter – 2

Precision Farming in India – Concepts and Approaches

S. K. Malhotra and Babita Singh

2.1 INTRODUCTION

With increasing population, urbanization and continuous depletion of natural resources, there has to be a paradigm shift in farmer's perception from production to productivity and to profitability. In the present scenario, the major challenges arising are shrinking land and depleting water and other related resources in agriculture. There is need for promoting farmer friendly location specific production system management technologies in a concerted manner to achieve a vertical growth in horticulture production duly ensuring quality of produce and better remuneration per unit of area with judicious use of natural resources. In this endeavour, precision farming aims to have efficient utilization of resources per unit of time and area for achieving targeted production of horticultural produce. With the revolution in information technology, now it has become possible to focus on the positioning of agricultural tools with high degree of accuracy and sophistication for maximizing returns. The combination of information technology and use of hi-tech equipments made it possible for the introduction of precision farming. It is a comprehensive system designed to optimize agriculture production through the application of crop information, advanced technology and management practices. With recent advances in geographic information systems, it has become possible to focus on site specific crop production technologies so as to harness the potential of horticulture through innovative technologies in a sustainable manner.

2.2 PRECISION FARMING : AN APPROACH

In the broad sense precision farming has been defined as application of a holistic management strategy that uses information technology to bring data from multiple sources to bear decision associated with agriculture production, marketing, finance and personnel. Precision Farming takes care of each and every plant in a farmers field throughout its crop cycle. Precision Farming is technology enabled information based and decision focused system. Information, technology and management are key elements of precision farming. The concept of precision farming has developed over the years by way of addressing various components of traditional crop management systems, starting with soil fertility. Precision farming, from a systems perspective, is similar to the early days of no-tillage crop production. The no-tillage was not adopted until the 1980s when the integration of appropriate technologies and public policies supported the dissemination to farmers. Certain technologies in the early days of precision agriculture allowed for the variable application of nutrients and pesticides, there did not exist a thorough understanding of how soil fertility and pests varied in space and time. But explanations were lacking on what specifically caused the variability so that appropriate inputs could be matched to site-specific conditions. These days, individual components of precision agriculture are being adopted on the farm but a distinctive precision farming system is required to be established for different situations.

2.3 ADVANCEMENT IN PRECISION FARMING APPLICATION

There has been two advancements in precision farming applications one is variable rate fertilizer and another application and yield mapping. In a traditional system of fertilizer application, the farmer obtains one or more soil samples, combines the samples, sends them off to a soil-testing laboratory. The farmer receives a fertilizer-mix recommendation and a suggested application rate based on the crop to be grown. The farmer applies the specified fertilizer mix at the recommended rate across the entire field. In this, the key concept has been uniformity. However, it is a well known fact that most of the fields are anything but uniform. With modern methods the precision farmers apply fertilizer at variable rate with the help of a soil sampling programme by demarcating grids on his plot. It is possible to identify the exact location of the plot with the help of one of the main precision farming technologies of Global Positioning System (GPS). For applying the fertilizer at variable rates, advanced equipments are available which carry the fertilizer elements in separate tanks to the field and mixes them on the fly just before dispersal. To accomplish this, the farmer must mount a GPS receiver on the fertilizer tank so that the equipment knows its location in the field. As a second component of precision farming practice, yield mapping has become quite popular all over the world. This is because information on crop yield is important to every farmer. Yield mapping has become possible with the use of GPS and sensors that are used to measure the flow rate of grain harvested through combines. As the combine works, an on-board computer gathers information from the GPS receiver and the sensors which are stored in a memory card and later downloaded to a PC equipped with Yield Mapping software.

2.4 VARIABILITY MANAGEMENT BASED ON SITE SPECIFIC INPUTS APPLICATION

Primarily, precision farming involves the measurement and understanding of variability over time and space. This system is based on use of information generated through surveys to manage this variability by matching inputs to conditions within fields using site-specific inputs. Finally, and most important, this system provides for the measurement and recording of the efficiency of these site-specific practices in order to assess value on and off the farm. The precision farming system being information based and technology enabled, is thus effiecient.

2.4.1 Enabling technologies

The enabling technologies of precision farming can be grouped into five major categories viz., computers, Global Position System (GPS), Geographic Information Systems (GIS), sensors and application control. Some of the enabling technologies were developed specifically for agriculture just about two decades back. It is the integration of these technologies that has enabled farmers and their service providers to do things not previously possible, at levels of detail never before obtainable, and, when done correctly, at level of quality never before achieved.

2.4.1.1 Computer system

Computers have helped to define precision farming in terms of a management strategy that uses information technologies for decision making. Precision farming requires the acquisition, management, analysis and output of large amounts of spatial and temporal data. The extent to which computer technology could be utilized in agriculture is important to the success of agriculture in general. However, the agricultural sector is lagging in the adoption of computer technologies on the farm as compared to other business sectors. However, in the long run, farmers will have to become as comfortable working with computers and their data as they are working with their farm machinery. Besides, computer hardware related software is equally important in precision farming. Computer software in precision agriculture has become better with time. For precision farming, the knowledge needed is that for managing variability on the farm, knowledge that is requisite for decision making. Therefore, in this system, computers and related softwares have become critical inputs.

2.4.1.2 Geographic Information System (GIS)

Geographic information system (GIS) is an important system which includes organized collection of computer hardware, software, geographic data and personnel designed to efficiently capture, store, update, manipulate, analyze and display all forms of geographically referenced information. The science of GIS has included data management and modeling, enabling a shift from mapping to spatial reasoning. The ability to perform spatial operations on data distinguishes a true GIS from the many software programme that do thematic mapping and database management. During the past few years, spatial operations have been added to mapping software programme. Workstation GIS software programme have spawned microcomputer versions with more limited GIS capabilities to fit desktop computer technologies and new microcomputer based GIS systems have emerged. There are many different mapping and GIS software programme that offer different GIS features. However none have captured the market for application in precision agriculture. Because, precision agriculture is concerned with spatial and temporal variability and because it is information based and decision focused, it is the spatial analysis capabilities of GIS that enable precision agriculture. This is particularly true because the value of precision agriculture is derived only when resulting information is turned into a management decision that increases profitability, benefits the environment or provides some other value to the farm.

2.4.1.3 Global Positioning Systems (GPS)

In order to assess spatial variability and for site-specific application control, location control is essential to precision farming. The GPS technology enables precision farming because all phases of precision agriculture require positioning information. GPS is able to provide the positioning in practical and efficient manner. High-precision differential GPS are also now available that achieve high accuracy, allowed for automated machinery guidance and kinetic mapping of topography and are useful in the creation of digital elevation models needed for terrain analysis. While the GPS signal is ubiquitous, there have been problems in making available GPS as needed precision for agriculture. The US Department of Defence implemented selective availability (SA) in March 1990, which limited accuracy of GPS to civilian from about 8 to 10m without SA to about 100 m of SA. This was done by varying the reported precise time of clocks on board the satellites and by providing incorrect orbital positioning data. The SA has been overcome by the use of differential corrections transmitted to GPS receivers (rovers) from GPS receivers at known fixed locations. Differential GPS (DGPS) involves the transmission of differential correction, i.e., the difference between actual and predicted position at the base GPS receiver, to rover GPS receivers, which then apply the corrections to receive GPS signals to solve for a more accurate position. A differential correction is desirable even without SA because it is needed to archive the accuracies needed in some aspects of precision agriculture, included navigation and guidance. Currently, only WADGPS provides national coverage, whereas all others are dependent on whether the rover is close enough to a base station to receive a signal consistency. Accurate guidance and navigation systems will allow for farming operations not currently in use, including field operations at night when wind speeds are low and more suitable for spraying and the use of night tillage to reduce the light-induced germination of certain weeds.

2.4.1.4 Sensors

Sensors are the devices which are capable of transmitting an impulse in response to a physical stimulus such as heat, light, magnetism, motion, pressure and sound. Computers help to record the sensor impulse a GPS to measure position and GIS to map and analyse the sensor data, any sensor output can also be mapped at very fine scales. Sensor technology currently lags behind other enabling technologies and the availability of sensors has been cited as the most critical factor preventing the wider implementation of precision agriculture. Sensors are critical to success in the development of precision agricultural system for three important reasons. Sensors have fixed costs, sensors can sample at very small scales of space and time, and sensors facilitate repeated measures. This means that the cost per sample is determined by the extent of sensor use, sample intensity is determined by the capability of the sensor and not the cost of difficulty in sampling associated with traditional physical sampling schemes, and sampling frequency is determined by accessibility of the target and not costs. Sensors are essential for use in precision agriculture. Sensors are needed in precision farming because such a system requires the collection, coordination and analysis of massive quantities of data, some for strategic surveys and inventories and some for use in real-time applications. Sensors can be contact or remote ground based or space based and direct or indirect. Sensors have been developed to measure machinery, soil, plants, pests, atmospheric properties and water by sensing motion, sound, pressure, stain, heat, light and magnetism and relating these to properties such as reflectance, resistance, absorbance, capacitance and conductance.

2.4.1.5 Remote sensing

Many companies now offer commercial products to agriculture from images obtained by these satellites of enhance digital products derived from them. Remote-sensing satellites collect image data actively by sending a known signal from the satellite to the earth and measuring the portion of the signal that is returned. Passive data collection occurs by measuring the incoming energy from the sun reflected by an object or heat energy emanated from an object. The electromagnetic energy emanated from an object varies in wavelengths as determined by the object's physical and chemical structure. Different images of an object can be constructed by combining different wavelengths, creating images far more revealing than images obtained from visible light alone. Remote-system vary in spatial resolution (meter to kilometers), spectral coverage (portion of the light spectrum covered), and temporal frequency (days to months). Different applications in agriculture will require different spatial resolutions, spectral coverages, or temporal frequencies. Remote sensing holds great promise for precision farming because of its potential for monitoring spatial variability over time at high resolution.

2.4.1.6 System control

It is an automated system governing application control, in which sensed information is used to influence the system's state in order to meet an objective. For precision farming control must be achieved in space and time for varying single or multiple inputs at different rates, at varying single or multiple inputs at different rates, at varying soil depth, and in a uniform and location – specific manner within fields. Because it is a required component, control technology has been the strength of precision farming since its inception. Simply stayed, if the needed accuracy cannot be achieved at the point of application of inputs, then precision farming cannot be successful. Application control completes the precision farming loop. The control system are currently available at varying degree of precision for variable seed and metering granular fertilizers and pesticides, changing verities on the anhydrous N application, sprayers, irrigation, manure application, and various tillage implements. The application control system which includes navigation and guidance, has been enabling precision farming since its inception and continues to improve. Farmers and their service providers have the capability to apply very precise application of inputs site specifically. Application control technology will continue to improve and support the need of precision agriculture. What needed is knowledge of what inputs are required where and when; intensity in a yield map in orders of magnitute higher than used in grid sampling and the exercise of grid sampling. The exercise of grid sampling a yield map would reveal the weaknesses of inherent interpolation of spatial sample points. Yield maps do not indicate the causes of the yield magnitute and its variability that need to be quantified if farmers are to adjust their management practices to specific conditions within a field at appropriate times during the growing season.

2.5 MANAGING VARIABILITY

The agronomic inputs must be matched to know condition employing management recommendations that are site specific and use accurate application control equipment. The potential for accurate application control on the farm although exists but the extent to which it has reached the farm is limited. High-precision application control technology is available in many areas commercially, but it is not generally installed on farmer-owned equipment. Application technology is not the only factor that can limit precision management. The agronomic feasibility of precision agriculture, i.e., whether precision agriculture improves crop performance is a relevant question. Precision farming has best management practices developed from decades of agronomic research are applicable at any scale of management and are required to be validated for different crops and conditions.

2.6 PRECISION MANAGEMENT OF SOIL FERTILITY

Soil do not supply nutrients in sufficient quantities naturally to meet nutrient demands of commercial crops, therefore use of nutrient input to crop production is important. Important inputs to crop production, fertilizers and manures are also identified as major sources of nutrient contamination of surface and ground water in agricultural areas. Soils vary in their ability to supply nutrient to plants, and crops vary in their demands for nutrients. Since soil supply and plant demand vary in space and time and nutrient losses through leaching, erosion, and runoff also vary temporally and spatially indicates that significant opportunities may exist for precision management of soil fertility.

2.6.1 Management of nitrogen

The precision management of nitrogen is applicable to situations in which the factors that control total nitrogen in soils and nitrogen availability to plants vary spatially.

However, precision N management is increasingly more difficult but may in fact have increasingly more environmental benefits as the temporal component of spatial variability of nitrogen availability increase. For this reason, precision nitrogen management is more complex than precision management of lime, P, and K but may have significantly more value. Both nitrogen deficiency and excess nitrogen availability create problems for production agriculture. Most fertility concerns are focused on deficiencies in nitrogen availability to plants because they reduce yield and/ or quality for some crops. Leaching, runoff, and denitrification are processes that result in loss of nitrogen from the soil-plant system creating the potential for nitrogen deficiency in crop and degradation of water and air quality. The environmental threat of nitrogen losses from soil is further compounded by the application of manures at rate in excess of crop nutrient needs. Therefore, where soils or landscape consistently regulate water availability, precision nitrogen management would have potential.

2.6.2 Management of phosphorus and potassium

Precision management of phosphorus and potash was an early focus of precision farming because there was an established basis for fertilizer recommendations in soil testing that could theoretically be applied at any scale. Spatial variability in phosphorus and potash was already know and was not difficult to measure within agricultural fields. The concept of VRT was very intuitive and easy to understand and implement. The temporal component of the spatial variability of phosphorus and potash is low, making it easy to soil test at a convenient time and requiring only periodic/ validation. The current basic for precision management of phosphorus and potash, therefore, is fertilizer recommendations based on traditional soil fertility tests using various sampling schemes to asses within-field variability.

2.6.3 Precision management of soil moisture

Water is the most important input in agriculture for crop production. It is essential to ensure that adequate quantity of soil moisture has been made with high water productivity. Depending upon the soil and land features, the availability of soil moisture vary. Besides the moisture requirement of the crop would vary depending upon the stage of growth. The fact that crop water requirements vary over time and space indicate the scope of precision management of soil moisture. Irrigation scheduling using VRT combines knowledge of spatially variable soil water holding capacity, spatial and temporal variability of applied water, system delivery specifications and weather-driven crop evapo-transpiration (ET) calculation. On the other hand under conventional system, a single value of crop ET is typically used for entire field for use in irrigation scheduling software.

2.7 PRECISION MANAGEMENT FOR DISEASES, INSECT-PEST AND WEED MANAGEMENT

Weeds, diseases and insects are costly management problem to crop production because significant infestations reduce crop yield and quality. Precision weed management is possible because weeds are spatially aggregated and not randomly distributed within most agronomic fields and because the efficacy, efficiency, and fate of weed control inputs vary spatially and to a varying extent, temporally. A common approach to precision weed management is site-specific weed control achieved by applying herbicides only where weeds are present or above economic threshold levels, termed intermittent herbicide application or patch spraying, varying herbicide application (type, formulation, or rate) according to soil physical and chemical properties or weed characteristics (spices, growth stage, and density), or some combination of the two approaches. For prevention or pre-emergence weed control, site specific application require prior knowledge of historical weed are visible at the time of application. This knowledge can be obtained by mapping weed aggregation in previous years.

The important concerns regarding the impacts of pesticide use include health risks related to food safety, water quality, and worker safety and concerns over wildlife and ecosystem health. Therefore, agricultural management practices that reduce pesticide use, improve pest management or reduce risks of pesticide to human and economic benefit of precision pest management is the reduction in chemical/ non-chemical pest management costs, crop damage or both to more efficacious or efficient application of pest control measures. A reduction in pesticide use, however, does not translate into profit if, the cost of obtaining information about pest populations and distribution exceeds the savings. Environmental benefits are presumed to result from a reduction in pesticide usage, particularly in sensitive environments. However, the result of the few studies on the potential of precision farming to provide environmental benefits have been inconclusive regarding its effect on pesticide use. Precision insect management has potential because distributions of insect populations are spatially variable, in part because insects are mobile during at least part of their life cycle and in part because during at least part of their life cycle and in part because during the relatively non-mobile stage insects cluster in response to environmental and behavioral responses. Therefore, precision insect management has the potential to reduce insecticide application and improve the efficacy of both prevention and intervention with the management of insects is that their populations are highly dynamic and prediction of insect density is difficult or uncertain.

2.8 SCOPE FOR PRECISION FARMING IN HORTICULTURE

Precision Farming involves the application of technologies and principles to manage spatial and temporal variability associated with all the aspects of horticultural production for improving crop performance and environment quality. Precision farming calls for efficient management of resources through location-specific hi-tech interventions. Hi-tech horticulture encompasses a variety of interventions such as micro irrigation, fertigation, protected/greenhouse cultivation, soil and leaf nutrient based fertilizer management, mulching for *in-situ* moisture conservation, micro propagation, biotechnology for germplasm, genetically modified crops, use of biofertilizers, vermiculture, high-density planting, hi-tech mechanization, green food, soil-less culture, biological control etc.

This technology has been applied for various horticultural crops like potato, banana and oil palm. Studies conducted at the Irrigated Agricultural Research in Extension

Centre, Wash, USA on using site specific crop management in potato has demonstrated promising results. The tools available commercially for the potato growers are yield monitoring and VRT fertilizer application equipment. All aspects of precision involving VRT water application, VRT insect and pest control can be applied to horticultural crops. The critical factor would be cost effectiveness. A Management Information System (MIS) was developed for coconut and cassava plantations comprising monthly weather data, harvest, cultural operations, pest and disease incident and treatment of pests data base. For precision farming of cassava, modified QUEFTS (Quantitative Evaluation of Fertility of Tropical Soils) model for fertilizer calculation, leaf colour chart (LCC) for real time N management and Chlorophyll Meter (SPAD-502) have been standardized. The Site Specific Nutrient Management (SSNM) approach compared to Farmers' practice gave good response. Precision farming application of fertilizers has been proved profitable than the recommendations based on package of practices. About 17 Precision Farming Development Centers (PFDC) have been established in different agro-climatic regions under NHM. Activities like green house construction, mulching, shade net and plastic tunnels are also being promoted and 195.2 ha have been already brought under protected cultivation. The crops where some of the components of precision farming have been practised are banana, grape, pomegranate, capsicum, tomato, chilli, cashew and selected flowers. The dissemination of technology on hi-tech horticulture and precision farming has to be addressed through training programs for the benefit of farmers as well as field functionaries.

2.9 RESEARCH AND DEVELOPMENT INITIATIVES

The development of precision farming as a package has received little attention, although some aspects have been developed. The precision farming has not been taken fully, but partially taken up for research under the National Agriculture System of ICAR through All India Co-ordinated Research Project on Water Management, Use of plastics in agriculture. The protected cultivation of horticultural crops has also received attention. In order to meet the increasing demand of horticultural produce for the domestic and export demand, the precision farming has been also considered important. Precision Farming Development Centre of Tamil Nadu Agriculture University and TANHODA, Chennai, conducted the training at the Agriculture Research Station, Bhavani Sagar. There was also a hands-on training as part of the programme. The initiative for promoting the precision farming concepts in India commenced in the year 2001 by way of re-designating 17 Plasticulture Development Centers into Precision Farming Development Centers (PFDC). Depending upon the location they have been assigned different crops to work on precision farming models. Besides, the Government of India have launched Hi-tech Horticulture and Precision Farming during the tenth plan period. The PFDC has been working on drip irrigation, sprinkler irrigation, fertigation, *in-situ* moisture conservation, anti-hail nets, anti-bird nets, shade nets, protected cultivation of fruits, vegetables and floricultural crops and LDPE lined water storage tanks. Under this centre drip irrigation studies are being undertaken on various commercially important fruits, vegetables and floricultural crops including crop geometry studies in vegetables with a view to minimize the cost of installation of drip irrigation. Recently, microirrigation mission has also been proposed. Under protected cultivation, studies on cultivation, mulching and propagation of fruit and vegetable crops under polyhouse condition, shading in ornamental crops and hail protection in fruit crops have been undertaken. The centre has come out with specific recommendations on drip irrigation and plastic mulching in peach, apricot plum, pea and tomato; crop geometry in pea, tomato, capsicum and cauliflower; fertigation in pea, tomato, apricot and chrysanthemum; propagation of apple kiwi fruit and carnation and varietal performance of tomato, cucumber and spinach under polyhouse; shading in lilium and carnation; anti-hail nets in peach, apricot and plum and cost effective/affordable rain water harvesting technology through LDPE lined water storage tanks. Precision Farming Development Centre has made 24 recommendations in crop geometry, plastic mulching, micro-irrigation, protected cultivation and anti-hail nets for commercially important fruits, vegetables and flower crops.

2.10 CONCLUSION

Precision farming basically integrates information technology with application system to enhance crop productivity with saving resources including time. The enabling technology or precision farming including computers, geographic information system, global positioning system, sensors and application control. It is a proven technology in many cases of horticultural crops such as banana, potato, tomato, capsicums, cucumber, gerbera and orchids. But in India, it has yet to come up at large scale. Some of the obvious reasons identified are small to medium size land holdings, variations in cropping systems, high cost involved in system, lack of technology and expertise in related areas and lacks for data on various aspects like soil characteristics, topographic features, crop requirements in terms of consumptive use and nutritional requirements. Moreover, the success depends largely on quick availability of new technology and further its adoption in the precision farming system. However, the application of precision farming as a package in the farmers field has received little attention, although some aspects has been practised. It is believed that once the system is adopted massively, there will be efficient utilization of resources per unit of time and area and further high production and productivity shall be achieved with maximum returns without polluting the environment. There is need to provide technical back up support to farmers to develop models for each horticulture crop, which can be replicated on a large scale.

2.11 REFERENCES

Gary T. Roberson, (2000). Precision Agriculture Technology for Horticulture Crop Production, *Horticulture Technology*, July-September, pp. 448-451.

Singh, H.P., Samuel. J.C and Singh, G.(2002). Precision farming- concepts and approaches. *Souvenir*, National seminar-cum workshop on hitech horticulture and precision farming, July 26-28, 2002 CISH, Lucknow.

Joan R. Davinport and Mary J. Hahendrof (2000). *Using site-specific Approaches to Advance Potato Management in Irrigated System*, pp. 452-457.

Pierre C. Robert (2000). Site-Specific Management for the Twenty first Century. *Horticultural Technology*, July-September, pp. 44-447.

❑❑❑

Chapter – 3

Precision Farming in Horticulture : A Scenario

A. V. Barad, Virendra Singh, L. K. Dashora, J. Shankara Swamy and Rita Markam

3.1 INTRODUCTION

Today's technological advancement has reached a level where a farmer can have access to information and tools to manage mechanized field operations. They can now measure, evaluate and deal variability within field (e.g. soil fertility, water availability and yield) that was known to exist previously but was not manageable, to his advantage. The ability to handle the variation in productivity in the field and maximize financial return, reduce waste and minimize impact on environment has always been objective of an enterprising farmer in horticultural sector, especially those who advocate sound horticultural practices. Since precision farming proposes to prescribe strategic farm management decision in the field operations to help to serve this purpose through mechanization, sensing and communication technology, automate data collection, documentation and utilization of this information. Such an approach in horticultural production management gives rise to what is now termed as precision horticulture, soil specific crop management (SSCM), spatially variable crop production (SVCP), smart farming (SF) *etc.*

3.2 PRECISION FARMING IN HORTICULTURE

Precision farming in horticulture involves the application of technologies to manage spatial and temporal variability associated with all aspects of horticultural production for improving crop performance and environment quality by using satellite, sensors and field or thematic maps. Precision farming calls for efficient management of resources through location-specific interventions. It encompasses a variety of interventions such as micro irrigation, fertigation and mulching for *in-situ* moisture conservation, soil and leaf nutrient based fertilizen management, protected cultivation and organic farming, *etc.*

3.2.1 Precision farming concept

Better yield management decisions that will lead to higher productivity, high profit by reduced cost of cultivation and reduced environmental impact because of the reduced the input addition of to the soil.

3.2.2 Objectives of precision farming

Increased production efficiency with improved product quality by more efficient chemical use, energy conservation and protection of soil and water.

3.2.3 Need for precision farming in india

The status of Indian environment shows that about 182 million ha of the country's total geographical area of 328.7 million ha is affected by land degradation. In this context, there is need to convert this green revolution into an evergreen revaluation. Since precision farming proposes better management practices, it can help to serve this purpose.

By catering to this variability, called precision farming, one can improve the productivity or reduce the cost of production and diminish the chance of environmental degradation caused by excess use of inputs.

3.2.4 Components of precision farming

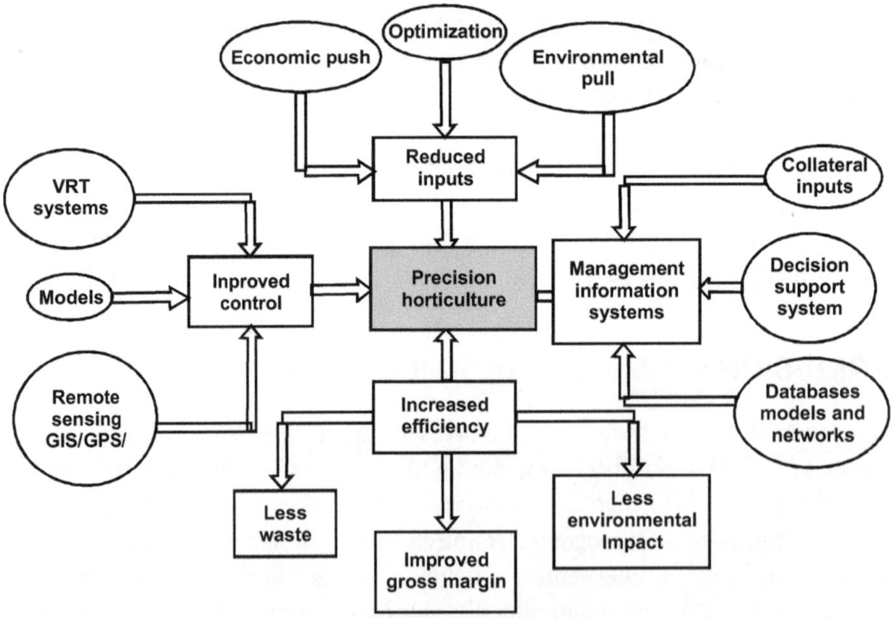

- Remote sensing (RS)
- Geographical Information System (GIS)
- Global Positioning System (GPS)
- Variable Rate Applicator (VRA)

3.2.4.1 Remote sensing

It is a group of techniques used to collect information about an object or area without actually being in contact with that object or area. This is for data acquisition of the farms to find the soil, vegetation and other parameters that are amenable for remote sensing. Remote sensing techniques play an important role in precision farming by providing continuous acquired data of agricultural crops. Remote sensors image vegetation, which is growing on different soil types with different water availability, substrate, impact of cultivation, and relief etc. These differences influence the state of the plants and cause heterogeneous regions within single fields. Hence, the heterogeneous vegetation acts as an interface between soil and remote sensing information, because vegetation parameters describing the state of the plants can be deduced from remote sensing imagery. The analysis of the variability occurring within the field was carried out by measuring soil and plant parameters through conventional methods as well as through spectral techniques using ground truth Spectroradiometer (350-1800 nm) and satellite data. Moran *et al.* (1997) reviewed potentials of remote sensing in providing information required for precision farming for Indian condition.

- Process of Remote Sensing
  1. Aerial photography
  2. Satellite imaging
- Aerial Photography
  1. **Aerial Photography :** It can be done with modern aircraft, using photography or electro-optical sensors based on the differences in colour within the image.

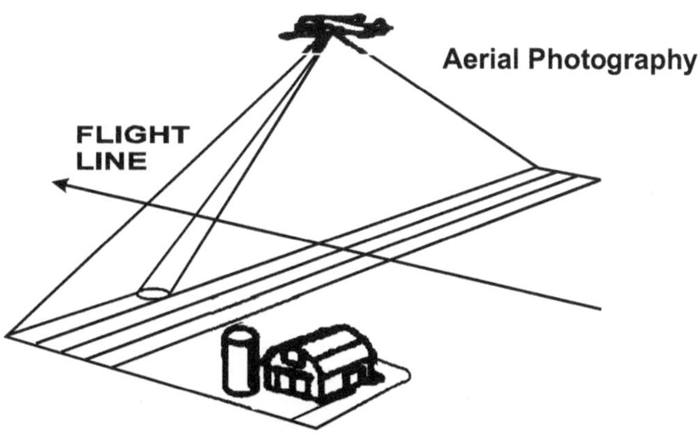

2. **Satellite Imaging :** This system uses satellites in space to gather detailed information on an area.

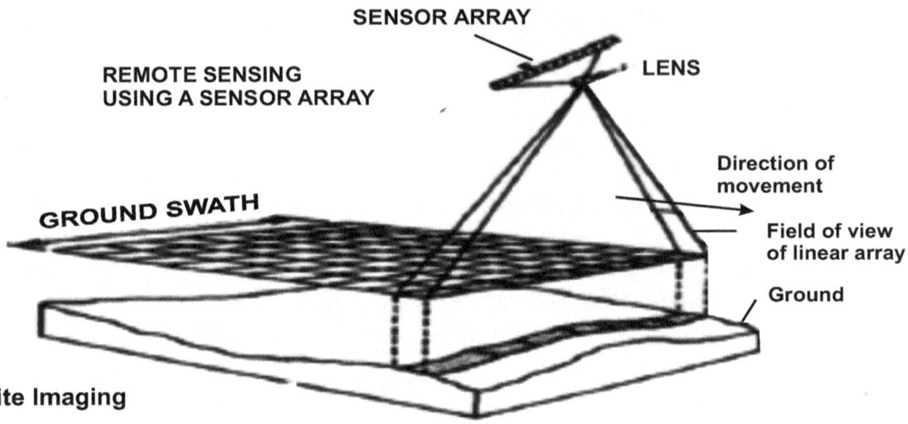

Satellite Imaging

- Use of Remote Sensing in Precision Farming
    - Pre harvest crop acreage estimation
    - Production estimation
    - Assessing and monitoring of water resources
    - Intensification and expansion of horticultural activities
    - To detect soil related variables and disease and pest incidence

The launch of resource sat in future will be a boon to monitor and assess the onion condition in Bhavnagar district of Gujarat state.

3.2.4.2 Geographic Information System (GIS)

Geographic information system is computer based management system used for computation, storage, retrieval, analysis and display of spatial data in the form of a map.

GIS contributes significantly to Precision Farming

- It contains base maps like topography, soil type, nutrient level, soil moisture, pH, fertility, weed and pest intensity maps etc.
- It can integrate all types of information and interface with other decision support tools.
- So these maps and information are used for application of recommended rates of nutrients or pesticides.

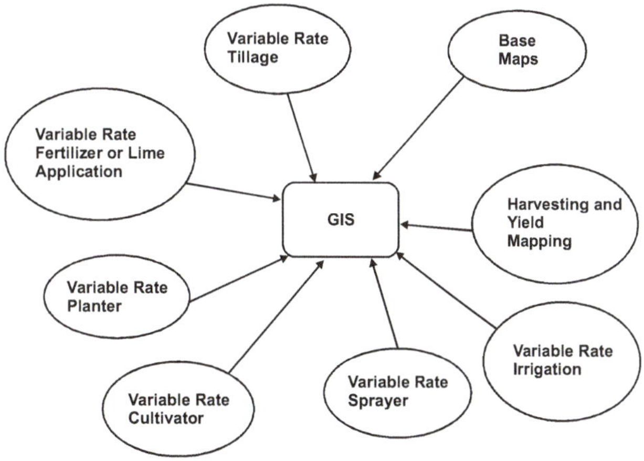

3.2.4.3 The Global Positioning System (GPS)

The Global Positioning System (GPS) is a location system based on a constellation of about 24 satellites orbiting the earth at altitudes of approximately 11,000 miles which provides accurate positioning system necessary for field implementation of variable rate technology. GPS receivers are designed for guidance in positioning of equipments (variable rate applicator, yield monitor, etc.).

Use of global positionong system in Precision farming

- Controlled application of inputs by equipments
- Identification the precise location of farm equipment within inches
- One can monitor and record the yield data of the field
- Fertilizers and pesticides can be prescribed according to the soil properties soil conditions and
- Tillage adjustment can be made as one finds various conditions

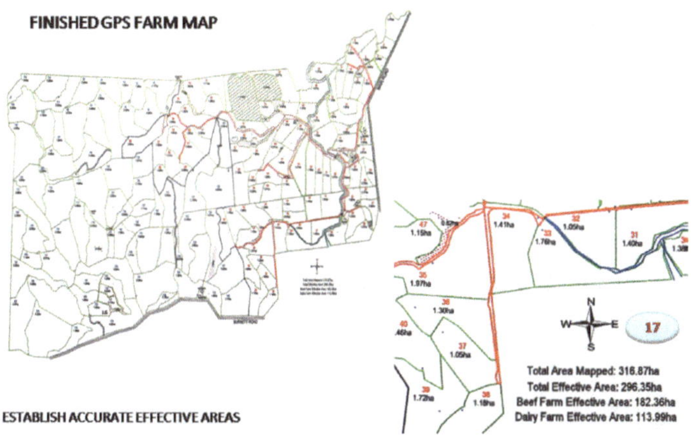

3.2.4.4 Variable rate technology

Variable rate technology is the advance component of precision farming technologies. A GPS receiver is mounted on the truck so field is easily recognized. An in-vehicle computer which contains input recommendation map, provide suitable input by comparing positional information received from GPS receiver.

Thus, mapping and analysis of within field variability is an essential input for precision crop management. Thus, precision farming involves acquiring the variations in crop or soil properties, mapping, analyzing the variations and adopting the suitable management techniques to maximize the yield. Farmers have been applying fertilizers based on recommendations emanating from research and field trials under specific agro-climatic conditions, which have been extrapolated to a regional level. Since soil nutrient characteristics vary not only between regions and between farms but also from plot to plot (Ladha *et al.*, 2000), and within a field or plot, there is a need to take into account such variability while applying fertilizers to a particular crop. Consideration of in-field/plot variations in soil fertility and crop conditions and matching the agricultural inputs like seed, fertilizer, irrigation, insecticide, pesticide, etc. in order to optimize the input or maximizing the crop yield from a given quantum of input, is referred to as precision farming or precision agriculture or precision crop management. The information for variability map can be obtained from soil tests for nutrient availability, yield monitors for crop yield, soil samples for organic matter content, information in soil maps, or ground conductivity meters for soil moisture (Mulla, 1997). Generally, the fields are manually sampled along a regular grid and the analyzed results of the samples are interpolated using geostatistical techniques. These techniques are time consuming, labour intensive and in many cases destructive especially, for agricultural situation in India. With small size of landholdings and low income of farmers, the adoption of this methodology in its present form is not feasible. Various workers (Hanson *et al.*, 1995, Moran *et al.*, 1997) have shown the advantages of using remote sensing technology to obtain spatially and temporally variable information for precision farming.

3.2.4.5 Variable rate applicator

The variable rate applicator has three components:

1. Control computer
2. Locator and
3. Actuator

The control computer coordinates the field operation. It has a map of desired activity as a function of geographic location. It receives the equipment's current location from the locator, which has a GPS in it, and decides what to do based upon the map in its memory or data storage. It then issues the command to the actuator, which does the input application (Ravi and Jagadeesha, 2002).

PRECISION FARMING IMPLEMENTS

Soil Sampler

Variable-Rate N Applicator

Soil Conductivity Measurements

Chisel plough

Variable-Rate Seeding

Laser leveler

Yield Monitoring Implement

Variable Rate Applicator

3.3 MAJOR CONCEPT OF PRECISION FARMING IN HORTICULTURE

Better management decisions that lead to higher productivity, high profit by reduced cost of cultivation and reduced environmental impact because of the reduced addition of the input to the soil.

Precision farming calls for efficient management of resources through location-specific hi-tech horticulture which encompasses a variety of interventions such as micro irrigation, fertigation, protected/greenhouse cultivation, soil and leaf nutrient based fertilizer management, mulching for *in-situ* moisture conservation, micro propagation, use of biofertilizers, vermiculture, and high-tech mechanization etc.

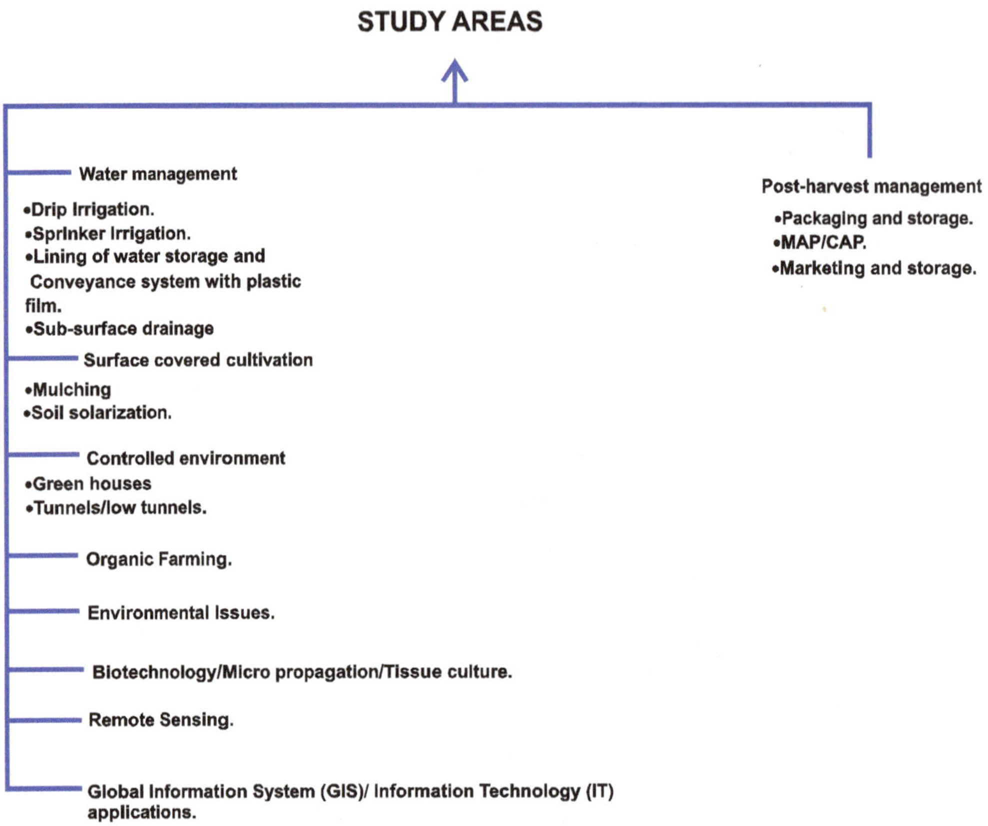

Precision farming development centers are established in different agro climatic Regions of India to carryout the research on above mentioned areas.

3.4 PRECISION FARMING VERSUS TRADITIONAL FARMING

Sr. No.	PRECISION FARMING	TRADITIONAL FARMING
1.	Farm field is broken into"management zones"	Whole field approach where field is treated as a homogeneous area
2.	Management decisions are based on requirements of each zone	Decisions are based on field averages
3	Precision farming tools (*e.g.* GPS/GIS) are used to control zone	Inputs are applied uniformly across a field

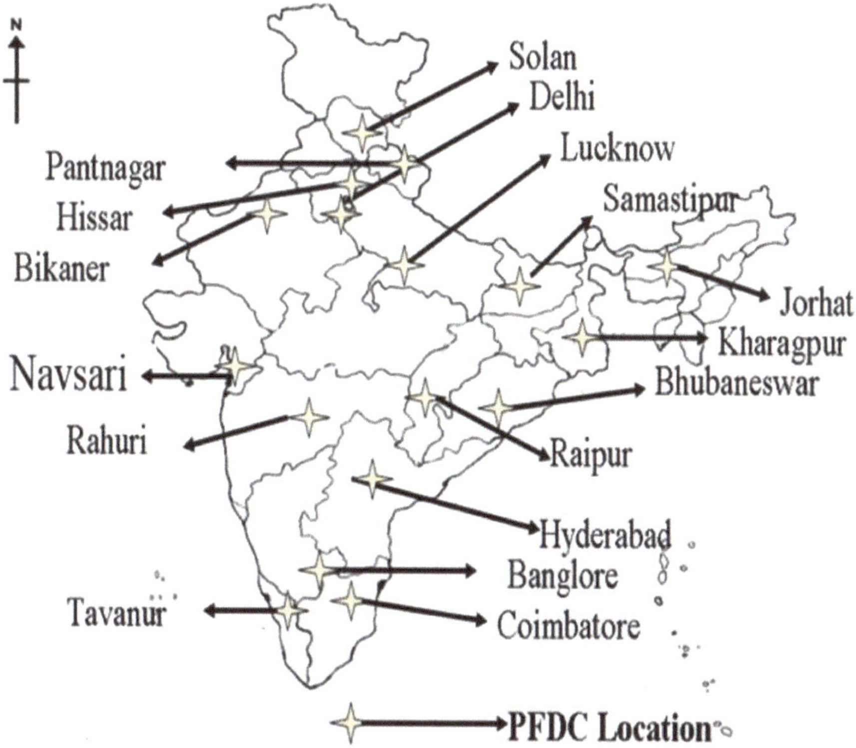

3.5 LIMITATIONS IN IMPLEMENTATIONS OF PRECISION FARMING UNDER INDIAN CONDITIONS

- Small farms size
- Heterogeneity of cropping systems
- High cost of obtaining site-specific data
- Complexity of tools and techniques requiring new skills
- Culture, attitude of farmers including resistance to adoption of new technologies and lack of awareness of environmental problems
- Precision farming as new story to Indian farmers needs demonstrated impacts on yields
- Lack of local technical expertise
- High initial investment
- Uncertainty on returns from investments on new equipment and information management system
- Knowledge and technological gaps

3.6 STEPS TO BE TAKEN FOR IMPLEMENTATION OF PRECISION HORTICULTURE IN INDIA

In the present existent situation, the potential of precision horticulture in India is limited by the lack of appropriate measurement and analysis techniques for agronomically important factors. High accuracy sensing and data management tools must be developed and validated to support both research and production. The limitation in data quality/availability has become a major obstacle in the demonstration and adoption of the precision technologies. The adoption of precision horticulture needs combined efforts on behalf of scientists, farmers and the government. The following methodology could be adopted in order to operationalise precision horticulture in the country :

1. Creation of multidisciplinary teams involving agricultural scientists from various fields, engineers, manufacturers and economists to study the overall scope of precision horticulture
2. Formation of farmer's co-operatives, since many of the precision horticultural tools are costly (GIS, GPS, RS, etc.)
3. Government legislation restraining farmers using indiscriminately farm inputs and thereby causing ecological/environmental imbalance would induce the farmer to go for alternative approach
4. Pilot study should be conducted on farmer's field to show the results of precision horticulture implementation
5. Creating awareness amongst farmers about consequences of applying imbalanced doses of farm inputs like irrigation, fertilizers, insecticides and pesticides in horticultural crops

3.7 CONCLUSION

Horticulture has continued to be the cornerstone of Indian economy. In the years to come, precision horticulture may help the Indian farmers to harvest the fruits of frontier technologies without compromising with the quality of land and produce. The adoption of such a novel technique would trigger a techno-golden revolution in India which the need of the hour.The major focus would be on technology development, adoption and its dissemination for all elements of the programme. In the overall perspective, with the introduction of innovative technologies, horticulture sector is expected to achieve a vertical growth

3.8 REFERENCES

Hanson, L.D., Robert, P.C. and Bauer, M., (1995). Mapping wild Oats infestation using digital imagery for site-specific management. In Proc. Site-Specific Management for Agricultural System. 227-230, March 1994, Minneapolis, MN, ASA-CSA-SSSA, Madison, WI, pp.495-503.

Ladha, J. K., Fischer, A. K., Hossain, M., Hobbs, P. R., Hardy, B., editors. (2000). Improving the productivity and sustainability of rice-wheat systems of the Indo-gangetic plains: a systematic synthesis of NARS-IRRI partnership research. IRRI Discussion Paper Series No.40.Makati City (Philippines): International Rice Research Institute.31p.

Mulla, D.J., (1997). Geostatistics, remote sensing and precision farming. In *Precision Agriculture: Spatial and Temporal Variability of Environmental Quality* (Eds. J.V. Lake, G.R.Bock and J.A. Goode) John Wiley & Sons, New York, pp.100-119.

Moran, M.S., Inoue, Y. and Barnes, E.M., (1997). Opportunities and limitations for image based remote sensing in precision crop management. *Remote Sensing Env.*, 61, pp.319-346

Ravi, N. and Jagadeesha, C.J., (2002) Precision Agriculture, Training course on Remote Sensing and GIS Applications in Agriculture, May 27th –7th June, 2002, RRSSC- Bangalore, pp: 225-228.

Chapter – 4

Precision Farming in Fruit Crops in India : Status and Prospects

M. K. Kaul

4.1 INTRODUCTION

Ever since the man appeared on the earth, he has been harnessing the natural resources to meet his basic requirements. Reference to soil, water and air as basic resources, their management and means to keep them pure are mentioned in the *Vedas*, *Upanishads* and in ancient *Hindu* literature. Agriculture began sometime during 10000 BC and during 1930 AD use of chemical came into vogue (Sarswat *et al.*, 2003). The phenomenal increase in population of both man and animal in the last century and fast growing industrialization and urbanization in last few decades have overstrained the natural resource base, which are getting degraded much faster than ever before. Thus, the attention of whole world is focused on how to increase production to feed the burgeoning population and the question uppermost in every ones mind is "how can we produce enough food without damage to the natural resource base, in a sustainable manner?" Precision farming is a promise in this regard. It appears to reduce risk of crop failure (De Boer, 1997).

4.2 ISSUES AND DEBATE

Despite all the natural advantages, India's productivity of food grains per hectare is no more than three-fourth of the world average and less than half of that in agriculturally advanced countries, per capita food grain availability even after the Green Revolution, has been less than two thirds of the world average. More than two third of the population lives in states that are still food-deficit. This requires transport of lac of tonnes of food grain, involving high costs and pilferage. Our effort should have been to make all the states self-sufficient with respect to food grains and if some disturbances occur due to unnoticed natural calamities the nation must be in an ever ready position to mitigate such challenging tasks. The status of Indian environment shows that, in India, about 182 million ha of the country's total geographical area of 328.7 million ha is affected by land degradation of this, 141.33 million ha are due to water erosion, 11.50 million ha due to wind erosion and 12.63 and 13.24 million ha due to water logging and chemical deterioration (Signalization and loss of nutrients) respectively. According to Human Development Report 2003, the percentage of the undernourished population in India, which was 21 per cent a few years ago, has now reached to 24 per cent.

To achieve the required growth will not be easy as some of the existing production systems are based on unsustainable use of the resources. The signs of fatigue in the natural resources have already appeared with serious concern to the planners, decision-makers and researchers alike.

4.3 THE BASIC COMPONENTS OF PRECISION FARMING

Precision farming would involve the measurement and understanding of variability over time and space. The system would use the information generated through surveys to manage this variability by matching inputs. This system must provide for the measurement and recording of the efficiency of the site specific practices in order to assess value on and off the farm. The precision farming is technology enabled, information based and decision focused. The enabling technology of precision farming can be grouped into five major categories: Computers, Global Positioning system (GPS), Geographic Information System (GIS), Sensors and application control. Availability of contiguous blocks of mono-crops and equipments needed for survey, recording and analysis on near real time basis has made the precision farming technology a reality in developed countries, where farms holdings is large and heavily equipment dependent.

Precision farming in Indian context is still in its infancy. A vast amount of data on soil characteristics, climatic parameters, topographic features, crop requirement in terms of consumptive use and nutritional requirements have been generated and instruments needed for recording are also available. Technology for fertigation has also been generated in specific crops. However, precision farming as package has received little attention at farmers field level, leaving besides a few aspects of precision farming. Use of fertigation in grape on petiole analysis coupled with bunch management has increased the yield and quality. There is an urgent need to develop a package based on knowledge of soil environment and crop needs to enhance the efficiency of inputs in given time frame.

The basic steps in precision farming are :

i) Assessing variation

ii) Managing variation and

iii) Evaluation

4.4 TECHNOLOGY TRANSITION

Precision agriculture is dependent on the existence of variability in either or both product quantity and quality. If this variability does not exist then a uniform management system is both the cheapest and the most effective management strategy and precision farming is redundant. Thus, in precision farming, "Variability of production and quality equals opportunity". Having said this, the nature of the variation is also important in determining the potential for precision farming in a system. For example the magnitude of the variability may be too small to be economically feasible to manage. Alternatively, the variability may be highly randomized across the production system making it impossible to manage with current technology. Finally the variability may be due to a constraint that is not manageable. Thus, the implementation of precision farming is limited by the ability of current variable rate technology (VRTmachinery/ technology that allows for differential management of a production system) to cope up with the highly variable sites and the economic inability to produce returns from sites with low variability using precision farming (VRT). Due to these constraints precision farming is at present operating on a zonal rather than a completely site-specific basis.

4.5 PRESENT SCENARIO

Before dwelling upon the precision farming, it is necessary to look into current scenario of horticulture which exhibited the growth rate of 6.9 per cent during the decade. The past interventions have proved beyond doubt that horticulture is one of the best options for improving land, generating employment, improving economic condition of farmers and the nutritional security. The programmes initiated during IX plan and continued in X plan with two new initiatives, including Hi-tech horticulture through precision farming. Precision farming is the application of technologies and principles to manage spatial and temporal variabilities associated with all aspects of horticultueal production for improving production, performance and environmental quality. The success in precision farming depends on the accurate assessment of the resources through location specific hi-tech interventions. Hi-tech horticulture encompasses a variety of interventions such as micro-irrigation, fertigation, protected/ green house cultivation, soil and leaf nutrient based fertilizer management, micopropagation, biology for germplasm, genetically modified crop, mulching for *in-situ* moisture conservation, use of bio-fertilizers, vermiculture, high density planting, hi-tech mechanization, green food, soil less culture, biological control etc. Utilization of these interventions put together with the aim to achieve higher output in given time period leads to precision farming that is mainly knowledge driven.

A centrally sponsored scheme on use of plastics in agriculture with an outlay of Rs. 250 crore was launched in VIII plan. Due to the focused attention and support, there has been a larger adoption of technologies and achievements have been higher than the targets (Table 1). On an average 4500 ha are being brought under drip irrigation annually under horticultural crops, as a result India has now emerged as one of the leading countries in using micro irrigation technology. In Sriganganagar, region of Rajasthan nearly 50 per cent of the total citrus area is under drip irrigation. However, the regional imbalance in use of drip is seen.

Table 1 : Coverage of Area under Drip irrigation

Period	Coverage (ha)	
	Target	**Achievement**
VIII Plan	107044	128444
!997-98	30425	45151
1998-99	43151	53017
1999-2000	38833	45676
2000-01	6476	13422
Total	**226200**	**285710**

4.6 PLASTIC MULCHING

Plastic mulching has proved its efficiency in moisture conservation which ultimately leads to increase in quality and yield. For promotion of plastic mulching 50 per cent subsid (maximum upto Rs. 7000/ per ha.) was provided by the Govt. of India. An area of more than 4000 ha has been covered under plastic mulching. Though the adoption is only for limited number of crops and needs to be promoted and extended in other crops, too.

4.7 FERTIGATION

It aims at maximizing profit of growers. The achievement depends upon factors such as input cost, crop growth, yield and quality of produce (Bar - Yosef, 1999). It is the technique of supplying dissolved fertilizers to crop through an irrigation system (Haynes, 1985.). For intensive and economical crop production fertigation has a potential. Fertigation not only provides essential elements directly to the active root zone but also minimizes the losses of expensive fertilizers. The experiments have clearly shown that fertigation saves the fertilizers up to 25-30 per cent and also saves time and labour. It has showed 90 - 95 per cent efficiency in N, 40 - 45 per cent in P and 70 - 80 per cent in K in comparison to 50 - 70 per cent, 10 - 20 per cent and < 50 per cent in surface flow system, respectively (Anand *et al.*, 2008). Further, it provides the opportunity to select the type of fertilizer most suited to a particular type of soil. Thus, soil pH and electric conductivity can be reduced/ increased accordingly to a level where the availability of nutrients including mico-nutrients can be increased. This has already been experienced by farmers in grape, pomegranate and banana. Fertigation is more ideally suited to hi-tech horticultural production system as it not only involves the efficient use of water and nutrients but also exploits the synergism of their simultaneous availability to plants. Significant yield enhancement can be expected only when drip is used along with fertilizer. In the past, the non availability of water soluble fertilizers was a limiting factor in its use but now a large range of fertilizers with varing range are available. Of course, such fertilizers are quite costly.

4.8 GREENHOUSE

Greenhouse provides protection to crop and create environment for growing crops out of season. Certain crops like grape, pomegranate, peaches, strawberry and lime etc. can be taken up in the areas where monsoons are limiting factor. Of course, its use has mostly been limited to vegetables and flowers. Package of practices for growing of different crops under greenhouse is required to be standardized.

4.9 BIO-TECHNOLOGY

The annual demand of *in-vitro* plants continue to grow at the rate of 15-20 per cent. The Govt. of India identified micro-propagation of plants as an industrial activity under the (D & R) Act of 1951, made effective in 1991 with many subsidies and incentives. Under this scheme, the tissue culture units established in the public sector has done a lot of work but more so of demonstrative work. Large scale multiplication has not been a reality except in banana. However, these interventions have created demand for *in-vitro* plants of large no. of crops. Recently under a scheme, Rajasthan state govt. has established two tissue cultural laboratories for developing virus free citus plans. One laboratory at the College of Horticulture,Jhalawar for virus free Nagpur *santra* and another at ARS, Sriganganagar for virus free Kinnow plants. In addition to it, large scale plantation of tissue culturally developed plants of Datepalm have been made in the field and to develop tissue culture developed Datepalm plants, a laboratory with the help of state govt. has been established at Jodhpur under private sector. Convincingly, *in-vitro* propagated plants in banana and strawberry *etc.* have become a commercial reality.

4.10 QUALITY SEED AND NURSERY PLANT

It determines the effectiveness of inputs and assumes significance when input investment is increasing. In addition to higher yield potential the cultivar must possess nutritional quality, resistance to diseases and proper shelf life. Thus, there is a need to have paradigm shift from production to productivity and the profitability. The solution lies in adopting newer technology to boost production under eco-friendly atmosphere. Propagation of plants through tissue culture, including meristem culture and molecular indexing of disease are of immense use in making available healthy propagules. Micro-propagation also has an added advantage of conservation and exchange of national and international germplasm, thus obviating quarantine related problems.

4.11 PROSPECTS

Precision farming, though in many cases a proven technology, is still mostly restricted to developed (American and European) countries. Except for a few there is not much literature to show the scope of its implementation in India. It is felt that, one of the major problems is the small field size. In India more than 57.8 per cent of operational holdings have size less than 1 ha. However, in the major agricultural states of Punjab, Rajastan, Haryana and Gujarat there are more than 20 per cent of agricultural lands have operational holding size of more than 4 ha. These are individual field sizes. However, when we consider contiguous field with same crop (mostly under similar management practices) the field (rather simulated field) sizes are large. Using aerial data, if has been found that in Patiala district of Punjab, more than 50 per cent of contiguous field sizes are larger than 15ha. These contiguous fields can be considered as a single field for the purpose of implementation of precision farming. However many horticultural crops in India, which are high profit making crops, offer wide scope for precision farming.

4.12 MISCONCEPTIONS ABOUT PRECISION AGRICULTURE

There are several mistaken preconceptions about precision agriculture.

4.12.1 Precision agriculture is a cropping rather than an agricultural concept

This is due to the fact that the cropping systems, in particular broad-acre cropping, is the face and driving force of Precision technology. However precision farming concepts are applicable to all agricultural sectors from animals to fisheries to forestry.

4.12.2 Precision agriculture in cropping equals yield mapping

Yield mapping is a crucial step and the wealth of information farmers are able to obtain from a yield map makes them very valuable. However these are only a stepping-stone in a precision farming management system. The bigger agronomic hurdle lies in retrieving the information in the yield map and using it to improve the production system.

4.12.3 Precision agriculture equals sustainable agriculture

Precision agriculture is a tool to make agriculture more sustainable. Precision farming aims at maximum production efficiency with minimum environmental impact. Precision farming will not fix problems such as erosion and salinity by itself, although, it will help to reduce such risk.

4.13 CONSTRAINTS

There are many constraints in the adoption of precision farming in developing countries in general and in India in particular. Some are common to those in other regions but the others are specific to Indian conditions which are as follows.

- Culture and perceptions of the users
- Small farm size
- Lack of success stories

- Heterogeneity of cropping systems and market imperfections
- Land ownership, infrastructure and institutional constraints
- Lack of local technical expertise
- Knowledge and technical gaps
- Data availability, quality and costs

4.14 CONCLUSION

Precision farming in many developing countries including India is in its infancy but there are numerous opportunities for adoption. It is believed that progressive Indian farmers, with guidance from the public and private sectors and agricultural associations, will adopt it on a limited scale as the technology shows potential for raising yields and economic or returns on fields with significant variability, and for minimizing environmental degradation. Although it is recognized that agriculture is a major polluter of the environment in many developing countries, farmers will not adopt precision farming unless it brings in more or at least similar profit as compared to traditional practice. The support from governments and the private sector during the initial stages of adoption is, therefore vital. It is necessary that not all elements of precision farming are relevant for each and every farm. For instance;

- Not all farms are suitable to implement precision farming. Some growers are likely to adopt it partially, adopting certain elements but not others
- Precision farming cannot be convincing if only environmental benefits are emphasized. On the other hand, its adoption would be improved if it can be shown to reduce the risk
- Do not oversell the technologies without providing adequate product support
- The adoption of precision farming also depends on product reliability, the support provided by manufacturers and the ability to show the benefits
- Effective coordination among the public and private sectors and growers is, therefore, essential for implementing new strategies to achieve fruitful success

4.15 REFERENCES

Anand, M.; Ponnusamy, V.; Vijayaraghavan, V. and Sumathi, T. (2008) Application of fertigation in horticultural crops. In : *Innovative Horticulture* (Eds. Arun Kumar *et al.*), NIPA, New Delhi, pp. 75-103

Bar - Yosef, B. (1999) Advances in fertigation. *Advances in Agronomy*, 65 : 1-69.

De Boer, J.L. (1997) Precision farming boosts risk management not profits. Ag Answers, Agric. Communication Service, Purdue University, USA, p.p.

Haynes, R.J. (1985) Priniciples of fertilizer use for trickle irrigated crops. *Fert. Res.*, 6 : 235 - 255.

Saraswat, V.N.; Bhan, B.M. and Yaduraju, N.T. (2003) Weed Management, ICAR, New Delhi, p.8.

❑❑❑

Chapter – 5

Fruit Based Diversified Cropping System for Arid Region

O. P. Awasthi, I. S. Singh and M. K. Verma

5.1 INTRODUCTION

Drought and rampant poverty, the double plague of arid region, present a scenario of misery and insecurity across the region where agriculture is the main livelihood and chief source for subsistence of the rural people. The conditions of low and erratic rainfall, extreme temperature conditions, high evapo-transpiration, scarce water resources, infertile and saline soils with ill-defined profile development threaten sustainable agriculture. Sustainable productivity is difficult to achieve by monoculture of either seasonal or forage crops. Thus cropping system approach for agricultural production is of far greater significance in these regions. Traditionally efforts have been made by the native people, obviously to achieve sustainable production ensuring ecological stability, to grow several species together, e.g. pearl millet, moth bean, cluster bean and gram between the natural stands of *khejri* (*Prosopis cineraria*), *bordi* (*Zizyphus rotundifolia*), *lasoda* (*Cordia myxa*), *pilu* (*Salvadora oleoides*), and *ker* (*Capparis decidua*). The perennial species also provide fodder for sheep and goats during scarcity periods.

Fruit based cropping system is now considered to be the most ideal strategy to provide food, nutrition and income security to the people (Chundawat, 1993, Chadha, 2002, Awasthi and Pareek, 2008). Integration of annual crops with fruit trees yields multiple outputs that ensure production and income generation (Randhawa, 1990 and Osman, 2003) in a sustainable manner. Natural resources (land, solar radiation, water, soil etc.) as well as socio-economic inputs (labour, credit, power, and market infrastructure) are efficiently utilized. Crop productivity under tree canopy is enhanced due to improved soil fertility (Young, 1989) and ameliorative influence of shade by reducing understorey temperature and evapotranspiration (Bunderson *et al.*, 1990). Incorporation of fruit trees into the cropping system is more remunerative. It increases the resilience of the system over time. It enables to i) maximize system productivity on annual basis, ii) utilize resources with high efficiency through due consideration of various interactions and direct, residual and cumulative effects occurring in soil-plant-atmosphere continuum, iii) intensify input use *vis-à-vis* quality of environment, and iv) impart sustainability of farm resources and environment in long term perspective. The type of horticulture based cropping system in arid region has been discussed in this paper considering these issues.

5.2 SUITABLE CROPS

5.2.1 Overstorey crops

The fruit trees selected should possess the following characters:

- Xeric characters such as deep tap root system, high bound water in tissues, sunken stomata, waxy coating and thick cuticle
- Ability to complete maximal vegetative and reproductive phase during the period of maximum moisture availability
- Root system and root growth of fruit tree species should be able to exploit different depth of soil layers that are different to those being tapped by the under and ground storey crops
- Branching habit should allow light penetration to the understorey crops
- Species should be strong coppicer and posses self-pruning properties or they should respond to pruning
- The rate of litter fall and litter decomposition should have positive effect on soil and fertility
- Resistance to pest and diseases

Several drought hardy fruit yielding tree species like *ker* (*Capparis decidua*), *pilu* (*Salvadora oleoides*), *lasora or goonda* (*Cordia dichotoma*), *goondi* (*Cordia gharaf*), *jharber* (*Ziziphus nummularia*), *bordi* (*Zizyphus rotundifolia*) and *ber* (*Zizyphus mauritiana*) are known to provide not only life support to the people living in the fragile ecosystem of arid region receiving an average annual rainfall of <300 mm but also provide their leaf fodder to the animals when other vegetation is scarce. The leaves of *khejri* (*Prosopis cineraria*) and *jharber* (*Z. nummularia*) contain about 75 per cent moisture even during the summer. Thus, animals foraging upon fresh leaves of these fruit trees can get considerable quantities of water embodied in these leaves. The fruit crops such as *aonla* (*Emblica officinalis*), pomegranate (*Punica granatum*), *bael* (*Aegle marmelos*), date palm (*Phoenix dactylifera*), *chironji* (*Buchanania lanzan*), tamarind (*Tamarindus indica*) etc. are grown successfully in rainfall zone of 350-500 mm or where irrigation facilities are available.

5.2.2 Ground storey crops

The selected crops should have the following characteristics.

- Crops should be selected according to their shade tolerance and amount of solar radiation available
- Should not grow too tall to compete with the perennial crop companion
- Should not be more susceptible than the main crop to the diseases they have in common
- Should not require harvesting or other operation that would damage the main crop or induce soil erosion or damage soil structure
- Should not have an economic life longer than the main crop
- Crops should be selected according to the soil type, rainfall pattern/ irrigation facilities and climatic conditions
- Desertic soils being low in N content, emphasis must be given to leguminous ground storey crops, which have nitrogen fixing capacity
- Crops to suit local market
- Processing facility and labour availability

5.2.2.1 Vegetables

Among the vegetable crops, brinjal (*Solanum melongena*), bottle gourd (*Lagenaria siceraria*), ridge gourd (*Luffa acutangula*), sponge gourd (*Luffa cylindrica*), *mateera* (*Citrullus lanatus*), round melon (*Citrullus lanatus* var. *fistulosus*), long melon (*Cucumis melo* var. *utillismus*), snap melon (*Cucumis melo* var momordica), *kachari* (*Cucumis callosus*), drumstick (*Moringa olerifera*), cluster bean (*Cyamopsis tetragonolobus*) and cowpea (*Vigna ungiculata*) are common.

5.2.2.2 Crops

Cereals and pulses constitute the major part of the diet and form major portion of calorie intake of people living in arid ecosystem. Pearl millet (*Pennisetum glaucum*) is the most important crop used as staple food by the inhabitants of *Thar* Desert. Plant stalks are fed to cattle or used for thatching purpose. There has been a rapid expansion in the area under wheat and groundnut during the last few years, particularly in command areas where irrigation facilities are available. *Kharif* pulses such as *moth* bean (*Vigna acontifolia*), *cluster bean* (*Cyamopsis tetragonolobus*), *urd* (*Phaseolus aureus*) and *chickpea* (*Cicer arietinum*) can survive under the extreme aridity. Rapeseed (*Brassica campestristoria*) and mustard (*Brassica junea*) are important oil seed crops.

5.2.2.3 Grasses

Sewan (*Lasiurus sindicus*), *anjan* (*Cenchrus ciliaris*), *dhaman* (*Cenchrus setigerus*) and *karad* (*Dichantimum annulatum*) grasses are naturally found in pasture land, *gochar*, *oran* and rangelands which flourish during monsoon season and require minimal care. *Sewan* is one of the important grasses of arid region and grows well in scarce rainfall zone upto 250 mm. *Anjan* and *Dhaman* have wide adaptability and grow in varying land forms and low to high rainfall zones. Halophytes like *Haloxylon salichornicum*, *Salsola baryosma* are preferred feed for camels and well adapted in saline soils (>20dsm^{1}).

5.2.2.4 Medicinal crops

Several medicinal plants grow in arid region, e.g. *senna* (*Cassia angustifolia*), *ashwagandha* (*Withania somnifera*), *guggal* (*Commifera wightii*), castor (*Ricinus communis*), *shankhpushpi* (*Evolvulous alsinoides*), *guarpatha* (*Aloe barbadensis*), *tumba* (*Citrullus colocynthis*), *dhatura* (*Datura sp.*) and *mehandi* (*Lawsonia inermis*).

5.2.2.5 Spice crops

Fenugreek (*Trigonella foenum-graceum*), Cumin (*Cuminum cyminum*), and Chilli (*Capsicum frutiscens*).

5.3 MAJOR CROPPING SYSTEMS

In practice, choice of crops and adoption of cropping system depend upon factors related to infrastructure, socio-economic factors, soil and climate, trade and marketing, all operating interactively at micro level. The economic returns or monetary gain per unit area and time is one of the major considerations for adoption of a cropping system. Based on the diversity of fruit, vegetable and other crops grown in arid region, the major cropping systems that emerge at the macro level are:

(i) Horti-Agri system

(ii) Horti-Pastoral system

(iii) Horti-Agri-Pastoral system

5.3.1 Horti-Agri system

This system provides improved nutrition and health standards to the farm family besides stability in income. Growing of vegetable crops, pearl millet, *moth* bean, cluster bean, and gram between *khejri*, *bordi*, lasoda, *pilu* and *ker* is a prevalent traditional Horti-Agri system in the arid region. Growing of fruit trees as an overstorey component and ground storey crops in association enforce the dual purpose of moisture and nutrient extraction in subsurface as well as above the soil surface. Singh (1980) reported that in arid condition of western Rajasthan about 25-40 per cent water penetrates into *murram* (hard layer of $CaCO_3$) sub-stratum during the rainfall or irrigation which is of no use for moderately rooted crops/grasses but inclusion of fruit trees under different alternate land use system due to their deep rooting characteristics, the water can be efficiently utilized by the fruit trees.

5.3.1.1 Ber based system

Growing of vegetable crops, pearl millet, *moth* bean, cluster bean, and gram between the trees of *Zizyphus nummularia* (*Jharber*), *Zizyphus rotundifolia* (*Bordi*) and *Z. mauritiana* is a prevalent traditional practice in the arid region. Crops like pearl millet, cluster bean and *moth* bean grow in association with *jharber* even on soil sediments in rocky plateaus near Jaisalmer and Jodhpur. Cluster bean has been found to be a good intercrop at all the locations. In agri-horticulture study involving ber plants and *mung* bean during subnormal years when rainfall was 51.3 per cent less than the long term average of 360 mm at Jodhpur, the yield of *mung* bean grown with *ber* was reduced by 44.2 per cent whereas under sole crop mung bean yield was reduced by 51 per cent. The inventory of the system shows that this agri-horticulture system can provide round the year supply of fodder for 5 goats/sheep/ha and fuel wood for family of 4 members besides efficient nutrient recycling and increase in economic stability (Faroda, *1998*). Gupta *et al.*, 2000 reported that three year old plantation of ber (*Z. mauritiana*) at a density of 400 plants/ha in association with green gram performed very well with a seasonal rainfall of 210mm. Intercropped green gram yielded only 160 Kg ha^{-1} as against 620 kg ha^{-1} from pure crop. The fruit yield from the intercropped system increased the net profit to Rs. 2,886. This shows that this Horti-agri system minimizes the risk in arid and semi arid regions and helps in imparting economic stability. This system is however, generally recommended for areas with rainfall more than 250-300mm.

Under hot arid eco-system of western Rajasthan (Bikaner), studies on Ber based Agri-Horti system was investigated by Saroj *et al.*, 2003. *In-situ* budded ber cv. Gola with three spacings (6x6m, 8x8m and 16x4m) as overstorey component and groundnut (*Arachis hypogia*), wheat (*Triticum aestivum*), cluster bean (*Cyamopsis tetragonolobus*), mustard (*Brassica juncea*) and indian aloe (*Aloe barbadensis*) as ground storey components were integrated into the system. During the initial stage of its establishment, there was no competition between groundstorey and overstorey components for resources. Instead, the inputs applied for the groundstorey crops enhanced the vegetative vigour of *ber* plants than sole plantation. Among the different crop combinations, the highest yield of *ber* was recorded with Indian aloe (13.55 q ha^{-1}) while with groundnut-wheat and cluster bean-mustard, the yield level was almost at par (9.62 and 9.80 q ha^{-1}) and minimum yield was recorded under sole plantation (8.09 q ha^{-1}). The average yield of ground storey crops over two years was recorded as 3.48, 50.97, 10.05, 11.49 and 20.48 q ha^{-1} in groundnut (dry pods), cluster bean (green pods), wheat and mustard (grain) and Indian aloe (green leaf pads) respectively. Conclusive results of the study indicated that cluster bean- mustard and Indian aloe can be integrated as a compatible ground storey component with *ber* as compared to ground nut-wheat, which is a dominant rotation in the irrigated hot arid ecosystem. To optimize productivity and profitability during pre-establishment phase of *ber* orchard under same agro eco-system, association of cluster bean-mustard and Indian aloe as a groundstorey component with *ber* cv. Gola have been recommended (Dhandar *et al., 2004*). These crops have been identified as low input and high returning crops giving net returns of 65, 802 and 26, 144 ha^{-1}, respectively.

Under arid and semi arid conditions, strong competition has however, been observed between ber trees and crops like wheat (Khan, 1993; Pateria *et al*., 2005) and chickpea (Khan, 1993). It appeared that the cost of growing *Zizyphus* trees was more than the value foregone by growing these crops. Thus such crops should not be grown in *ber* orchards. Cover cropping with *kulthi* (*Dolichos biflorus*) was found to increase water holding capacity of light soils as a result of increased organic carbon content in arid region in the tropics (Pareek and Nath, 1996).

5.3.1.2 Aonla-based system

Aonla is one of the important fruit crops of arid region. However, despite its adaptability in arid climate, productivity and economic return of the crop are affected by low temperature (-2°C) and frost (Anon., 2006) which cause economic loss to the growers. In order to mitigate the risk of total crop failure, suitable multistorey crop combinations were evaluated at Central Institute for Arid Horticulture during 2004-2008 in the interspace of *aonla* orchard which could generate extra income, improve productivity, ameliorate and improve ecological situations (Awasthi *et al*., 2009). *Mothbean* grown during kharif season was a common ground storey crop grown in rotation with *rabi* crops i.e., fenugreek, chickpea, mustard and cumin. Growth parameters in terms of plant height, stem girth, canopy spread and canopy volume of aonla was recorded to be significantly more with intercrops compared with its sole plantation. Higher grain and straw yield were recorded in *mothbean*-chickpea (497, 1250 kg/ha) and *mothbean*-fenugreek (465, 1161 kg/ha) crop sequence. Amongst the rabi crops, grain yield of fenugreek, chickpea, mustard and cumin were higher by 28.05, 38.11, 19.96 and 36.05 per cent, respectively, when grown in association with aonla compared to its sole crops. The highest net profit (Rs 28260/ha) was obtained from *mothbean*-cumin cropping system, followed by *mothbean*-chickpea (25024/ha) cropping system. *Mothbean*-chickpea intercropping with *aonla* supplemented 22.01, 5.00 and 27.90 kg/ha nitrogen, phosphorus and potassium through crop residues, followed by *mothbean*-fenugreek crop sequence. In a similar study, Arya *et al*., 2010 recorded significant differences in yield levels of perennial as well as ground storey components in multi species cropping models as compared to sole cropping. The growth and yield in perennial components were more under the multispecies cropping system, i.e. *aonla-ber-karonda*-clusterbean-brinjal and *aonla-ber-karonda*-clusterbean-fallow. Minimum yield was recorded in sole perennial crops. Plant height, number of branches/plant, pods/silique/fruits/plants and yield was found to be superior in multispecies cropping systems of *aonla-ber-karonda*-clusterbean-brinjal and *aonla-ber-karonda-mothbean*-indian mustard as compared to sole cropping except mustard where a reverse trend was observed.

5.3.1.3 Khejri based system

Prosopis cineraria, commonly known as *khejri* is considered as a "*Kalpvriksha*" of the Indian desert. It is one the chief indigenous trees of the Indian north-western plains and also of gently undulating ravine lands. The trees have monolayered canopy and deep root system and provide nutritious edible pods and fuel wood to the human beings, nutritious forage to the animals and fertility to the poor soils. Farmers usually grow jowar, pearl millet, moth bean and cluster bean between the scattered *khejri* trees in the arid region. Barley (*Hordeum vulgare*), pearl millet, *toria* (*Brassica tournefortii*) and chickpea are the common intercrops grown in Haryana. Puri *et al.* (1992, 1994) observed 86 per cent higher grain yield of barley in association with *khejri*. Besides, the yield of chickpea also improved and resulted in higher soil fertility and moisture conservation. Maximum net return (15,197/ha) and benefit-cost ratio (3.73) were obtained when pearl millet in *kharif* was followed by *toria* in *rabi* between the *khejri* trees (Kaushik and Kumar, 2003). Yield of fodder crops during both *kharif* and *rabi* seasons was higher in association with khejri trees as compared to sole cropping.

5.3.2 Horti-pastoral system

A combination of fruit trees and pasture species, commonly known as "Horti-pastoral system" is one of the several ways to satisfy human needs and alleviate cattle hunger. The system involves growing of fruit trees and grasses in combination during the initial and later years of trees growth as per the suitability of the species. Fruit trees usually form the 1st tier whereas grasses are grown as ground storey crop. Such system sustains the requirement of rural masses for fodder through leaves, grasses and fruits, which is a much-needed component in human diet. In such system it is however, important to ensure that forage crops do not affect the growth and development of the main crop. Trees such as *Jharber* or *bordi* are commonly grown in the rural areas and is an important top feed species. Even in drought years when crop fails, the top feed of this species is obtained which fulfills the need of the farmers. The trees are heavily lopped during winter for its leaf fodder known as "*pala*". One-hectare area of *jharber* may yield about 125 kg of *pala*. The perennial grasses such as sewan, *anjan* and *dhaman besides karad* (*D. annulatum*), *P. antidotale* and *Schima nervosum* are found in natural pasturelands, gochar, oran and range lands and give high forage yield under rainfed conditions.

Suitable *ber* based pastoral combinations for arid regions are *ber-anjan*; *jharber-anjan*; *jharber-sewan* and *boradi-anjan*. Results of growing different grass species between productive and unproductive *ber* trees with varying degree of reports have been reported. *Ber* based-pastoral studies on sandy rangelands of Rajasthan revealed that *Cenchrus ciliaris- Zizyphus mauritiana* system produced 1.2t/ha forage and did not affect the fruit yield (Sharma and Diwakar, 1989). Sharma *et al.*, 1994 reported that the productivity of wasteland in Western Rajasthan could be increased by planting top feed species *Z. nummularia* in combination with grasses such as *C.ciliaris*. Valuable information on paddocks has been gathered which may provide positive clues on management in western Rajasthan.

5.3.3. Horti-Agri-Pastoral system

Horti-Agri-Pastoral system refers to the combinations of fruit crops along with other agricultural or fodder crops in a manner that farmers get some return throughout the year. This is however, variable and depends on the demand of a farmer and agro-climatic situations. For instance if a farmers approach is livestock oriented, he can diversify his crops such as *Ber-Ker*-Grass or *Ber-Pilu-Lana* (*H. salichornicum*) or *Ber*-Drumstick-Grass, Through these systems the farmers can obtain fodder through *ber*, *pilu*, *lana* (a preferred feed for camels) and moringa leaves as top feed whereas fodder through grasses. *Lana* (*Haloxylon salichornicum*) is very common and preferred feed by the camels. It is naturally found in saline patches of arid region. The crop models have to be critically adjusted seeing into the rooting behaviour, fruiting period and food cum fodder demand particularly during the lean period. The crop-pasture components selected in the models can be made to interact synergistically to increase the productivity and generate higher net returns. While growing agricultural and fodder crops in association with fruit trees due consideration should be given on tree morphology and other characteristic so that there is a symbiotic association between over storey and ground storey components. Research in this important field is lacking.

Harsh *et al*., (1992) reported that when fodder grasses were grown in combination with *ber* plants as well as *Acacia tortilis* tree, the fodder yield and seed production were higher than the yields obtained when grass alone was grown. Similarly when *A tortilis*, *ber* plants and pearl millet and cluster bean were grown on same unit of land, the fodder and grain production under the canopy of woody component were not much affected.

5.4 CONCLUSION

Mounting pressure on natural resources due to rocketing population rise has ushered in large-scale degradation of environment and ecosystem thus calling for immediate attention for seeking newer approaches in cropping system to meet the food, fibre, firewood and timber requirement for the country. For sustainable management of arid lands and enhancing the economic viability of the cropping system, partial shift from the existing high input requiring rotation to low input requiring system is the need of the hour.

5.5 REFERENCES

Anonymous (2006) Annual Report (2005-2006), CIAH, Bikaner, pp. 7-8

Arya, R., Awasthi, O. P., Singh, J. and Arya, C. K. (2010) Comparison of fruit based multi-species cropping system under arid region of Rajasthan. *Indian Journal of Agricultural Sciences,* 80 (5):423-26.

Awasthi, O. P. and Pareek, O. P. (2008) Horticulture-based cropping system for arid regions-A review. *Range Management and Agroforestry*, 29 (2): 67-74.

Awasthi, O. P., Singh, I. S. and More, T. A. (2009) Performance of intercrops during establishment phase of aonla (*Emblica officinalis*) orchard. *Indian Journal of Agricultural Sciences*, 79 (8):587-91.

Bunderson, W.T., Wakeel, A.El., Saad, Z., Hashim, I. (1990) Agroforestry practices in Western Sudan. In: Budd, W., *et al.* (eds.). *Planning for Agroforestry,* New York, Elsevier Science Publisher.

Chadha, K.L. (2002) Diversification to horticulture for food, nutrition and economic security. *Indian Journal of Horticulture*, 59 (3): 209-229.

Chundawat, B.S. (1993) Intercropping in orchards. *Advances in Hort*., 2(2):763-775.

Dhandar D G, Saroj, P. L., Awasthi, O. P. and Sharma, B. D. (2004) Crop Diversification for sustainable Production in irrigated Hot Arid Eco-System of Rajasthan. *Journal of Arid Land Studies* (*Special Issue*) 148: 37-40.

Faroda, A.S. (1998) Arid zone research: an overview. (In) Faroda, A.S. and Manjeet Singh Ed. *Fifty Years of arid Zone research in India*. CAZRI, Jodhpur.pp: 1-16.

Gupta, J.P., Joshi, D.C. and Singh, G. B. (2000) Management of arid agrosystem: In: National resource management for agriculture production in India (Eds.) (J.S.P.Yadava and G.B.Singh). International conference on managing natural resource for sustainable agriculture production in the 21st Century.

Harsh, L.N. and Tewari, J. C. (1993) Tree crop interactions in agroforestry practices. In: *Agroforestry for Rural Needs* (Vol II) (Eds. D.K. Khurana and P.K. Khosla). Indian Society of Tree Scientists, Solan. (H.P.), India. pp. 535-541.

Kaushik, N., Kumar Virendra. (2003) *Khejri* (*Prosopis cineraria*) based agroforestry system for arid Haryana, India. *Journal of Arid Environment*. 55: 433-450.

Khan, F. S. (1993) Bioeconomic interaction of single scattered trees on farm lands. *Pakistan Journal of Forestry*, 10 (12): 47-49.

Osman, M. (2003) Alternate landuse systems for sustainable production in rainfed areas. In: *Agroforestry-potential and opportunities* (eds. P.S. Pathak and Ram Newaj): pp. 177-181.

Pareek, O. P. and Nath. (1996) *Ber*. In Coordinated Fruit Research in Indian Arid Zone- A two decades profile (1976-1995). National Research Centre for Arid Horticulture, Bikaner, India: 9-30.

Pateria Dinesh Kumar, Jaggi Seema, Batra P K and Gill A S. (2005) Modelling the impact of fruit trees on crop productivity. *Indian Journal of Agricultural Sciences,* 75 (4):222-24.

Randhawa, K. S. (1990) Pulse based cropping system in juvenile orchards. International Symposium on Natural Resource Management for Sustainable Agriculture.

Saroj, P.L., Dhandar, D. G., Sharma, B. D. Bhargava, R. and Purohit, C. K. (2003) *Ber* (*Zizyphus mauritiana* L.) based Agri-Horti System: A sustainable Land Use for Arid Ecosystem. *Indian J. Agroforestry*, 5 (1&2):30-35.

Sharma, S. K. and Diwakar, G. D. (1989) Economic evaluation of hortipastrol system on arid region of western Rajasthan. *Indian Journal of Range Management,* 10 (2): 119-122.

Sharma, S. K., Singh, R. S., Tewari, J. C. and Burman, U. (1994) Silvipastoral studies in arid and semi-arid degraded lands of Western Rajasthan. In: *Agroforestry systems for degraded lands.* Vol.II.Pp749-754.

Singh, H. P. (1980) Improving the moisture storage in sandy desert soil by subsurface moisture barrier. *Arid Zone Research and Development*: 245-252.

Young, A. (1989) *Agroforestry for soil conservation*. International Council for Research in Agroforestry, Nairobi (Kenya).

Chapter – 6

Production System Management in Precision Farming of Fruit Based Cropping Systems under Hot Arid Ecosystem

P. L. Saroj

6.1 INTRODUCTION

Presently, pressure on productive land is greater than ever before and will be further more in years ahead. Per capita arable land is projected to decline from about 0.23 ha in 2000 to about 0.15 ha by 2050 (Lal, 1991). The global demand for food is projected to increase by 1.5 to 2 times due to ever increasing population and also due to demand for richer diets. On the other hand, volatility in the cost of agricultural inputs and the income generated from farm products leads to instability in the farm economy. Moreover, in conventional system, agriculture is practised for uniform application of fertilizer, herbicide, insecticides, fungicides, and irrigation, without considering spatial variability. In some cases, there is excessive use of nitrogenous fertilizer and pesticides, owing to air and water pollution but there are certain cases, where farmers hardly use any chemicals. Hence, to alleviate the ill effects of over and under usage of inputs, the scenario calls for the introduction of modern technologies to improve crop yield, provide information to enable better in-field management decisions, reduce chemical and fertilizer costs through efficient application, permit more accurate farm records, increase profit margin and reduce pollution.

Precision agriculture is a production system that promotes variable management practices within a field, according to site conditions. This system is based on new tools and sources of information provided by modern technologies. These include the global positioning system (GPS), geographic information systems (GIS), yield monitoring devices, soil, plant and pest sensors, remote sensing, and variable-rate technologies for applicators of inputs. Precision farming is concerned with the management of variability in the dimensions of both space and time. Aspects of precision farming, therefore, encompass a broad array of topics, including variability of soil resource base, weather, plant types, crop diversity, machinery performance and most of the physical, chemical and biological inputs used in production of a crop. These are closely linked to the socio-economic aspects of production system. Success in precision framing is directly related to how well it can be applied to manage the space-time continuum in production system. Thus precision farming is technology enabled, information based, and decision focused. It is the integration of these technologies that has enabled farmers and their service providers to do things not previously possible, at level of detail never before obtainable, and, when done correctly, at level of quality never achieved before. In our country, vast data on various aspects like soil characteristics, climatic parameters, topographic features, crop requirement in terms of consumptive use and nutritional requirements have been generated and instruments needed for recording these parameters are also available. However, applications of precision farming as a package in the farmer's field have received little attention, although some aspects of precision farming have been prasticed. This has been primarily due to lack of awareness about the potential for increasing productivity and improving the quality of produce with minimum use of inputs.

Horticulture is expected to have accelerated growth rate above 7 per cent for achieving overall growth of agriculture to the tune of 4 per cent. In the task of achieving the higher growth rate, hi-tech interventions like precision farming in horticulture is essentially required, as horticultural crops, whether it be a fruit, vegetable, flower, spice, medicinal and aromatic plants or plantation crops, respond to technologies like micro-propagation, micro-irrigation, fertigation etc. Horticulture is expected to have accelerated growth rate above 7 per cent for achieving overall growth of agriculture to the tune of 4 per cent. In the task of achieving the higher growth rate, hi-tech interventions like precision farming in horticulture is essentially required, as horticultural crops, respond to technologies like micro-propagation, micro-irrigation, fertigation etc. Use of *In-vitro* plants, fertigation and nutrient management based on soil analysis in banana have shown increase in yield manifold and improvement in the quality. Use of fertigation in grape coupled with nutrient application based on petiole analysis added with bunch management has increased the yield and quality. Use of high yielding tropical varieties coupled with proper irrigation and nutrition produced significantly higher yield in potato, tomato, onion, cassava etc. High density planting and canopy management proved economical in guava. There are many other examples wherein a few components of precision farming have been adopted to greater advantages in increasing the returns from the land. Therefore, there is an urgent need to develop a package based on knowledge of soil environment and crop needs to enhance the efficiency of inputs to get higher output in given time frame.

6.2 REQUIREMENTS FOR PRECISION FARMING

a) **Crop related :** Type of crops, stage of the crop, crop health, special care, nutrient requirement, canopy architecture etc.

b) **Soil and topography related :** Physical and chemical properties, depth, texture, nutrient status, salinity and toxicity, soil temperature, productivity potential, field shape and size, slope, drainage etc.

c) **Climate related :** Temperature, rainfall, humidity, wind direction and speed, evaporation, evapotranspiration etc.

d) **Irrigation related :** Availability of surface water and ground water, irrigation method, levels of irrigation, interval etc.

e) **Precision machinery :** Tillage machinery, intercultural machinery, crop specific machinery, harvesting machinery etc.

f) **Other planning inputs :** Electricity, road and transport, market, labour, agri-inputs, secondary services, etc.

6.3 FRUIT BASED CROPPING SYSTEMS

It is demand driven approach by integrating multi enterprise based on resource situation on a land management unit for sustainability towards eco-balance, multiple and optimum outputs, employment generation, risk coverage and ultimately improving living standard of the people. The approach cater the diverse need of farmers/ entrepreneurs for food, fruit, fodder, fuel, fibre etc, by utilization of both natural resources i.e. soil, water, vegetation, solar energy etc. and farm inputs i.e. seed fertilizers, chemicals, labour etc. However, the approach markedly influenced by power supply, credit support system, storage and transport facility, processing units, market infrastructure etc. The additional advantage of fruit based cropping systems are; use of labour force, ecologically/economically synergistic influences , protective influences against strong wind, frost, hail storm etc., establishment of secondary enterprises like bee keeping, fisheries/duck rearing, dairy, goat/ sheep rearing, poultry, wicker works, processing unit etc. besides other intangible benefits.

The region may be characterized as very low and erratic rainfall, extremes of temperate, high evapo-transpiation, scarce water resources, sandy soil with poor organic matter, saline soil and water, problem of high wind velocity, frost etc. problem of wild life, poor input availability, poor marketing network, insufficient processing units, poor storage facility etc. These biophysical farming constraints are main limiting factor of agriculture in the region. The traditional farming, sometimes, poorer uneconomical especially due to hostile climate and diverse riskprone situation coupled with high human and livestock pressure. Though, the vast land resources, surplus family labours, increasing canal command area, less problem of pest and diseases, sufficient solar and wind energy, increasing infrastructural facilities are the strength of the region the traditional farming is less economical. The famine and poverty are associated with the rural mass where agriculture is the major source of subsistence. Thus, there is a need of system approach by which even a subsistence farmer can get at least some yield in the worst climate year (Saroj *et al.*, 2004a).

6.3.1 Components of fruit based cropping systems

6.3.1.1 Agriculture component

6.3.1.1.1 Perennials

Group	Crop
Fruit crops	*Ber*, Date palm, *Bael*, *Lasoda*, Citrus, Mulberry, *Karounda*, *Aonla*, *Ker*, *Pilu*, *Phalsa*.
Forest tree	*Khejri*, *Kumut*, *Babul*, *Rohida*, *Neem*, *Mopane*, *Sheesham*
Shrubs	*Phog*, *Kheep*, *Euphorbia*, *Bhui*.
Grasses	*Sevan*, *Dhaman*, *Gorda*

6.3.1.1.2 Annuals

Group	Crops
Food grains	Wheat, Jawar, Bajra, Barely
Vegetables	Cluster bean, Kachri, Mateera, Tinda, Snap melon, Brinjal, Chillies, Radish, Beans, Spinach.
Pulses	*Moth*, Gram, *Moong*
Oilseed	Mustard, Groundnut, *Taramira*
Seed spices	Coriander, Fenugreek, Cumin, Fennel, Chillies

6.3.1.1.3 Medicinal and aromatic

Group	Crops
Shrubs	*Aak*, *Dhatura*, *Gokhuru*, *Gugal*, *Guarpatha*, *Periwinkle*, *Tumba*, *Menhdi*, *Ker*, *Ashwagandha*, *Senna*, *Guarpatha*
Tree	*Aonla*, *Bael*, *Jamun*, *Neem*, *Amaltas*

6.3.1.1.4 Ornamental

Group	Crops
Perennial	*Bougainvillea*, *Lantana*, *Menhdi*, *Clerodendran*, *Cactus*, *Kaner*
Annual	Marigold, Annual Chrysanthemum, Gomphrena, Hollyhock

6.3.1.2 Animals component

Purpose	Animals
Milk	Cow, Goat, Camel
Meat	Goat, Sheep, Poultry, Fish
Draft	Camel, Bull, Horse, Donkey

Wool	Sheep
Eggs	Poultry, Duck

6.3.1.3 Other component

Group	Enterprises
Processing unit	Dehydration, Pickling, Canning
Miscellaneous	Bee keeping, Vermiculture, Mat Making, Furniture
Secondary enterprises	Wicker works, Packing materials, Labeling, Transportation, Storage facility.

6.4 PLANNING OF FRUIT BASED CROPPING SYSTEMS

(i) **Selection criteria :** There are various considerations, which one has to look into while working with fruit based cropping systems.

6.4.1 Over storey component

i) The perennial components should have xeric characters such as deep tap root system, high bound water in tissue, sunken stomata, waxy coating, thick cuticle and pubescence

ii) Ability to complete maximum vegetative and reproductive phase during the period of moisture availability

iii) Root system should be able to exploit deeper soil layers than those tapped by the under and groundstorey crops

iv) Species with sparse foliage and branching habit that allows light penetration to the understorey crops

v) The species should respond to pruning

vi) It may be deciduous in nature particularly litter fall during the growth period of the ground storey crops

vii) Leaf litter should not have allelopathic influence

viii) Specific quality attributes

ix) Resistance to various biotic and abiotic stresses

6.4.2 Ground storey component

i) The annual crops must be suited to agroclimatic situations and need based

ii) It should be shade tolerant and can respond to available solar radiation of the region

iii) It should not compete with the perennial component for resource pool

iv) It should be short duration and early maturing

v) Leguminous groundstorey crops that have nitrogen fixing capacity should be preferred

vi) It should not require harvesting or other operation that would damage the main crop or induce soil erosion

6.4.2.1 Selection of varieties

While selecting varieties of perennial components based on their growth behaviour, production potential, quality attributes, resistant/ tolerant to pest and diseases etc., proper attention should be given. The verities of fruit crops have been identified for specific purpose (Table. 1)

6.4.2.2 Provision of windbreaks

After delineation of the boundary, live fences with staggered system of close planting of wind breaks by integrating perennial shrubs and trees should be done. The trees like *lasoda, bael, jamun, khejri, sheesham, neem* etc. should be planted outer side while shrubs like *bordi, karonda, ker* etc, in the inner side. In general, it is suggested that the wind breaks must be established before inside plantation. For protection during establishment of these windbreaks, dried thorny twigs of *khejri, bordi, babool* etc. should be used. These fast growing wind breaks not only protect the inside plantation from hot winds and wild life but also an additional source of income after some time, if managed properly.

6.4.2.3 Use of rootstocks

In arid region, least emphasis has been given to exploit the potential of important rootstocks of various fruit crops. Some important rootstocks which can be utilized for establishment of orchards under arid or semi arid region are given in Table. 2

Table 1 : Varieties of fruit crops for specific purpose

Fruit crop	Variety	Purpose
Datepalm	Medzool, Shamran, Khadrawy, Chip-Chap	For dehydration (*chhuhara*)
Aonla	NA-6, NA-7, Chakaiya, Banarasi Kanchan, Krishna, NA-10, BSR-1	For preserve pickle and shreds medicinal purpose
Pomegranate	Amlidana, Goma Khatta	For dehydration (*Anardana*)
Ber	Umran Gola, Seb, Mundia, Banarasi Karaka Seb, Umran, Mundia	For dehydration, preserve
Fig	Poona, Deanna, Excel	For dehydration
Bael	NB-9	For preparation of candy

Table 2 : Influence of rootstocks in arid fruit crops

Fruit crop	Rootstock	Influence
Ber	*Z.mauritiana* var. Tikdi	Vigorous growth
	Z. mauritiana var. Shukhawani	Vigorous growth
	Z. mauritiana var. *rotundifolia*	Vigorous growth and drought hardy
Citrus	*Poincirus trifoliataCitrus karna*, Rangpur lime	Drought hardyand sodic soils
Guava	*Psidium cujavillis, P.Molle*	Tolerant to drought and sodic soils
Sapota	Khirni	Drought hardy
Grapes	Salt Creek, Dogridge, St. George , 110 R	Drought and salinity tolerant

6.4.2.4 Proper soil working

Proper soil working is essential for sandy soils of arid region as poor water holding capacity of soil due to high infiltration rate, high rate of evaporation, poor soil physical condition and deficient in several nutrients are the limiting factors of plant establishment. Based on the plant types, optimum pit size and spacing should be decided. The pits should be prepared and left open during summer at least for 15 days and thereafter, filled with filling mixture (1: 1: 1, soil; FYM, pond silt). Before filling, chloropyiphos or mean cake should be applied in the pits for control of termites. Planting should be done during monsoon in well saturated pits. Proper compaction of soil while planting is also important.

6.4.2.5 In- situ establishment of orchard

Due to limited and erratic rainfall and saline underground water resource, there is poor establishment of orchard in arid region. During summer, hot winds coupled with sand deposition and biotic interference are other limiting factors. Therefore, *in-situ* orchard establishment is suggested. For hot arid region, it is suggested that the seeds of *ber, bael, aonla, lasoda* should be sown in polybags/tubes of 25 x 15 cm size in the nursery in the month of February. Irrigate them regularly till the germination, there after at 3-6 days intervals either by sprinkler or by watering cans. These seedlings should be shifted in the well prepared pits at desired spacing during the month of July-August followed by irrigation, while sifting these seedlings, the polybags should be removed. It is also pertinent to mention that in no case seedling should be more that 9 months old, otherwise there is every chance of root coiling. During subsequent monsoon season, budding should be done with desired scion cultivars.

6.5 FRUIT BASED SYSTEMS

Over centuries, traditional wisdom of integrating agriculture with livestock farming, efficient water harvesting systems, utilizing perennial trees and grasses evolved as most favoured practice in the arid region. Many of such farmers' wisdom have high relevance even today, which should be given due emphasis while planning and execution of fruit based cropping system in the region. Some important systems of arid region are mentioned below;

6.5.1 Khejri based

Khejri (Prosopis cineraria) 'Kalp Vriksh of Desert' is found growing naturally, though at different plant densities in hot arid ecosystem. Growing of crops like pearl millet, *moth bean*, cluster bean, gram, mustard with Khejri and retaining of *ker*, *pilu* and *bordi* is an age old tradition in the arid region. In fact the deeper root system, tiny leaves, spines, prone to cent per cent lopping, leguminous nature, multiple outputs (loong/fodder, fuel/fencing material, green and dry *sangria* value added products, timber etc.) made this plant most compatible agroforestry component under arid region.

Harsh *et al.* (1992) documented the growing of the trees, shrubs and annual crops integrated with *Prosopis cineraria* on farmer's field both under rainfed and irrigated conditions in different district of Rajasthan. The effect of scattered trees of *P. cineraria* at farmers' field in the arid region of Haryana was assessed on the productivity of barley. The yield of barley was enhanced by 68 per cent over control. The soil under tree canopy was rich in organic carbon content, moisture availability and nutrient status (N, P and K). The positive response of *P. cineraria* on the yield of *chickpea* was reported by Puri *et al.* (1994). They further suggested that increase in growth and yield of *chickpea* was due to improvement in soil fertility and conservation of moisture. Kaushik and Kumar (2003) found higher benefit cost ratio (3.73), when pearl millet in *kharif* and *toria* in *rabi* season was growth under khejri trees.

6.5.2 Lasoda based

The *lasoda* plant (*Cordia dicotoma*) is fast growing and the fruits having good nutritive value. The tender fruits are used for making vegetable and pickles. The plant is now being grown on boundaries of irrigated crops as wind breaks. The tree is suitable to integrate with various crops. In fact, the characteristics like leathery leaves, sunken stomata etc. have made the *lasoda* plant to survive under arid region. The ground storey crops grown with *lasoda* on farmers field is given in table 3. However, it was found that growing of *lasoda* as over storey component has adverse effect on the pad yield of Aole crop (Saroj *et al.,* 2004b).

Table 3 : Lasoda based cropping systems in different districts of Rajasthan.

Growing areas (District)	Ground storey crops	
	Irrigated conditions	**Rainfed conditions**
Bikaner, Barmer, Jaisalmer	Vegetables	Bajra, Grasses
Pali,Sirohi, Jalore	Vegetable, Spices	Wheat, Mustard, Taramira, Sesame, Bajra, Cluster bean,
Jodhpur	Mustard	Bajra, Cluster bean

6.5.3 Ber based

Ber (*Zizyphus mauritiana*) based cropping system is very common in arid region both under irrigated and raifed conditions. The drought hardy crops like cluster bean, mustard, green gram, *kachri*, *mateera*, pearl millet and grasses like *sewan*, *dhaman* etc. are grown since long back at the farmers' field. While working with *ber* based agri-hortisystem, Gupta *et al.* (2000) suggested that the system minimizes the risk in arid region and helps in imparting the economic stability of the farmers'

Under limited irrigation, Saroj *et al.* (2003) reported that various ground storey crops can be grown successfully with *ber* without any adverse effect on over storey component, rather growing of crops in initial years had better response in term of plant survival, growth and yield the integration of different component into the system led to multiple benefits e.g. from groundnut dry pods, from cluster bean green pods (for vegetable), from wheat and mustard grain yield, from Indian aloe green leaf pad (for vegetable) and from *ber* trees besides fruit; fuel and fodder can be obtained. The phytomas of different ground storey components was recycled in the same plot from where it was harvested in order to improve fertility status of the growing site. Overall, highest economic return was obtained with integration of *ber* and + Indian aloe but from resource management point of view, *ber* + cluster bean + mustard is very compatible and profitable combination suggested for this region. Under rained conditions of Jhansi, Kumar *et al.* (2006) found that *ber* based horti-pastoral systems are very promising and suggested that Guinea grass + *Stylo* is a compatible combination with ber.

6.5.4 Aonla based

Under arid region, while working with *Aonla* (*Emblica officinalis*) based cropping system, Awasthi *et. al.* (2007) established various models like; *Aonla* + *ber* + Brinjal + *Moth bean* + Fenugrrek, *Aonla* + *bael* + *Karounda* + *Moth bean* + Gram, *Aonla* +

Khejri + *Saji* + *Moth bean* + Mustard, *Aonla* + Drumstick + *Senna* + *Moth bean* + Cumin and Control. In the first year itself, good economic return was obtained from the crops when perennial trees were in the establishment phase. In the second year, the economic return was further improved. Thus, it is found that in the juvenile phase of perennials, there are ample opportunities of raising annual crops, which in tern can meet the diversified need of the farmers. However, under extreme arid situations, raising of *aonla* plants needs protection against frost.

6.5.5 Date palm based

The date palm (*Phoenix dactylifera* L) is an index crop of arid region. Being a tall growing, single stem, grown under irrigated condition and widely spaced plant; there is enough scope for intercropping. However, removal of suckers and leaf pruning is essential for introduction of ground storey components. So far no systematic and long term experiment has been conducted to identify compatible intercrops with date palm. In Punjab, kinnow has been grown in between two rows of date palm. Farmers are also growing paddy as intercrop with date palm. In Rajasthan, vegetables were grown in interspaces of date palm. Whereas, in *Kachchh* region of Gujarat; *lucern* and wheat in *rabi* and cotton, sorghum, *guar* and pearl millet in *Kharif* are the popular intercrops on farmers field.

6.6 PRODUCTION SYSTEM MANAGEMENT

Desired output can only be obtained with proper managements. Management of fruit based cropping systems is not very easy as there are multiple components integrated on the same land. It is also suggested that in first year no-intercrops should be grown and full attention should be given to establish perennial components. The suckers arising from the rootstock portion should also be removed. The requirement of individual component also varies greatly. The sowing/planting, training, purning, irrigation, nutrition, fruiting, flowering, ripening, harvesting, processing, marketing etc. requirements also vary. Thus, from very beginning intensive management is required as per need of the component for sustainable production without impairing the fertility of the growing site.

6.6.1 Runoff management

For *in-situ* moisture conservation, inward basin, staggered trenches, circular trenches, ridge and furrow system of cultivation etc. are very useful in arid and semi arid conditions. The ridge-ditch method of moisture conservation is adopted in arid region since long back. For optimum water harvesting in *ber* orchard, a catchment area of 31.5m^2 with 5 per cent slope was found adequate for harvesting bulk of rain water within the deeper soil profile. At 200-250 mm per annum rainfall, 1m soil profile maintained 40-60 per cent higher soil moisture storage than control (Sharma *et al,* 1982). In semi arid region there is good scope for rain water harvesting, which can be effectively utilized for establishment of young plantation, life saving irrigation, irrigation during critical stage of crop growth, irrigation for intercrops etc. owing to better return.

6.6.2 Mulching and antitranspirants

Mulching is an important practice to (i) reduce evaporation from soil surface, (ii) breaking supply of heat energy to the evaporating site and (iii) decreasing thermal conductivity of the soil. There are various types of mulch materials like dry grasses, leaves, crop residues, saw dust, black or colourless polythene etc., which are commonly used. The locally available organic mulch materials also improve the hydraulic conductivity and field capacity, besides 'conservation of soil erosion and improving site conditions after decomposition by reducing the ECe, ESP, cataion and improving the organic carbon content. In *aonla* based cropping systems, use of organic mulches curtailed soil temperature by 1.1 to 5.6°C during summer and increased by 0.6 – 3.2 °C during winter months (Awasthi *et al*; 2007).

Similarly, the transpiration loss from the foliage can be minimized by using radiation reflectants, waxing, plastic films, stomata closing chemicals etc. spraying of *kaolin* (4-6 per cent), liquid paraffin (0.5-1.0 per cent) and power oil (1.5 per cent) in low rainfall areas. The chemicals such as decinyl succinic acid (DSA) and abscisic acid (ABA) are also useful in reducing the water loss through stomata closure. However, these antitranspirants/ chemicals have not been exploited at commercial venture so far by the farmers.

6.6.3 Irrigation management

Every drop of water should be utilized judiciously and effectively, as water is a precious commodity in arid region. Besides, little precipitation, there are two sources of irrigation in arid region i.e. one underground resource and second canal water. Underground water resource is very limited and difficult to develop deep well by an ordinary farmer, as it is very costly. The canal irrigation water is good and per unit irrigation cost is reasonable. However, it is difficult to cover each and every field of farmer's by the canal network. The irregular and interrupted supply of canal water is another limiting factor. Thus, if we have to develop any fruit based cropping systems, we must think to develop a water collection body in order to avoid uncertainity of water supply, to laid out an effective micro irrigation system, to integrate lesser water requiring crop combinations and to fallow other moisture conservation practices.

Under sole plantations, sub-surface irrigation by drip system avoids losses of moisture through both evaporation, and gravitation. In deep rooted perennial fruit trees, the concept of irrigation scheduling based on plant water contents rather than the ET alone would perhaps be ideal. Because of high infiltration rate and poor water holding capacity of sandy soil, frequency of irrigation and critical stage of irrigation should also be taken into consideration. Being a multicomponent system, sometimes, with holding of water is also required. After six year of experimentation in *ber* based cropping system, it is found that ber + cluster bean + mustard can be successfully grown just in one-third of water than either groundnut alone or ber + groundnut system under sprinkler irrigation. The study further indicated that if such system is adopted, two-third more area can be covered under same amount of available irrigation water, with better tangible and intangible benefits.

6.6.4 Nutrition management and site productivity aspect

Proper nutrition in fruit based cropping systems needs special attention because of two reasons i.e. (i) fertility status of inherent soils in arid region is very poor and (ii) in a system where there is always two or more than two components of drivers nature; there is a competition for resource pool in general. However, complimentary and synergistic influences are also found if compatible components were interrelated and managed properly. If interspaces are utilized for growing crops, then there is no need to apply recommended dose of fertilizers to perennial plants after full canopy development, rather it should be curtailed gradually. However, annual application of manures and fertilizers near the root zone of tree coincides either with rainfall or moisture availability gives good result. Moreover, in annual crops, fertilizers should be applied as per recommendation. The split application of nitrogenous fertilizers is recommended in order to avoid leaching losses.

It is advisable that leguminous crops must find place in the rotation of fruit based cropping systems. Application of organic manures, recycling of crop residue and green manuring should be considered as essential practice to improve the physico-chemical condition of the soil. Arid soils are generally deficient in nitrogen, zinc and iron; thus foliar feeding should also be done. In fact, in fruit based agroforestry system; hence canopy cover by the perennial component becomes permanent feature, which creates better microclimatic conditions for decomposition of organic matter. The work done at CIAH, Bikaner indicated that in *ber* based agroforestry system and *aonla* based multistorey systems both, the soil of growing site got improved in term of organic carbon, available nitrogen, phosphorus and potassium content. The physical condition of soil was also improved considerably.

6.6.5 Pest and diseases management

Pest and disease management is crucial in fruit based cropping systems. Based on threshhold value proper management strategy should be adopted. Sometimes, if perennials plants left un-pruned, they not only interfere with the groundstorey components but also act as host for many pest and diseases, Similarly, poor light penetration and over humidity buildup may also cause of higher infestation of diseases. Therefore, all possible undesirable interaction should be taken care of timely. Regular monitoring of pest and diseases should be done. Proper sanitation of the field, regular intercultural operations, proper canopy management, soil solarization, use of well rotten FYM / well decomposed crop residues, use of botanicals and bio-agents are useful to keep the pest population below economic threshold. The use of chemical insecticides / fungicides should be avoided to the possible extent but should be applied under severe conditions.

6.7 REFERENCES

Awasthi, O. P., Saroj, P. L., Singh, I. S. and More, T. A, (2007). Technical Bulletin, *Fruit based diversified cropping system for arid regions,* CIAH, Biakner.

Gupta, J. P., Joshi, D. C. and Singh, G. B. (2000) *Management of arid ecosystem*,. International Conference on Managing Natural Resources for Sustainable Agriculture Production for 21st centaury, New Delhi.

Harsh, L. N., Tiwari, J. C., Burman, U. and Sharma, S. K. (1992). Agroforestry in arid regions . *Indian Farming,* 45:32-37.

Kaushik, N. and Kumar, V. (2003). *Khejri* (*Prosopis cineraria*) based agroforestry systems for arid Haryana, *India Journal of Arid Enviroment*, 55: 433-450.

Kumar, S., Singh, J. B. and Ram, S. N. (2006). Influence of pruning and weather on productivity of ber based horti-pastoral system in rainfed conditions., *Indian J. Hort*., 63(3): 260-263.

Lal, R. (1991). Soil structure and sustainability. *Journal of Sustainable Agriculture*, 4:67-92

Puri, S., Anil Kumar and Singh, S. (1994). Productivity of *Cicer arietinum* under a *Prosopis cineraria* agroforestry system in arid region of India. *Journal of Arid Environment*, 27 (1): 85-98.

Saroj, P. L., Dhandar, D. G. and Singh, R. S. (2004b). *Indian Aloe, Technical Bulletin* CIAH, Bikaner.

Saroj, P. L., Vashishtha, B. B. and Dhandar, D. G. (2004a). *Advances in Arid Horticulture*, Vol. I, International Book Distributing Comp., Lucknow.

Sharma, K. D., Pareek, O. P. and Singh, H. P. (1982). Micro-catchment water harvesting for raising jujube orchards in an arid climate. *Transaction of American Society of Agricultural Engineers*, 29 (1): 112-118.

Chapter – 7

Prospects of Temperate Fruit Production in Subtropical Climate

M. K. Verma, O. P. Awasthi and R. K. Giri

7.1 INTRODUCTION

Production of temperate-zone fruits in subtropics may seem fanciful to be initiated, but it is a practice that has existed in localized regions for generations. Seedling peaches are grown in Venezuela, Northern Thailand, and Southern Mexico. No one knows precisely when these species were introduced, but local selection has produced cultivars adapted to areas with little or no chilling temperatures. Rest either does not occur or is sufficiently shallow to be broken by stress induced by defoliation or drought. Commercial cultivations of temperate-zone fruits are well-established in subtropical areas, where winter temperatures allow partial breaking of the rest period. In areas of Northern Mexico, Southern Africa, and Israel, for example, chemical treatments are regularly applied to supplement the effects of limited chilling. Because temperature declines with increasing elevations, many microclimates exist in the subtropics where temperate-zone fruits are not currently grown, but could be. Subtropical regions where temperate fruits are grown include the southern past of California (Mediterranean type), the low deserts of the Southwest USA (hot arid type), the Gulf Coast and most of Florida (humid type), the Southern Mediterranean and Northern Sahara, Northern India, Southeast China, the middle part of South America, and much of Australia. Similar potential climatic conditions are also available in subtropics of India.

- India : North Indian plains, parts of Eastern states, Eastern and Western *ghats*, Nilgiri hills.
- Russia : Sochi
- Turkey : Mersin, Antalya, Rize
- Australia : Brisbane and Sydney

- Middle East : Jerusalem, Baghdad, Kabul, and Islamabad
- Europe : Lisbon, Barcelona, Rome, Dubrovnik, Athens and Sochi
- United States : Houston, Texas, Charleston, South Carolina, Orlando, Florida, Tucson, Arizona
- Brazil : São Paulo, Curitiba, Florianópolis and Porto Alegre
- North Africa : Cairo
- South Africa : Durban and East London
- New Zealand : Auckland
- Uruguay : Bella Unión
- China : Hong Kong, Macau and Shanghai

The subtropical region in India lies between latitude 15° and 30° North of equator covering north Indian plains, the Gangetic basins, large parts of eastern states, eastern and western *ghats*, Nilgiri region etc. In this zone spring is generally frost free and chilling availability ranges from 50 to 500 hours. This region is thus not considered suitable for temperate fruits but grows several low chilling varieties of peach, plum, apricot, almond, pear and nuts. The area and production of stone fruits in this region is estimated to be much higher than in the temperate zone in India. These areas being closer to the markets, offers better production potential than those in higher hills. The winters are relatively warm, but not as hot as the summer season. These climates rarely - if ever - see frost or snow, and plants such as palm, citrus and many broadleaf evergreens flourish, in contrast to the hardier deciduous and coniferous trees which dominate mid latitude climates. As one moves toward the tropical side the slight winter cool season disappears altogether, while at the polar threshold of the subtropics the winters become much cooler. Rainfall patterns vary widely throughout the subtropic. In India, North Indian plains, Gangetic basins, parts of Eastern States, Eastern and Western Ghats, Nilgiri region receive varied quantum of rainfall.

7.2 SIGNIFICANCE OF CHILLING

Dennis (1997) stated that several systems have been devised for growing temperate zone fruits in warm climates, where chilling is insufficient to break bud dormancy in "standard" cultivars. These are: using chemicals to break dormancy, grafting with low chilling scions, moving the trees to cold storage or high elevations in containers, evaporative cooling applications, preventing the induction of dormancy by defoliation. Countries, such as Argentina, Australia, Brazil, Chile, Egypt, France, Reunion Islands, India, Indonesia, Israel, Italy, New Zealand, Mexico, Morocco, Pakistan, Philippines, Portugal, Spain, South Africa, Sri Lanka, Thailand, Tunisia, Turkey, Zimbabwe, etc. have restricted conditions to grow temperate zone fruits. Since the temperatures are too high during winter periods, low chilling requiring cultivars are preferred and some preventations should be taken.

The temperate fruits, nuts and berry crops require an exposure to low, but above freezing, temperatures, for resumption of growth and fruiting, called the chilling requirement. In general, their commercial cultivation is limited by high latitudes, where too much cold kills plant tissues and by the low altitudes, where, the winters are not cold enough to fulfil the chilling requirement. Hutchins (1932) reported that the chilling requirement is met, only, if the plants are exposed to less than 7°C temperature. However, later on, it was found that low-chill cultivars/species of these fruits can partially meet the chilling requirements even at higher temperatures. In order to find out the suitability of non-temperate regions for growing temperate fruits, various models/formulae were proposed to predict the number of available chilling hours (Chill unit) for a specific location. The model of Sharpe (1969) and Weinberger (1956), based on the mean temperature of the coldest month, is very simple and practical, as mean maximum and mean minimum temperatures are available for most of the locations. Sharpe (1969) further suggested that Florida peaches can meet their chilling requirements at higher temperature, so his allotment of chill-units is higher than that of Weinberger (1956), who does not assign any chilling unit to areas with mean monthly temperatures higher than 14.4°C. The states of Punjab, Haryana, Northern Rajasthan, plains of U.P., Bihar and West Bengal falling in the subtropical zone there is possibility of growing less chilling requiring temperate fruits. The areas being contiguous to Himalayas, the cool climate prevails during November to February (winter season), with January being the coldest month, when mercury may touch -1°C, at some places, for few hours for a few days. The ground frost/fog may occur from end December to early February. The mean monthly temperature of coldest month (January) at various locations in these plains varies from 11.6°C to 18°C at various places located in this zone. Therefore, using the model of Sharpe (1969) the chilling units available at Agra, Aligarh, Alwar, Amritsar, Bhatinda, Churu, Dehradoon, Ferozpur, Meerut, Delhi, Sriganganagar, Ludhiana, and Hissar are ranging from 415 to 685 chilling units. So there exists the possibility of growing low-chilling requiring cultivars of temperate fruits. However, these area experience very hot summer, with 40-45°C air temperatures prevailing for 2-3 weeks during May-June, when sunscald occurs in some fruits. The highest temperature recorded is 50°C at Sriganganagar, which also receives about 523 chill units. Therefore, in addition to low-chilling requirement these cultivars should also be heat tolerant. The other abiotic stresses to be experienced in these areas include slightly alkaline/sodic saline soils, flooding during monsoons and high winds velocity. The region being near to major consumption centres and with well-developed physical connectivity, through network of rural roads, the returns from these fruits are comparable to that in higher hilly areas and other subtropical fruits. Many medium to large plantations of temperate fruits especially that of pear and peach exists in Punjab, Haryana and Western U.P. etc. and total area under stone fruits is more than in the hills.

Table 1 : The climatological data of various places of subtropical climate (1951-1980)

Station	Altitude(m)	latitude	Longitude	Mean Temperature (°C)	Temperature (°F)	Chilling accumulation (chilling hours)
Agra	169	27.1	78.02	14.65	58	415
Aligarh	187	27.53	78.04	14.2	58	415
Allahabad	98	25.77	81.44	16.15	61	253
Alwar	271	27.3	76.35	14.65	58	415
Amritsar	234	31.38	74.52	11.6	53	685
Bahraich	124	27.34	81.36	15.7	60	307
Bareilly	173	28.22	79.24	15.2	59	361
Barmer	194	25.45	71.23	18	64	121
Bhatinda	**	30.1	74.35	12.5	55	577
Bhopal	**	23.15	69.4	17.85	64	91
Chandigarh	347	30.44	76.53	13.25	56	523
Churu	291	28.15	74.55	13.25	56	523
Dehradun	682	30.19	78.02	12.6	55	577
Ferozepur		30.55	74.4	12	54	631
Jaipur	390	26.49	75.48	15.15	59	361
Kanpur (UP)	126	26.26	80.22	15.6	60	307
Lucknow	128	26.52	80.56	14.75	59	361
Meerut	222	29.01	77.58	14.2	58	415
Delhi (Safderjung)	216	28.35	77.12	14.2	58	415
Delhi (Palam)	233	28.34	77.07	13.5	56	523
Varanasi	85	25.27	82.52	15.85	61	253
Sriganganagar	177	29.55	73.53	13.2	56	523
Ludhiana	247	30.56	75.52	12.95	55	577
Jhalawar (Rajasthan)	321	24.32	76.1	17.1	63	145
Hisar (Haryana)	221	29.1	75.44	13.45	56	523

(*Source:* IMD)

7.3 CHILLING ZONES

Much of the Southeast Asia region is within the tropics and thus receives no chilling temperatures. Beginning with about 20° latitude, there is a positive accumulation of chilling which increases with the latitude. Generally more than 1000 chilling units are accumulated above 26° latitude whereas in Australia low- to medium-chilling areas are found at latitudes of up to 35° south (Table 2). The accumulation of chilling unit is affected by the ocean currents and local geography, especially the altitude in tropical and subtropical zones. With cooler temperatures at the higher altitudes, it was estimated in Taiwan that with each increase in altitude of 100m there is an increase in chilling accumulation of 27 chilling units. When using the expected temperature decrease with altitude of 1°F or 0.56°C per 100m, the estimated chilling increase was 54–61 units per 100m altitude.

Table 2 : Estimated mean chilling accumulation in Southeast Asia and Australia as a function of latitude.

Latitude Southeast Asia	Chilling range	Australia
< 20	None	
20–23.9	0–630	0–700
24–27.9	100–1500	50–500
28–31.9	> 1000	300–1100
32–35.9	> 1000	450–1300
36–39.9	> 1000	> 850

Chilling accumulation estimated by using the following equation:
CU = 3547 – 53 (mean coldest month temperature in degrees F.

Although the mean chilling received is useful, it is the year-to-year fluctuations that are critical to examine when choosing varieties that would be well adapted to the region. Thus, the most appropriate chilling range (and most consistent in production) would be varieties requiring between 450 and 600 CU. If cold weather during bloom was not a problem, lower chill varieties could also be used. Based on chill requirement of commercial cultivars mild winter regions of the world are classified into three major adaptation zones: the transitional, medium and low-chill zones :

i) The transitional-chilling zone includes regions in which most commercial cultivars have a chilling requirement of 650–950 chilling units (CU). This zone experiences years in which the higher chill varieties produce poorly due to mild winters. This is true of the southern San Joaquin valley in California on the West Coast and central Georgia and the coastal plain of South Carolina on the East Coast of the USA, some regions of southern Australia, and the southern and higher altitude regions of South Africa.

ii) The medium-chill zone includes areas where cultivars with a chill requirement of about 400–650 CU can be grown consistently. This would include areas such as Southern Georgia, Northern Florida, Southern Louisiana, South central Texas, Southern Spain, the highlands in Mexico, Northern parts of subtropical India, and lower altitude areas of South Africa.

iii) The low-chill zone is characterised by frequently receiving less than 400 CU. This includes almost all parts of the subtropical areas lies in India.

7.4 SELECTION OF TEMPERATE FRUITS

Although no temperate fruit is indigenous to subtropical region, some apple (crabapple), peach, plum, pear and apricots have got established here. Some of these fruits may have come with invaders, traders, slaves and soldiers and later by missionaries. For instance, pear and peaches were planted by Chinese hostages, during the period of King *Kanishka*, near Amritsar (Punjab) in the first century AD. While reviewing horticulture in ancient India, Firminger (1863) mentioned that an apple, like White Joanneting variety had flower and fruits there at Ferozepur (Punjab). He also recorded that the apple and pears grew into big trees near Kolkatta (WB), had profusely flowering but the entire fruits drop, but a dark kind of plum, eatable after cooking had its fruting near Kolkatta. The systematic efforts to introduce temperate fruits in the northern plains were initiated in the later half of the 19th century with the establishment of Agro-Horticulture Society of India and later on by the Agricultural colleges at Pune, Coimbatore, Faisalabad (Pak), Nagpur, Sabour and Kanpur. The introduced fruits varieties were planted at Govt. Botanical Gardens (Company Baghs) at Kolkata, Saharanpur and at other places. In the first half of the 20th century large introductions were made at Palampur, Punjab and Kandghat (Patiala State).

There are a large number of temperate fruits which are grown in the temperate climate but all these fruits cannot be grown in the subtropical climate. Only those temperate fruits can be grown in tropical/subtropical climate which have comparatively low chilling requirement and have the ability to tolerate high summer temperature. Proper selection of temperate fruits should be done which can be successfully grown under subtropical climatic conditions. For example, crops such as pear (*Pyrus communis*), peach (*Prunus persica*) and strawberry (*Fragaria ananassa*) are ideal crops for subtropical climate followed by fruits such as plum (*Prunus salicina*), apricot (*Prunus armeniaca*), almonds (*Prunus dulcis*), olive (*Olea europaea*) and persimmon (*Diospyros kaki*) which can be grown in a wide range of subtropical and warm temperate climate. To some extent the cultivation of apple (*Malus domestica*) can also be tried.

7.5 SELECTION OF CULTIVARS

Selection of cultivars of the temperate fruits to be grown under subtropical climate is one of the key requirements for their successful cultivation in the subtropical regions. Only those cultivars which have low chilling requirement and ability to tolerate high summer temperature should be selected. Breeding programes have provided new cultivars selected for their short chilling requirements and tolerance to high temperature. One of the most successful cultivars is 'Anna' apple, developed by A. Stein in Israel. This is now extensively grown in areas with limited chilling. The introduction and the selection of superior types have remained major research activities at various institutes, as a result of which cultivars like Sharbati peach, Titron and Kala Amritsari plum, Patharnakh, LeConte and Baghugosha pears became commercially popular in some parts of the subtropical India.

Table 3 : Temperate fruits and their cultivars suitable for cultivation under subtropical climate

Sl. No.	Fruit crop	Cultivars (*Chilling hours required)
1	Peach (*Prunus persica*)	Okinawa (50), Red Ceylon (50), UF Sun (100), Florda Grande (100), Flordabelle (110), Flordared (110), Flordawon (110), FlordaPrince (150), FlordaGlo (150), Tropic Beauty (150), UF Beauty (200), Sunred (210), Flordabest (250), UFO (250), UF 2000 (300), UF Blaze (300), Flordadawn (300), Early Amber (310), Flordasun (310), UF Sharp (325), FlordaKing (350), Flordacrest (350), Gulfking (350), Desert Gold (350), Jewel (350), Gulfcrimson (400), Gulfprince (400), Sun Ripe (400), Florda Home (400), La Festival (400), La Pecker (400), Rio Grande (400), Tejan (420), Rechan (450), Tex Star (450), Whiterobin (500), Bonita (500), Sunfire (500), Ventura (500), May Gold Su (500), Gulfcrest (525), Flordaqueen (540), Sungold (540), Springtime (650), June Gold (650), Maygold (660), Junegold (660), Springtime (660), Armgold (660), Suwannee (660), Tropic Snow, Double Flowering, Parbhat, Pratap, Khurmani, Sharbati, Safeda Early Cream, Saharanpur Prabhat, Shan-e-Punjab, Shaharanpur No. 6, Ranjit Bagh Early, Safeda (LR Brothers), Shaharanpur Hybrid 3, China Flat, Babcock,
2	Nectarine (*P. persica* var. *nucipersica)*	Sunbest (225), Sunraycer (250), UF Royal (250), UF Queen (250), Sunmist (275), Sundollar, (350), Suncoat (375), Sunred
3	Apple (*Malus domestica)*	Dorsett Golden (250), Anna (300), Tropic Mac (300), Tropic Sweet (300), 88-20 (375), Ein Scheimer (400), 60-39 (400), Tamma, Neomi, Tropic Beauty, Gallia Beauty, Winter Banana, Tame, Vered
4	Pear (*Pyrus communis* and *P. pyrifolia*)	Patharnakh, Gola, Leconte, Keiffer, Smith, Baghugosha, China Pear, Pineapple, Baldwin, Tenn, Flordahome, Ayers Hood, Orient, Carnea, Tsu Li, Ya Li, *P. calleryana* (rootstock requires 400 chilling hours)
5	Strawberry (*Fragaria ananassa*)	Chandler, Tioga, Torrey, Selva, Belrubi, Fern, Pajaro
6	Plum (*Prunus salicina*)	Satluj Purple, Kala Amritsari, Jamuni Meeruti, Titron, Aloo Bokhara, Alucha Black, Titron, Howe, Gulfruby, Gulfbeauty, Gulfblaze, Gulfrose
7	Apricot (*Prunus armeniaca*)	New Castle, Early Shipley, St. Ambroise, Benazir, NJ-13
8	Almond (*Prunus dulcis*)	California, Papershell, Hybrid 15, Pathick.s Wonder, JKS-55, H-98, Achak (266), Desmayo Largueta (309), Ramillete (326), Marcona (435), Marta (478), Antoneta (514), Ferragnes (558)
9	Sweet Chery (*Prunus avium*)	Stella (200-250), Cristobalina, Temprona del Sot, Precoce de Bernard, Sunburst, Lapin, (Rootstocks : Gisela-5, Mahaleb)
10	Persimmon (*Diospyros kaki*)	Hachiya, Fuiju, Jiro, Hyakuma
11	Olive (*Olea europaea*)	Arbequina, Barnea, Frantoio, Koroneiki, Leccino, Picual, Coratina, Picholine
12	Blue berry *(Vaccinium angustifolium)*	Sharpblue (150), Emerald (200), Jewel (200), Windsor (225), Springhigh (225), Chaucer (400), Woodard (400), Brightwell (400), Climax (450), Tifblue (550), Powderblue (550)

*Values in parenthesis are chilling requirement of that cultivar

Sunbest nectarine

Sunmist nectarine

Suncoat nectarine

7.5.1 Peach

A large number of promising hybrids of peach and plum were introduced at Ludhiana from Gainesville, Florida (USA) and out of these introductions peaches like, 'Flordasun' (obsolete now, because of shorter shelf life), '16-33', 'Partap', 'FlordaPrince', 'EarliGrande' etc. are gaining popularity. The other important cultivars also found promising are :

Parbhat : It is an early ripening, white-fleshed peach hybrid was released from Saharanpur. Its fruit size is smaller and shelf-life shorter, but absence of acidity in the flesh suites Indian palate.

Shan-e-Punjab : It is the most popular cultivar of peach because of larger size, yellow flesh and the skin colour and better keepability. It has proved to be potential variety to grow under Ludhiana conditions produced yield 8.5 t/ha of better quality in terms of fruit weight (86g/fruit) and 12.23^0 brix when planted at spacing of 6 x 1.5 m accommodating 1111 trees per hectare (Kanwar and Singh, 2004).

Early Grande : This is an introduced variety, which performs very well under Punjab conditions. Tree is productive (95.0kg/tree). Fruit is large (91.0g) with red blush on the surface. Pulp yellow, firm (firmness of 10.9lbs/inch^2) and some red colouration next to the pit and semi-free from stone when fully ripe. Fruit ripens in the 1st week of May and possesses excellent shipping quality.

Flordaprince : This is an introduced variety which suits well to the Punjab conditions. Fruit is medium to large (76.0g), round with little or no tip, red blush on yellow ground, pulp melting, yellow with some red colouration, semi cling having good fruit firmness (10.0 lbs/inch^2). It is a high yielding (100kg/tree) variety. Fruit ripens in the last week of April.

Pant Peach-1: This is a chance seedling selection from the population of cultiver. Sharbati under Pantnagar conditions. It ripens about one week prior to Sharbati but fruit quality is similar to *Sharbati*.

Pratap : This variety was selected for subtropical conditions in Punjab. Tree is good yielder (70 kg/tree). Fruit is medium to large (66.0g), yellow with red blush and pulp is firm (fruit firmness of 15.0lbs/inch^2) and yellow with red colouration. Fruit matures one week earlier (3rd week of April) than that of Flordasun.

Florda Prince Peach

Earli Grande

Babcock Peach

7.5.2 Pear

In pear, 'Patharnakh', is predominant cultivar, because of longer tree life (more than 50 years) unless killed by ganoderma, heavy bearing capacity and shelf-life (about 3 weeks, at ambient conditions). However, it has little local demand as immediately after harvest the fruit is of unattractive colour and hard texture. It has been found that the ripening treatments like, exposing to 0°C for 3 days or dip in 1500 ppm Ethephon, followed by ripening at 20°C results in development of aroma, attractive colour, softer flesh and increase in soluble solids. The semi-soft pears, 'LeConte' and 'Baghugosha' are erratic and poor croppers. Under Pantnagar conditions, 10-year old 'Baghugosa' plantings gave good yield 4.0 t/ha, 4.7 t/ha, 7.2 t/ha, 5.4 t/ha and 9.8 t/ha in square (5 m x 5m), paired (within the pair 2.5 m and between the pairs 5.2m), hedgerow (within the hedgerow 2.5 m and between the hedgerows 5 m), cluster (within the cluster 2.5 m and between the clusters 5 m) and double hedgerow (within the double hedgerow 2.5 m and between the double hedgerows 5 m) systems respectively (Bist and Yadav 2004). However, the selection of regular bearing and superior quality pears, like, Punjab Beauty, is likely to revive pear industry of northern plains. The salient features of other important cultivars/ varieties are as under :

Baghugosha : Tree is upright and vigorous cultivar. Fruit greenish yellow, stem-end tapering. Pulp white or cream coloured, sweet somewhat gritty. It ripens in August and stands transportation well. Average yield is 100kg per tree.

Patharnakh : Its tree is spreading and vigorous. Fruit medium, round, green with prominent dots. Pulp gritty, crisp and juicy. Keeping quality good. It stands transportation well. It ripens in the last week of July. Average yield is 150 kg per tree.

Punjab Beauty : Its tree is upright, medium in vigour, precocious, regular in bearing and good yielder (100kg/tree). Fruit is medium to large (125.5g) and firm (10.4lbs/inch^2 firmness). The ground colour of the fruit is yellow with red blush, pulp whitish, and good in taste (total soluble solids 14.0° B, acid content 0.30 per cent). Fruit becomes less gritty when mature and mellowed when ripe. It matures in the 3rd week of July.

7.5.3 Plum

Plum is also one of the potential fruit crops suitable to grow under subtropical conditions. In recent past, a number of successful results revealed that under Amritsar (Punjab) conditions the selection named as 'Alu Bokhara Peshawari' showed its promise by producing large sized fruits (23.50g) of attractive red colour having high TSS: acid ratio (11.08) coupled with some special merits like free stone, pink coloured flesh with uniform sweetness (TSS 13.3^0 brix) and thus is absolutely free from sourness near the pit, the serious most drawback of Indian plums. The maturity period is also quite earlier (2nd week of June) as compared to those of temperate plums (Bal, 2004). The other important varieties/ cultivars are as under :

Kala Amritsari : This is the most popular cultivar, particularly in Punjab. Fruit is medium (13g), round and dark brown at maturity. Pulp is yellow, moderately juicy and excellent for jam making. This is self-fruitful but yield improves if pollinated with Titron. It ripens in middle of May and average yield is 40-45 kg per tree.

Pant Plum-1: This is a selection from the seedling population raised at the Department of Horticulture, GBPUA&T, Pantnagar (Uttarakhand). It is dwarf and bears yellow and sub-acidic fruits. It can be a dwarfing rootstock for plum cultivars.

Satluj Purple : Tree is self-incompatible and must be inter-planted with 'Kala Amritsari' as pollenizer. Average yield of fruits is 30kg/tree. Fruit is quite big (28.0g), crimson, pulp thick, excellent in shipping quality and ripens in the 2nd week of May.

7.5.4 Apple

Winter chilling requirements for most varieties are 500 to 1000 hr. below 45° F (7° C). There are relatively few varieties adapted to mild winter climates (chill hours below 500). To ensure successful apple production in mild winter zone, select from the following varieties that need less than 300 hr. of chilling: 'Beverly Hills', 'Gordon', 'Tropical Beauty', 'Anna', 'Dorsett Golden', and 'Ein Sheimer'. Under mid-hill conditions of Solan, 'Tropic Beauty', 'Parkin's Beauty', 'Red Baron', and 'Aziza' are found promising (Sharma, *et al.*, 2004)

Dorsett Golden Apple

Anna Apple

Tropic Sweet Apple

7.5.5 Almonds

With the discovery of a peach x almonds hybrid in Nawanshahr (Punjab), and undue publicity to it, the almond improvement work was started at Ludhiana, under an adhoc project of ICAR in early 70's. Some useful introductions – 'California Paper Shell', 'Pathich's Wonder'; 'Selections – 'J.K. 55' and Hybrids 'H-15', 'H-258' were identified, but with the termination of the project in 1978 the cultural practices for almond cultivation could not be standardized as a result of which this promising nut could not establish in plains. The salient features of some other suitable cultivars are as under :

California Papershell : It ripens in the last week of June. The thin-shelled nuts are bold, weighing about 2.1g each and the average yield is 3.8kg/tree.

Hybrid 15 : The thin-shelled variety with bold nuts ripens in the first week of July. On an average, a nut weighs 1.7g and the average yield is 3.7kg/tree.

Pethick's Wonder : Thin shelled nuts of this variety are bigger than all other varieties. On an average, a nut weighs 2.5g, the average yield being 3.7kg/tree.

JKS 55 : This is an earliest ripening variety. Its nuts are harvested in the first week of June. The nuts are semi-hard and small and the average yield is 3.5kg dry nuts/tree.

7.6 FLOWERING AND FRUIT MATURATION

The floral biology, pollination and fruit set of temperate fruits under subtropical region have been studied by many researchers, possibly because of the fact that it is most easily studied aspect of development, but it has been of little practical use in the absence of any systematic breeding programmes. The availability of less-than required chilling exposures and abrupt rise of 3-5°C in daily temperatures and gusty winds in Feb-March results in gynoecial abnormalities and very poor fruit set in almond and in medium or high-chill varieties of plums and 'Baghugosha' variety of pear. 'Santa Rosa' plum blooms around mid-March at Ludhiana and ripens by mid-June but yield is poor because of stigmatic desiccation. The flowering is advanced by GA_3/Thiourea treatments. However, the findings require pilot scale testing. The post-fruit set availability of more heat units and soil moisture leads to faster rates of development and maturation of temperate fruits in the subtropical region. Some peaches and plums mature as early as the 3rd week of April, about 9-10 weeks after full bloom. The ripening precedes from eastern UP to northern Punjab in the same cultivars fruit. The pear requires 135-145 DAFB (Days after full bloom) to ripen. Because of faster rate of maturation and temperatures above 40°C in May, the keepability of the temperate fruits, except 'Patharnakh' pear, is shorter and with the existing marketing infrastructure the stone fruits remain in marketable condition only for 2-3 days of the harvest, that too with some physical damages. However, 'Patharnakh' pear have 10-15 days of post harvest life even at ambient temperatures, can easily be transported to distant markets of Kolkata, Bangalore, Chennai or exported to South-East Asian countries.

7.7 PEST AND DISEASES

The tree life of stone fruits is 10-12 years in the sub-tropical region because of wind damage, limb breakage, due flat head borer attack and gummosis or bacterial canker infections. No germplasm selection and effective biological or chemical control measure is available to manage fruit fly damage for want of which superior quality June ripening peaches cannot be grown and tree storage of pear from August onward which is physiologically feasible is not possible. Many sources of tolerant/resistant genes have been identified in these crops world over, but without dedicated breeding programmes for sub-tropical peach, plum and pears these can be put to little use only. Without solving some of these problems, the area and production of these fruits is likely to remain static or follow vicious cycle of aggressive planting and desperate uprooting following price crashes during gluts.

7.8 PROBLEMS

- Majority of the temperate fruit trees require climates with sufficient winter chilling to break internal bud dormancy. This condition generally can not be fulfilled in subtropical climate
- Temperate-zone cultivars have not been commercially productive in subtropics; however, some poor-quality fruits could be produced
- Lack of suitable cultivars for subtropical climate having higher yields and good quality.
- Lack of quality planting material of scion cultivars and rootstocks
- Poor fruit quality.
- Lack of technology to release the dormancy in all the fruit crops and cultivars
- Inconsistent dormancy breaking through use of chemicals
- Insufficient breeding work to support and development of low chilling, high yielding and better quality fruit crops
- Incidence of pest and diseases

- Lack of awareness
- Lack of financial support

7.9 FUTURE SCOPE

There are several success stories highlighting that temperate fruits could be cultivated under subtropical climate as in Venezuela, Israel, Vietnam, Australia and Mexico hence, these can be attempted in Indian conditions too. A collaborative effort among researchers, extension workers and development department could lead to make a breakthrough for their commercialization. This require a systematic approach which includes, identifications of zones according to chilling accumulation and high summer temperature, selection of appropriate fruit crop, selection of appropriate cultivar/ hybrid, awareness campaign by extension workers, and financial support by the financial institutions.

- India has a huge domestic demand of temperate-zone fruits
- Growing Indian economy making consumers more economically stronger
- Availability of market infrastructure for export-import of horticultural commodities, processing and value addition
- Recent attempts have led to the development of new cultivars of apple, peaches, and nectarines. These cultivars can be grown successfully under subtropical conditions. Locally adapted or developed cultivars/ varieties of peach and some plum cultivars produce crops of high quality in many parts of Punjab, Uttarakhand and Uttar Pradesh
- The use of dormancy-breaking agents made it possible to grow cultivars with moderate chilling requirements in areas having mild winters
- Due to increasing population pressure as well development of economy are forcing for higher demands. As a result a large amount of foreign exchange is required for importing the fruits. Since the majority of the countries in this area are still developing, these imports are not favourable to the national economy

7.10 IMPORTANT RECOMMENDATIONS OF 7TH INTERNATIONAL SYMPOSIUM ON TEMPERATE FRUITS IN THE SUBTROPICS, HELD AT DR YS PARMAR UNIVERSITY OF HORTICULTURE AND FORESTRY NAUNI, SOLAN, HP IN 2004

- There should be free exchange of germplasm and information among the scientists of different states and countries working on temperate zone fruits in the tropics and subtropics
- Priority should be given to conservation of wild relatives of temperate fruits and under utilized temperate fruits

- Concerted efforts need to be made for conserving the valuable temperate fruit germplasm suitable for growing in the tropics and subtropics.
- Apart from *in-situ* and on-farm conservation of low chilling requiring germplasm, *in-vitro* and cryogenic methods of conservation need to be developed
- Reports on evaluation and utilization of germplasm should be documented Passport data based on descriptors should be collected
- Research should be undertaken to workout the chilling requirements of different temperate fruit varieties, developing low chilling scab resistant apple varieties and walnut selections for varietal diversification
- There is a need for new low chilling requiring pome and stone fruit varieties having high yield and good quality should be developed for cultivation in the tropics and subtropics
- The *Utah* model was not found suitable under South African conditions due to high degree of chilling but daily positive chill unit accumulation gives good results. Evaporative cooling was found to be beneficial in supplementing winter chilling. Dormax (hydrogen cynamide) sprays in late winter had synergistic effects with evaporating cooling
- Total and partial lack of chilling affected the rate of bud break. The availability of carbohydrate was not the key factor but the capacity to utilize the soluble sugar seems to be more crucial. The capacity to synthesize starch could be used as a marker for growth capacity
- Insufficient winter chilling of apple tree in Brazil resulted in abnormalities like smaller flower, leaves and fruit size, small anthers with less pollen, low fruit set and flat fruit development.
- Estimated values of CU and GDH°C were more accurate than the suggested values (*Utah model*) for predicting full bloom in Delicious apple under Indian conditions
- In arid zone in the subtropical area in India like Rajasthan there exists a vast scope of growing low chilling cultivars of temperate fruits like peach, plum, apricot, strawberry and almond
- Multivariant analysis of temperature, intensity of light and rain indicated that rain and radiation can influence the overcoming of dormancy when these are linked to temperature
- Biotechnological approaches should supplement the conventional breeding tools for crop improvement
- In apple cultivar Red Fuji, chip budding in March proved to be superior over tongue grafting in terms of highest percentage of success, canopy spread,

length of plants and proportion of saleable plants under mid hill condition of India. This method of budding is required to be popularized for the commercial multiplication of all the temperate zone fruits grown in the tropics and subtropics

- In Silver Spur and Oregon Spur cultivars of apple grafted on EMLA 7a rootstock recorded an yield of 89.40t/ha and 91.17t/ha, respectively under valley areas and mid hill conditions
- In apple understanding of carbohydrate synthesis, assimilation and translocation may be helpful in manipulating flowering, fruit set and growth. Better understanding of abscission process may be useful in regulating fruit set and drop. Application of GA_3 and PP333 improved the fruit set
- M7 clonal rootstock is recommended for commercial cultivation of apple in warmer areas.
- Higher yields from Baghugosha trees during the initial years can be obtained by planting them at higher densities in double hedge row, hedge row and cluster system under subtropical conditions.
- The post-harvest treatments of $CaCl_2$ (4 and 8%) were recommended for retaining storage quality of Asian pear (Patharnakh) and semi-soft pear (Punjab Beauty), respectively. For increasing Ca uptake, it was suggested that Ca infiltration under vacuum should be tried. Waxing of Punjab Beauty pear with Citrashine before storage is also recommended for enhancing the storage/shelf life of fruit
- The cultivars like 'Madernassa' of pear which is being grown in Italy without any pesticide sprays should be introduced and evaluated in India
- Peach trees planted at 6 × 1.5 m and trained to 'Y' shaped structure intercepted higher radiation and improved its distribution within the tree canopy as compared to those planted at 6 × 6 m and 3 × 3 m and trained to modified leader system under subtropical conditions
- In peach girdling at full bloom stage enhanced fruit maturity by 13 days in Shan-e-Punjab cultivar. Thinning and girdling produced better quality fruit
- There is a distinct possibility of planting peach in a high density system under subtropical conditions with proper tree training. Trees planted at 6 × 1.5 m and trained to 'Y' shaped structure resulted in high yield and productivity and better fruit quality.
- Application of $Ca(NO_3)_2$ (0.2 to 0.4 per cent), three weeks before harvest was recommended for extending the storage life of Satluj Purple cv. of Plum. However for post-harvest treatments the use of $CaCl_2$ (4 %) dip for 5 minutes was also recommended for enhancing the storage/shelf life of Satluj Purple plum fruits.

- The cherry cultivars can be adopted for growing in tropics and subtropics as in case of peach and nectarines. The cultivars which are early season have good market potential. The cultivars like 'Cristobalina', 'Temprana de Sot', 'Precoce de Bernard', 'Sunburst', 'Lapins', 'Cordia' and 'Larian' which require low chilling are recommended for growing in the subtropics. Rootstocks like 'Giesla 5' and 'Mahaleb' are in use in Turkey and can be tried elsewhere also
- In cherry GA_3+NAA application at petal fall stage gave highest fruit set in 'Stella' cultivar
- The cultivars 'Chandler' and 'Etna' of strawberry which produces large sized fruits can be successfully grown in the subtropics and tropics
- Zinc sulphate (0.4 per cent) and ferrous sulphate (0.2 per cent) improved vegetative growth, yield and quality considerably in Chandler cv. of strawberry grown in subtropical areas of India
- For retaining the shelf life of strawberry, dipping of fruits in 0.05 per cent $CaCl_2$ at 10°C for 5 minutes has shown good results
- Early blooming almond strains should be tried for successful cultivation in warm areas
- Temperate fruits grown in subtropical and tropical areas are facing more severe pest and disease problems due to congenial and favourable environment for their development and spread. Hence, these fruits require intensive plant protection measures
- Ecofriendly biocontrol agents should be a major component of integrated disease and pest management so as to minimize the pesticide load in the orchards
- Soil solarization has been found to be effective against soil borne pathogens, which is cost effective and ecofriendly approach. This technique should be used to raise the healthy nursery relatively free from pathogens and nematodes. Effectiveness of soil solarization may be supplemented by the use of biocontrol agents and soil organic amendments
- Proper and regular monitoring of pests and diseases in relation to environmental changes should be made so as to record any change in pest status, as witnessed in India with respect to phytophagous mites and pre mature leaf fall of apple. There is a need to develop predictive models for pest and disease forecasting

7.11 REFERENCES

Bal, J.S. (2004) Genetic resources of plum under subtropical conditions of Punjab, India. *Acta Hort.*, 662: 147-150.

Bist, L.D. and Yadav, Ashish (2004) Long term studies on planting systems-cum-density of planting in low-chill soft pear under subtropical conditions. *Acta Hort.,* 662: 217-220.

Dennis F.G. Jr. (1997) Temperate zone fruits in out-of the way places- what does the future holds. *Acta Hort*, 441: 23-27

Dhatt, A.S. (2004) Prospectus of temperate fruit cultivation in Ind-Gangetic plains. *Acta Hort,* 662: 107-109.

Hutchins, (1932) Unpublished paper presented at the 1932 meeting of *Amer. Soc. Hort. Sci.*

Kanwar, J.S and Singh Harminder (2004) Scope of high density plantings of peach in the subtropics of India. *Acta Hort.*, 662: 221-224.

Mohammed Shafat (2004) Present status and future scope of growing temperate fruits in Rajasthan. *Acta Hort.,* 662: 101-105.

Sharma, G., Chua, G.D. and Sharma, O.C. (2004) Studies on evaluation and variability parameters in low chilling apple (*Malus x domestica* Borkh.). *Acta Hort,* 662: 157-162.

Sharpe, (1969) Subtropical peaches and nectarines. *Proc. Fla. State Hort. Soc,* 82:302-306.

Weinberger, J.H. (1956) Prolonged dormancy trouble in peaches in the southeast in relation to winter temperature. *Proc. Amer. Soc.,* 67:107-120.

Chapter – 8

Canopy Architecture Management for Precision Farming in Fruits

Vishal Nath and Amrendra Kumar

8.1 INTRODUCTION

In developing countries like India, sustainable development of horticulture is a biggest challenge. There is need to formulate strategies for their management through "state of art" technologies for harnessing the naturally available resources in a sustainable manner to meet the ever growing demands of fruits. Canopy architecture has its role in influencing the productivity of fruits. It has been well highlighted by Halle and Olderman, (1970) and Halle *et al.* (1978). The emerging concerns in the area of canopy architecture and its management are the existence of large number of old senile orchards among the fruits crops and the development of a rejuvenation technology to restructure their canopy and enhance productive and remunerative plant life. Actually, canopy management "designing the plant as per need, using inherent plant characteristics in accordance with given set of conditions and resources to make the plant to perform maximum. As a result of canopy management, the trees acquire their natural shape which very often is not ideal for quality production. A majority of the trees attain tall, upright and curved growth structure and the canopy is marked with criss-cross branches leading to a highly dense vegetative mass with very poor penetration of active solar radiation. Such conditions not only affect the photosynthetic rate but also facilitate proliferation of pests as they prefer shady conditions. Consequently with pressure of pest abundance and poor photosynthetic efficiency over the years in want of due care of tree canopy architecture, the trees turn senile and unproductive and uneconomical.

Existence of such unproductive senile orchards has a telling impact on the socio-economic as well as livelihood status of the farmers and consumers are equally affected with low poor production and poor quality of produce is the biggest problem with old orchard. Uprooting such orchards and undertaking new plantation may not be a wise option considering the long gestation period of orchard establishment and resulting impact on livelihood of farm families, environment health as well as the investment cost. Mass scale uprooting of trees for replacement of the orchard will also have very adverse impact on production scenario at the regional and national level.

Considering the enormity of the problem of the high acreage of senile orchards and slow adoption of rejuvenation technology (especially in mango and litchi), emphasis is laid upon canopy management and tree architectural engineering and management of orchards right from the establishment stage. The advantage with canopy management technology is that every individual tree in the orchard is subjected to proper care for development of ideal architecture with outward spread of canopy to facilitate greater penetration of active solar radiation in the canopy and thus enhancing the photosynthetic efficiency. Besides, the tree height is regulated at a manageable distance making it amenable for available machines for pesticide spray, training and pruning of branches, intercultural operations and also for harvesting of fruits. A sound architecture of the trees with open and outward spread canopy is critical for healthy plantation and sustainable production. Based on numerous research investigations, it could be said that fruiting potential of the tree is largely governed by its architecture, canopy density and photosynthetic efficiency.

The productivity of fruit crops depends upon several agronomical/ horticultural factors. Poor management of canopy architecture among them is the most important one. In fact fruiting potential of the trees is largely governed by their architecture, canopy density and photosynthetic efficiency. Canopy management in fruit crops deals with the development and maintenance of their structure in relation to the size and shape for maximum productivity with quality fruits. The tree architecture or canopy management, especially size control, has become a priority for reducing production cost and increasing yield and quality. Canopy design and shape influence light interception with assured higher monetary returns to fruit growers. Therefore, early height control and trees canopy management are important techniques which should be practised in fruit crops for higher returns. Trees have inherent plant characteristics viz. growth behaviour, flushing pattern, dormancy pattern and bearing behaviour on which the productivity of the trees lies and for yield maximization is governed a lot by :

planting system, spacing, solar radiation, wind velocity, rainfall pattern and light distribution etc.

8.2 CANOPY ARCHITECTURE MANAGEMENT

A series of operations aiming maximized production of quality fruits per unit canopy area is carried out towards proper canopy architecture management of the tree. An appropriate canopy architecture management encompasses the development of :

- Strong frame of primary branches
- Wider crotches in scaffold branches
- Healthy and well distributed secondary and tertiary branches

- Sufficient fruiting terminals in most productive areas
- Healthy foliage with high photosynthetic efficiency to maximize the solar radiation use efficiency
- Enough space for air circulation inside canopy

8.3 NEED OF CANOPY ARCHITECTURE MANAGEMENT

Poor canopy architecture management is one of the most important production constraint confronting the fruit industry in India. Indeed, canopy management is the manipulation of trees canopy to optimize its production potential with excellent quality fruits. In many fruit crops, increase in production with enhanced fruit quality is obtained by managing canopies for short-statured trees. Srivastava (2007) reported the influence or pruning on canopy environment and fruit colour intensity. Small trees capture and convert sunlight into fruit production in a better way than larger ones. Fruit production involves the capturing and conversion of sunlight into production of fruit biomass (dry-matter content). Kumar *et. al.* (2010) observed that medium pruning treatment had the highest yield in Flordasun peach. The main controlling factors in canopy management are amount of incoming radiation and percentage of that radiation which is intercepted by energy into fruits. In general, the question always arises as to why the plants may not be left as such to be governed by the nature itself. The answer is that the nature's purposes are not always the same as ours. Nature intends that a plant should live merely in order that it may reproduce itself in as great quantities as possible. Nature does not care how many good sized, high quality fruits are to be born by trees. For economical fruit growing, the natural form and shape of fruit trees have to be modified through the practices of pruning to achieve the target of more yield with better quality fruits. The fruit growers desire to regulate the growth and development of their trees in a specific manner. This is because it is not always wise to allow a plant to develop naturally, since unwanted portions may develop at the expense of those which are essential from the cultivator's point of view. The removal of unwanted plant parts and having a balance between vegetative and reproductive growth is thus known as pruning. The practice of pruning, as and when performed in an appropriate manner, keeps the plant in such a shape and condition so as to yield the fruits of desired quality. Climate has significant bearing on pruning. The fruit crops grown in different climate zones (tropical, subtropical and temperate zones) are required to be trained and pruned accordingly depending upon the type of plant and specific objectives of the fruit growers. Canopy management depends upon the choice of trees size, number of trees/ ha and pruning techniques to get optimum yield. Pruning should be done when trees are young, to establish their basic shapes and sizes, and continues as tree mature.

8.4 PRINCIPLES OF CANOPY ARCHITECTURE MANAGEMENT

- Facilitating aeration inside the canopy
- Maximizing light penetration in and outside of canopy

- Facilitating conversion of sunlight into chemical energy by plant leaf
- Controlled plant structure including stature
- Improved fruit colour, quality and yield

Canopy architecture management is the manipulation of tree canopies to optimize the production of quality fruits. The canopy management includes tree training and pruning which decides the quantity of light intercepted by trees, as tree shape determines the presentation of leaf area to incoming radiation. An ideal training strategy centres around the arrangement of plant parts especially to develop a structures that optimizes the utilization of sunlight and promotes productivity.

Light is critical to trees and their fruit growth and development. The green leaves harvest the sunlight to produce carbohydrates and sugars which are transported to the site where they are needed- buds, flowers and fruits. Improving light penetration into the tree canopy improves tree growth, productivity, yield and fruit quality. High density planting and orientation of planting is also a problem for light penetration in an orchard. Generally, in close planting, quicker shading as becomes a problem. An east-west row orientation results in more shading as compared to western and southern orientation of trees. Strong bearing branches tend to produce larger fruits. They also transport nutrients and water more efficiently throughout the trees. Pruning aims to encourage new and strong growth of all branches.

Some of the basic principles of the canopy management are maximum utilization of light, avoidance of build-up of micro climate congenial for disease and pest infestation, convenience in carrying the cultural operations, maximizing the productivity with quality fruit production, economy in obtaining the required canopy architecture *etc.*.

8.5 UNDERSTANDING ARCHITECTURAL BEHAVIOR OF PLANTS

Before, going for canopy management, understanding of plant nature and behaviour i.e. plant architecture is essential because :

- The architecture of a plant depends on the nature and relative arrangement of its each part.
- The expression of an equilibrium between endogenous growth process and exogenous constraints exerted by the environment

8.6 COURSE OF ACTION FOR CANOPY ARCHITECTURE MANAGEMENT

1. Analysis of architectural engineering aspects in plant, such as angling of branches and pattern of branching
2. Understanding the endogenous growth process like flushing pattern and action of growth hormones

3. Understanding the level of influence of environment on architecture at a point of time/location
4. Objectives for which canopy is being managed

8.7 BASIC OBJECTIVES OF CANOPY ARCHITECTURE MANAGEMENT

- To restore root/shoot ratio
- To obtain specific form of the plant
- To develop specific geometry within the plant and its overall topology
- To maintain optimum balance between vegetative and reproductive parts
- Plants respond differentially in different environment
- Development of shoot and root system in the plant has symbiotic relationship
- Weak crotches in the branches and direction of the branches can be corrected by training and pruning
- Pruning for dwarfing and promotion of maximum shoots/fruiting points
- Vegetative growth and fruiting/yield have antagonistic effects
- Apical dominance is a natural phenomenon

8.8 STEPS OF CANOPY MANAGEMENT

a. Selecting dwarf cultivars
b. Selecting appropriate planting system
c. Use of dwarfing rootstock and inter-stocks
d. Initial frame development
e. Training the plants with open centre
f. Allowing the canopy to grow horizontally
g. Centre opening of grown up trees
h. Pruning of the shoots and excessive growth

8.9 PRUNING AND TRAINING

Pruning refers to removal of parts of a tree, especially shoots, roots, limbs, buds or nipping away of the terminal parts. It is done to make a tree more productive and amenable to bear quality fruits, increase longevity of the tree, make the tree into more manageable shape and to get maximum returns from the orchard. In ornamentals, pruning gives a more aesthetic look to plant, but it is being given the lowest priority in commercial production. Meyers (1988) reported that heavy pruning by removal of potential leaf area and development of excessive growth, occurs at the expense of root growth and total tree growth. Heavy pruning early in the tree life decreases size and root growth as compared to the light pruning.

8.10 DIFFERENT TYPES OF PRUNING

There are two main types of pruning i.e. heading back and thinning out. In heading back, the terminal part of the shoot is removed to the desired level, for example, one-fourth or two-third of the shoots may be headed back. Such pruning encourages development of secondary branches. In thinning out, the entire branch is cut right from the base. It allows more of sunlight to enter the interior of the tree because of its opening effect.

Another type of pruning is dehorning. It consists of removal of all the wood after leaving 7-10 cm thick stub all over the tree. Such severe treatment should normally not be given except forms the last option to commercial bearing.

8.11 STRATEGY FOR MANAGEMENT

8.11.1 Light management

- Light exposure to all parts of the canopy
- Maximize bearing area in a minimum of time
- Sustain maximum bearing area for maximal time

8.11.1.1 Mango

An initial branching height between 60-70 cm is appropriate. Train the plants and allow 3-4 horizontal branches to grow. Remove the upright growing branches and select horizontal shoots form the main scaffold limbs of the trees. Allow the branches to grow and during the second flush the tips should be pinched for maturation of the shoots. This procedure is to continue during the 2-3 year also. If selected shoots grow vertically this should be removed. Thus the horizontal scaffold limbs are stronger and flower and fruit earlier. Short branches within tree canopy produce a complex, compact and strong structure for fruiting shoots after third year.

For rejuvenation senile orchards and also the bearing behaviour of the free is identified. The branches which are to be removed are marked and also those branches which are to be kept for development of canopy in future and heading back operation to be done during November – December are also marked. All the cut branches and twigs are removed from the orchard and surface of the land is cleaned for making the basin for irrigation, manures and fertilizer application. Profuse emergence of new shoots starts after few months (March-April) on pruned branches which require light thinning of excessive shoots. About 8-10 healthy shoots with outwardly growth per branches are retained and remaining are removed. The recommended intercultural operation for control of insects and diseases are followed. The properly managed trees start bearing from third year and onwards with quality fruits.

8.11.1.2 Litchi

In general, litchi trees grow continuously and develop into larger canopy. But if we grow systematically promotion of large bearing area, accommodation of large no of plants/ per unit area, tree architecture management particularly tree height, canopy and vigorous shoots concentration to provide more fruiting twigs and penetration of light to inner portion become more relevant. Thus giving a proper shape to the litchi plants from initial year of growth and pruning of branches after harvesting in bearing trees is more essential. Since more sprouting takes place in young plants, proper shape is quite convenient. Single stem air layered plants (40-50cm) should be transplanted in field and allowed to grow straight up to 60-70cm, then 3-4 branches spread in all direction are allowed to grow from 50-55cm height on stem and all other branches in initial stage are removed. It is necessary to continue shaping the plant by removing those branches which form narrow crotches with main branches. To develop good and compact canopy, 15-20cm fruit bearing shoots at the time of harvesting are removed. In this way 2-3 new terminals develop which consequently develop into fruiting branches next season.

Unproductive trees are pruned heavily to develop new fruitful shoots. In such cases, heavy reiterative pruning, usually up to limbs at a height of 3-4m is commonly followed, supplemented with heavy application of nutrients and these shoots starts fruiting 2-3 years after pruning. Thereafter, general pruning is followed to maintain ideal vigour and productivity of trees.

(a) (b)

Fig. 1: Hedgerow cropping in (a) apple and (b) guava

8.12 REJUVENATION

This to be done in old senile unproductive orchard during dormancy at a height of 2.5-4m depending upon the tree structure. After topping 3-4 main branches with outward growth are kept and rest is to be removed. All the cut branches need to be removed from the field for making the basin and application of manures and fertilizer. The new shoots start to emerge within November and onwards, and a large number of the new shoots emerges and cover the entire branches. Thinning up the new shoots are important and allow 5-6 outward shoots to mature and form the canopy in each branches as semi circular or open umbrella type canopy. There should be access of light and air in all parts of trees.

8.13 FRONTIER CONSIDERATIONS OF CANOPY MANAGEMENT

- Conservation of stored food
- Maintaining root : shoot : fruit by pinching rather pruning
- Pruning time and severity
- Provision of more light and air inside the canopy for stereo fruiting, colour development
- Single canopy with multiple root system concept
- Boosting up of economical canopy formation
- Exploring the possibilities of mechanization in pruning

8.14 CONCLUSION

Canopy architecture management is therefore, an essential tree management operation which starts from the first year of plant establishment and species and environments but enables the plant to produce high quality and quantity by providing proper framework, more fruiting area and utilization of available resources.

8.15 REFERENCES

Halle, F. and Olderman R.A.A. (1970). *Essai Sui I'architecture et al. dynamique de croissance des arbes tropicaux*. Masson et Cie, Paris.

Halle, F., and Olderman R.A.A. and Tomlinson, P.B. (1978).*Tropical trees and forests*. Springer - Verlag, Berlin.

Kumar, M.; Rawat, V.; Rawat, I.W.S and Tomar, Y.K. (2010) Effect of pruning intensity on peach yield and fruit Quality. *Scientia Horticulture*, 125 : 218-221.

Meyers, S.C. (1988) Basics in open-centre peach tree training. In the *Peach* (Eds. childers NF. and Sherman WB), Somerset Press, Inc. Somerville, New Jersey, pp. 389-403.

Srivastava, K.K (2007) *Canopy management of fruit crops*. IBDCO, Lucknow.

Chapter – 9

Biotechnological Interventions in the Improvement of Fruit Crops

Krishan Kumar, Kuldeep Singh and M.I.S. Gill

9.1 INTRODUCTION

India is the second largest producer of fruit crops with annual production of 68.4 million metric tonnes from an area of 6.1million ha (Anon 2010). The national horticulture mission commenced in 2005 had fixed a target to double the horticultural production by 2011-12. According to the available production data of horticultural crops, it seems that we are very close to achieve our target. But, if the loss caused by biotic and abiotic stresses is accounted, a sound conclusion can be drawn that still there is a huge scope for productivity improvement in fruit crops.

Among biotic factors bacteria, fungi, nematodes, viruses and insect pests, diseases and other pests are widely recognized as the biggest threat to the sustainable production of fruit crops. Every fruit crop is infected with one or other pathogen/pest, which causes a significant reduction in yield. The disease alone both in field and storage accounts for 25-30 per cent losses (Rangaswami and Mahadevan 2002). Along with diseases, insect-pests also cause reduction in the yield. About 600 insect and mite pests have been reported in temperate fruits alone in India (Sharma and Singh 2006).

Beside biotic factors, abiotic factors (salt, drought, low and high temperature) also limit the potential yield of fruit crops. It is estimated that nearly 10 million ha is affected by salinity/alkalinity in India (Singh 2006). Moreover, fruit cultivars are also to be modified for yield and traits of consumers preference e.g. seedlessness in grapes and citrus.

Genetic improvement can be achieved by following the conventional breeding approaches, when the gene of interest is present in related cultivars/species and the crop is sexually crossable. But due to long juvenility and genetically prevalent problems like incompatibility, sterility in most of the fruit crops and apomixis in citrus, mango and apple, conventional techniques seem to have limited role in the improvement of fruit crops.

Various biotechnological approaches either alone or in combination with conventional breeding approaches can provide a direct solution for the genetic improvement of fruit crops.

9.2 ROLE OF BIOTECHNOLOGICAL TOOLS

The role of biotechnological tools for the improvement of fruit crops has been discussed under the following sub-heads :

1. Somaclonal variation
2. *In vitro* mutagenesis
3. Somatic hybridization
4. Embryo rescue
5. Genetic modification of fruit crops
6. Markers aided breeding

9.2.1 Somaclonal variation

Variation in the somatic tissue generated by the use of tissue culture-cycle is called somaclonal variation (Larkin and Scowcroft 1981). The plants regenerated from cell/ tissue culture differ in a number of morphological traits like precocity, yield and reactions to pathogens, insects and abiotic stresses. Variability is observed in plants regenerated from various culture sources such as callus, protoplast, embryo and even meristem. It is a major tool for creating variability in the population, when it is not present naturally. The actual cause of somaclonal variation is not known but, it is thought that it may be due to one of the reasons like : chromosomal changes, enhanced amplification of certain genes, mitotic crossing over, cytoplasmic genetic changes, methylation of genes etc.

The somaclones and other plant propagules can be effectively selected against a particular stress under in vitro conditions. Advantages include : minimized environmental influence, the potential to handle large numbers of individuals in a very small space and accelerated growth and development (Wenzel and Foroughi-Wehr 1993). For screening of somaclones against biotic stresses, they are grown on medium containing (cultural filtrates of bacteria, fungus etc.) while for abiotic stresses (like salt, drought *etc.*) in medium with higher salt concentrations and high molecular weight compound like polyethylene glycol (PEG). A strong correlation has been reported between the resistance/ tolerance expressed in cell cultures and enhanced level of resistance in the whole plants.

The potential of the technique has already been exploited for resistance to diseases in fruit crops *viz*., peach (Hammerschlag and Ognjanov 1990; Hammerschlag *et al.*, 1994), apple (Donovan *et al.,* 1994), strawberry (Takahashi, 1993), banana (Hwang and Ko, 1992) *etc*. The somaclones regenerated against different stresses in fruit crops are presented in Table1a.

Table 1a : Somaclonal variants for stresses resistance/tolerance in different fruit crops

Fruit crop/cultivar	Objectives	Somaclones	Remarks	Reference
I. BIOTIC STRESS				
a) Bacterial disease				
Prunus persica cvs. Sunhigh and Redhaven	Leaf spot (*Xanthomonas campestris* pv. *pruni*) and canker (*Pseudomonas syringae*)	S61, S156 (Sunhigh) S 122, 31 and 46 (Redhaven)	S156 R* and S122 HR to leaf spot, MR to canker	Hammerschlag and Ognjanov 1990
Apple cv. Green Sleeves	Fire blight (*Erwinia amylovora* strain T)	16 somaclones	21% less symptoms than Green Sleeves	Donovan *et al.* 1994
b) Fungal diseases				
Apple rootstocks (M26, MM106)	*Phytophthora cactorum*	-	S-2 (M26) performed better	Rosati and Predieri, 1990
Giant Cavendisha Banana[20%]	*Fusarium* wilt (race-4)	10 somaclones	GCTCV215-1 for commercial planting	Hwang and Ko, 1992
Williams banana	Yellow and black sigatoka	CIEN BTA-03	High yield and resistance to yellow and black sigatoka	Unai *et al.*, 2004; Giménez *et al.*, 2008
Strawberry cv. Morioka	Black spot (*Alternaria alternata*)	M16AR-1, 2& 3	-	Takahashi, 1993
II. ABIOTIC STRESSES				
Quince BA29 somaclones	High pH and lime soils	-	7-A (High pH) and 6-A, 5-7-A (In line soils performed better)	Marino *et al.*, 2000
Apple rootstock M_9, M_{111} & M_7	Al toxicity	Somaclones	-	Dantas *et al*, 2001
Actinidia deliciosa cv. Tamuri	NaCl (85.5 mm)	5 somaclones	Tolerant	Caboni *et al.*, 2003

[20%]Except others (calli) in this case meristem derived plantelts were used for screening

*R - Resistant, HR – Highly resistant, MR – Moderately resistant

However, some variation in the somaclones may also be due to epigenetic changes that puts a question on the reliability of the technique. Keeping this in mind, the field stability of the somaclones has also been studied by various researchers and examples are presented in Table 1b.

Table 1b : Stability of somaclones to biotic stresses under field conditions

Fruit crop/cultivar	Somaclonal variant	Resistant/ tolerant to	Field performance	Reference
Prunus persica	S19-1 and S-156 (Sunhigh) S122-1 (Redhaven)	Bacterial leaf spot	*HR after 3 yrs in field	Hammerschlag *et al.*, 1994
	S122-1	*Pseudomonas syringae*	HR compared to Redhaven	
Strawberry cv. Marioka 16	M16AR-1, M16AR-2 and M16AR-3	Alternaria leaf spot	H.R., M16AR-2 registered as Akita Berry after 3 yrs	Takahashi *et al.*, 1993
Apple cv. Green Sleeves	-	Fire blight (*Erwinia amlylovora* strain T)	60% less symptoms as compared to parent cultivar	Donovan *et al.*, 1994
C. limon	FS 01 and FS 11	Mal Secco (*Phoma tracheiphilla*)	Tolerance level comparable to Monachello (T)	Gentile *et al.*, 2000
Agbagba (False Horn plantain)	AO 2B2-2	Black sigatoka disease	Higher tolerance as compared to Agbagba	Nwauzoma *et al.*, 2002
Banana *cv.* William	CIEN BTA-03	resistance to yellow and black sigatoka	Similar performance even after 5 years of field trials	Unai *et al.*, 2004 ; Giménez *et al.*, 2008

*R - Resistant, HR – Highly resistant, MR – Moderately resistant

Beside selection of the somaclones, *in vitro* screening technique has also been utilized for evaluation of resistances of genotypes against various stresses *viz*., virus resistance to plum pox virus in apricot (Martinez-Gomez and Dicentia 2000), high pH tolerance in blue berry (Finn *et al.,* 1991), drought water stress in olive (*Olea europea* ssp. *maderensis*) (Brito *et al.,* 2002).

9.2.2 *in vitro* mutagenesis

Since the discovery of X-rays by H J Muller in 1927, the mutagens have been used for fruit breeding routinely. The mutagens have been used in two ways in fruit crops : treatment of seeds and budwood with LD_{50} dose of the mutagen and their subsequent phenotypic evaluation in the field (*in vivo* mutation breeding) and combining the mutation breeding with *in vitro* regeneration system (*In vitro* mutagenesis). The use of *in vivo* mutation breeding has assisted in the production of as many as 50 cultivars in fruit crops and some notable examples include pear cv. Gold Nijisseiki is resistant to black spot (Yoshokita *et al.,* 1999), dwarf growth habbit of Pusa Nanha papaya (Ram and Majumdar 1981) and various seedless citrus cultivars. Under *in vitro* mutagenesis, any plant part, which can regenerate into complete plantlets are used as explants. *In vitro* mode of mutagenesis can prove even more advantageous in fruit crop improvement than its counterpart *in vivo* mutation breeding due to the following points :

i) Pre and post mutagen treatment of source material to obtain a large population in a small space

ii) When the mutagens are used on undifferentiated cell/callus or somatic embryos or tissues without preformed axillary buds either prior to regeneration or at different stages of differentiation, somaclonal variation effect is added to the mutagenesis.

iii) Use of single cell or few cells helps in direct identification of mutants without undergoing 3-4 cycles of vegetative propagation after mutation induction i.e. M_1V_4

iv) Mutants can be easily screened against the selective agent

Both physical (Gamma rays, X- rays, UV rays, thermal neutrons and Heavy ion beam) and chemical (EMS, MMS MNH *etc.*) mutagens are being used for mutation induction. The advantages of physical mutagens are accurate dosimetry, reasonable reproducibility and uniform penetration of multicellular tissues. The benefit of using chemical mutagens is a high mutation rate and predominantly point mutations (Jain, 2005). For fruit crops, EMS and gamma rays have been widely used mutagens for mutation induction. The mutagens create variations by causing alterations at genomic level. Gamma rays induce breaks in the genomic DNA and mutation is produced by counter reaction of the genome, when the DNA starts operating its repair mechanism. EMS causes mutation by adding ethyl group to N or O at 7th and 6th positions of guanine base, respectively. These changes impose higher affinity of the altered guanine to bind with thymine instead of cytosine. Thus, GC is subsequently converted into TA. Among the physical mutagens category, Heavy ion beam (HIB) is the latest addition, and it can induce more and complex DNA breaks and thereby, possibility of mutation frequency is increased (Jain, 2010). The heavy ion beam can be generated by ^{14}C, N, Ne or Fe ions. The potential this mutagen has been exploited in the improvement of ornamental crops but there is no report regarding its use in fruit crops.

The basic steps of *in vitro* mutagenesis are : mutation induction and estimating the LD_{50} (the dose at which 50 % growth reduction or plants survived), treatment with LD_{50} dose of the mutagen, mutant isolation by *in vitro* screening and regeneration of the mutants. The research regarding the estimation of LD_{50} dose of various mutagens for various explants in fruit crops and subsequent screening against selective agents is presented in Table 2.

Table 2 : Optimum doses of mutagens for different types of explants in fruit crops their utilization towards fruit crop improvement

Fruit crop	Explant	Mutagenesis source and dose	Improved trait	Reference
Citrus sinensis	-	37.5 Gy gamma rays	-	Goldman and Ando, 1990
Citrus aurantifolia	Callus	Gamma rays (5-10 Gy) and EMS (0.1-0.4 %)	Mutants resistant to citrus canker	Srivastava *et al.*, 1997
Citrus	Callus	50-80 Gy gamma rays and EMS (0.3-0.5 %)		Wan *et al.*, 1991
Banana cv. High Gate (AAA)	Shoot apices	2.3 mM NaN_3 for 30 min., 25 mM Diethyl sulphate (DES) for 60 min. 200mM EMS for 30 min.	-	Bhagwat and Duncan, 1998
Grand Naine	Shoot tips	80 Gy gamma rays	Novaria- 10 week early flowering and high keeping quality	Mac *et al.*, 1996
Grand Naine	Shoot tips	Gamma rays	DPM 25 – with high yield, large fruit size and resistance to black sigatoka disease	Smith *et al.*, 2006
Apple cv. Indo	-	80 Gy gamma rays	Mutants resistant to *Alternaria alternate*	Masuda and Yoshika, 1997
Apple	Nodal segments	10 Gy X- rays	Tolerance to *Podosphora leutricha*	Schum *et al.*, 1999
Pear cultivars	Leaves	Gamma rays at 20-50 Gy	Mutants resistant to *Erwinia amylovora*	Pinet Leblay *et al.*, 1992
Pineapple cvs. Sugarloaf and Smooth Cayenne	Shoot tips	Gamma rays at 45 Gy	Induction of mutants for drought tolerance	Loko and Amoatey, 2001
Avocado cvs. Duke and Hass	Zygotic embryos	Gamma rays : 28 and 27 Gy for Duke and Hass	-	Fuentes *et al.*, 2004
Datepalm cv. Deglet Noor	Embryogenic cultures	Gamma rays at 20 Gy	-	Jain, 2007
Mango cvs. Tomy Atkins, Keitt and Hindi be Sennara	Embryogenic cultures	Gamma rays at100 Gy	Resistance to anthracnose	Litz, 2009
Guava cv. Safeda	Shoot tips	Gamma rays at 45 Gy	-	Zamir *et al.*, 2009

Although the optimum dose of the mutagens has been determined for a number of explants in different fruit crops (Table 2). However, *in vitro* mutagenesis has led to varietal development only in banana [(Mac *et al.,* 1996) and DPM 25 (Smith *et al.*, 2006)].

9.2.3 Somatic hybridization

Somatic hybridization is a technique of fusion of protoplast from two diverse parents to combine the nuclear and cytoplasmic genome of the organisms. Somatic cell fusion has the potential applications for crop genetic improvement by overcoming sexual incompatibility or reproductive barriers and by creating novel combinations of nuclear and or cytoplasmic genomes (Liu *et al.,* 2005).

Related or distant species/genera are the reservoir of many desirable genes. Gene transfer from these species through conventional breeding is difficult owing to post and pre-zygotic sexual incompatibilities in fruit crops. For Somatic hybridization, citrus has served as the model crop. In citrus, besides the above-mentioned barriers, nucellar embryony is also the major hurdle in its genetic improvement through sexual hybridization. Since the production of first somatic hybrid in 1985 (Ohgawara *et al.,* 1985), more than 300 somatic hybrids (Intergeneric and Interspecific) have been produced in citrus to date (Pavan *et al.,* 2007). Out of these 300, more than 150 somatic hybrids are the contribution of Jude W. Grosser group (*Citrus Research and Education Center, IFAS, University of Florida*) while rest is the combined contribution of X X Deng group (*National Key Laboratory of Crop Genetic Improvement, National Center for Crop Molecular Breeding, Huazong Agricultural University, Wuhan, China*) and groups from Japan, France and Israel. The somatic hybridization potential has been exploited for rootstock and scion improvement in citrus (Grosser *et al.,* 2004). The possible applications of somatic hybrids in citrus are as follows :

i) Direct use as a rootstock/scion,

ii) To utilize the somatic hybrid in the breeding programes as pollen parent to impart seedlessness through triploidy,

iii) Partially combining the genome from the donor parent into the recipient through irradiation/ asymmetric fusion, and

iv) Cybrids production to introduce seedlessness in the commercial cultivars.

9.2.4 Embryo rescue

This is the technique of aseptically culturing the developing embryo after excising from the ovule, if possible and even in some cases culturing the ovule or ovary to save the embryo, which could otherwise be lost due to one of the following factors :

i) Due to incomplete development in early ripening varieties of grapes and peaches (flesh in early ripening cultivars mature before seed maturity),

ii) High abscission of fruits e.g. avocado,

iii) In crosses involving distant parents,

iv) Post zygotic barriers e.g. stenospermocarpic grapes.

This is a classical technique which supplements well the conventional breeding. The first use of embryo culture dates back to the work of E. Hanning in 1904, who successfully cultured the mature embryos of two crucifers on artificial medium (Narayanaswami and Norstog, 1964). In fruit crops, it was Tukey (1933), who first cultured cherry embryos on the artificial medium. The success of cherry was replicated in peach and grapes. The success of embryo culture depends upon the stage of embryo development at the time of excision and the culture medium composition (Sharma *et al.,* 1996). Generally, smaller is the embryo, more complex is the medium requirement. The exact age of embryo excision has to be standardized for successful conversion of embryo into a plantlet. If embryo aborts at an early stage, the medium composition has to be modified in a manner so as to allow the proper development of the embryo before its germination into a plantlet. The culture medium generally contains all the constituents, which an endosperm provides to the developing embryo. Still, the culture medium requirement varies from crop to crop (for further reading, see Sharma *et al.,* 1996). The embryo rescue has benefited the fruit crop breeding in following ways :

i) Early ripening peach varieties can be employed as female in the breeding programme.

ii) By using embryo rescue, seedless (A) x seedless (B) strategy can be followed in grapes, which helps in getting rid of using intermediate seeded female parent in the crossing programme to combine the desired traits from two seedless varieties in a hybrid and one additional generation and at least 5 year time (Ramming, 1990)

iii) Transferring genes of disease resistance from wild relatives .

The use of embryo rescue along with its achievements in fruit crops is listed in Table 3.

Table 3 : Embryo culture in the rescue of fruit hybrids

Fruit crop	Crosses	Objective	Embryo excision time	Achievements	Reference
Grapes	Muscat Hamburg (*Vitis vinifera*) x *V. amurensis*	Resistance to downy mildew and anthracnose	-	Released as variety 'Beichun' with high resistance to powdery mildew and anthracnose	Li *et al.,* 1983
	Bharat early x Beauty Seedless, Beauty Seedless x Perlette		Beauty Seedless x Perlette, best age was 24 days after pollination		Midani *et al.,* 2002
	Red Pearl (Tetraploid) x Muscat Bailey (Diploid)	Triploids	Embryos larger than 1mm in their length		Hiramatsu *et al.,* 2003
	Thompson Seedless x Concord			Blue black seedless grape released as Thomcord in 2003	Anon., 2004
	cross between complex seedless hybrids (C33-30 X C103-41)	Seedless grape with uniform red colour and muscat flavour		Released as Sweet Scarlet in USA in 2004	Anon, 2004
	Thompson Seedless × 15 different lines with resistance to downy mildew	Downy mildew resistance	In ovul embryo culture 6-10 weeks after pollination		Sahijram *et al.,* 2005
	Diploid × tetraploid cultivars	-	6-9 weeks after pollination (WAP) in early varieties, 7-10 WAP in mid and 9-12 WAP in late varieties		Xu *et al* 2005
	Emerald Seedless x Beichun	Downey mildew and anthracnose resistance	3 days after veraision	Two accessions showed less severity index	Tian and Wang, 2008
Citrus	Allote traploids somatic hybrids × diploid (Shatian pummelo , HB pummelo and Guoqing satsuma)	Triploid seedless		147 triploids from the cross with Shatain pummel	Song *et al.,* 2005
	Diploid × tetraploid	Triploid seedless		11-65% developed into normal plants	Viloria *et al.,* 2005
Papaya	*Carica papaya* x *Vasconella quercifolia*	Resistance to PRSV			Drew *et al.,* 2006
Peach and nectarine				'May Fire' nectarine and 'Gold Crest'peach resulted from embryo culture	Ramming and Tanner 1987a, 1987b

9.2.5 Genetic modification of fruit crops

The technology of introduction of artificial genes in the host genome is called genetic transformation. The process of introduction, integration and expression of *trans* (foreign) or *cis* (genes from same species/ genera in the undisturbed state,) gene(s) in the host is called *trans/cis* genesis. Genetic improvement through transgenesis is an old concept. The first transgenic through transgenesis was produced in tobacco (1983), closely followed by a fruit crop (papaya), which was later commercialized in 1998. The production of transformed plants with plant genes (cis genesis) is relatively a new concept (Jacobsen and Schouten, 2007) and genetically modified fruit crop through this mean are yet to come.

Several approaches have been used for the development of transgenics but *Agrobacterium* and particle bombardment is the methods of choice for the genetic transformation. Although the commercialized transgenic fruit tree (papaya) was produced with the aid of particle bombardment but most researchers have relied upon *Agrobacterium* for gene transfer in fruit crops. In citrus, the modifications in the existing *Agrobacterium* method were also tested (Oliveira *et al.,* 2009). They compared the Sonicated assisted *Agrobacterium* mediated transformation (SAAT), vaccum infiltration, a combination of the two and the traditional *Agrobacterium* mediated transformation. The higher transformation efficiency was obtained with the combined approach : a combination of SAAT for 2 seconds followed by 10 min of vacuum infiltration. Table 4 gives details of the experimental transgenic produced in field crops.

Table 4 : Transgenic fruit plants with improved horticultural traits

Particulars	Gene used	Fruit crop	Reference(s)
Virus Resistance			
Arabic mosaic virus (ArmV)	Coat protein (*cp*) gene	Grape#	
Citrus tristeza virus (CTV)	Coat protein (*cp*) gene	Citrus#	
	Coat protein (*cp*) and Galanthus nivalis agglutinin (*gna*) gene	Citrus#	
Citrus psorosis virus	*cp* gene	*Citrus sinenesis*	Zanek *et al.,* 2008
	RdRp (anti sense and untranslatable version of Replicase derived RNA polymerase)	*Citrus paradisi*	Cevik *et al.,* 2006
	GLRa V-3 gene	Grape#	

Contd.

Grapevine leaf roll virus (GLRV)	*GLRa V-2* and *GLRa V-3* gene	Grape#	
Grapevine virus A (GVA)	Coat protein (*cp*) gene	Grape#	
Grapevine virus B (GVB)	Movement protein (*mp*) gene	Grape#	
Grapevine fan leaf virus (GFLV)	Coat protein (*cp*) gene	Grape#	
Grapevine chrome mosaic virus	Coat protein (*cp*) gene	Grape#	
Papaya ring spot virus (PRSV)	Coat protein (*cp*) gene	Papaya#	
	PRSV replicase (*rp*) gene	Papaya#	
	PRSV TK (*cp*) gene	Papaya#	
PRSV Papaya leaf-distortion mosaic virus (PLDMV)	*cp* gene	Papaya	Kung *et al.*, 2010
Plum pox virus (PPV)	Coat protein (*cp*) gene Coat protein (*cp*) gene	Apricot# Plum#	
Bacterial Resistance			
	Shiva-1 (Insect cecropin)	Apple#	
Bacterial resistance	Antibacterial AP-D (synthetic antibacterial peptide D) gene	Citrus#	
Crown gall resistance	dsRNA from *iaa*M and *ipt* sequences	Apple#	
Black heart resistance	PPO (poly phenol oxidase gene)	Pineapple#	Graham *et al.,* 2000
Fire blight	*Attacin E* gene	Apple#	
	Cecropin MB-39 gene (modified SB-37 gene)	Apple#	Liu *et al.,* 1999, 2001
	Shiva-1	Apple#	
	Shiva-1	Grape#	
Pierce's disease	**pPGIP* (Pear) Polygalacturonase inhibiting protein	Grape	Aguero *et al.,* 2005

Contd.

Citrus canker	*Attacin* A gene	Citrus	Boscariol *et al.,* 2006, Cardoso *et al.,* 2010
	**Xa* 21 from rice	Early Gold sweet orange	Guo and Grosser, 2004
	*hrp*N genes from *Erwinina amylovora*	*Citrus sinensis* cv. Hamlin	Mendes *et al.,* 2009
Citrus blight	Expansins related P12	Carrizo citrange	Kayim *et al.,* 2004
Fungal Resistance			
Apple scab	Chitinase gene	Apple#	
	Ech 42 (endochitinase) and *Nagl* (exochitinase) genes	Apple#	
	ThEN-42 (endochitinase) gene	Apple#	
	AMP (antimicrobial peptide) gene	Apple#	
Fungal/microbial resistance	PD (plant defensin) gene-cystine rich antimicrobial peptide	Apple, Pear#	
Fungal resistance	Stilbene sysnthase (*Vstl*), PGIP genes	Apple#	
Powdery mildew and anthracnose	*RCC 2 (rice chitinase gene)	Grape#	
Phytophthora palmivora	*Stillbene synthase gene from grapevine (*Vst1*)	Papaya	Zhu *et al.,* 2004
Botrytis rot	**p*PGIP	Grape	Aguero *et al.,* 2005
Fusarium wilt	AMP gene	Banana# (AAA)	
Verticillium wilt	**pcht 28* (Chitinase gene)	Strawberry#	
Grey mold	**fa*PGIP (*Fragaria ananasa*)	Strawberry	Schaart, 2004
Collectotrichum acutatum	Acetolactate synthase gene	Mexican key lime, USA	Chung *et al.,* 2002
Insect Resistance			
Insect resistance (BT)	Cry IA(c) gene	Walnut#	Mc Granahan *et al.,* 1988; Dandekar *et al.,* 1989, 1994
	Cry IA (c) and ICP gene	Apple#	Dandekar *et al.,* 1991
	Btk-ICP gene (*B. thuringenesis* var kurstak gene)	Juneberry#	Hajela *et al.,* 1993

Contd.

Abiotic Stress Tolerance			
Cold/freezing tolerance	Antifreezing gene CBFI gene	Grape# Strawberry#	
Salinity tolerance	HAL 2 (halotolerance) gene MdNHX1 *cod*A gene & cDNA S6PDH	Citrus# Apple rootstock M 26 Persimon#	 Li *et al.*, 2010
Aluminum tolerance	Chimeric citrate synthase	Papaya#	
Other Traits			
Precocious bearing	*AP1* gene from *Arabidopsis* for floral meristems	Carrizo citrange (Citrus) *Fortunella crassifolia*	Cervera *et al.*, 2009 Duan *et al.*, 2010

*Plant genes, #In : Gill *et al.*, (2004)

9.2.6 Marker aided breeding

The role of molecular markers for improving fruit crops is presented under the following subheads :

A. Markers assisted selection

B. Map based gene cloning

C. Genome sequencing

9.2.6.1 Markers assisted selection

It is an indirect selection approach, in which a particular individual for a trait is selected on the basis of the linkage of molecular marker with gene of interest. The molecular markers used in this way increases selection efficiency. The pre-requisites to follow marker assisted selection are :

i) Generation of molecular linkage map

ii) Gene mapping and tagging

iii) High resolution mapping

9.2.6.1.1 Generation of molecular linkage map

Linkage map indicates the position and relative genetic distance between the markers along chromosomes, which is analogous to signs or landmarks where gene are the houses (Collard *et al.*, 2005). Construction of linkage map is the first step towards locating the gene or QTLs of importance on a particular linkage group. The production of linkage map is based on the principle that genes (markers/loci) segregate via chromosome recombination during meiosis. During this process, the markers located on the different chromosomes assort independently (unlinked) while the markers present on the same chromosome tend to remain linked and form linkage group. As during meiosis, two non sister chromatids are involved in strand exchange, recombination does take place between the DNA segments. The chance of a cross over producing recombination between genes/markers is related to the genetic distance between them. Closer they are situated, lesser is the chance of cross over between them (Semagn *et al.*, 2006). The production of linkage map has the following requirements :

a) Development of mapping population

b) Identification of poylymorphism between the parents and marker genotyping of the individuals

c) Linkage analyses and map construction

9.2.6.1.1.1 Development of mapping population

The parents used for developing mapping populations should show genetic difference for the desirable traits at least and maximum number of traits ideally. The ideal mapping populations are recombinant inbred lines (RILs), doubled haploids (DH), near isogenic lines (NILs) and F_2 etc. But it is difficult to develop these populations in out crossing species such as fruit crops, as these crops can not tolerate inbreeding and will require a longer time to develop these. Two way pseudo test cross (Rosa *et al.*, 2003, Mehlenbacher *et al.*, 2006, Gisbert *et al.*, 2009, Gulsen *et al.*, 2010), F_2 (Dillon *et al.*, 2006, Sargent *et al.*, 2006, Blas *et al.*, 2009), half and full sib derived populations (Riaz *et al.*, 2008) have been used in fruit crops for mapping purpose (Table 5). The size of the population may vary from 50 to 250 for initial map development, but for high resolution mapping, larger populations are required.

9.2.6.1.1.2 Identification of polymorphism between the parents

Available DNA markers are screened against the genomes of the parents to find out the polymorphic ones (that can differentiate the two parents). Generally, different authors have used different markers *viz*., RFLP, RAPD, AFLP, SSRs SNPs *etc.* for map development in fruit crops. The markers used should have higher information content, co-dominant inheritance, reproducibility, easy transferability and ease of automation. SSRs represent all these advantages (Gupta *et al.*, 1999). The polymorphic markers are screened against the individuals of the population, and data is recorded according to its similarity to the banding pattern of the parents.

9.2.6.1.1.3 Linkage analyses and map construction

Although linkage analyses for the few markers can performed manually, it is not feasible to manually analyze and determine linkages between large numbers of markers.

Linkage analysis and the respective distance between the markers is generally calculated by computer programs such as MAPMAKER (Lander *et al.,* 1987), JoinMap (Stam *et al.,* 1993), GMENDEL (Echt *et al.,* 1992) *etc.* In these computer programs, lankage between the markers is based on the Log of likelihood (LOD) scores. LOD score of e" 3 is considered ideal (LOD 3 : linkage between the concerned markers is 1000 times more likely than no linkage). The genetic distance between them is calculated on the basis of mapping functions such as Haldane and Kosambi mapping function.

The linkage maps based on molecular markers in fruit crops are listed in Table 5. Highly saturated maps are available in fruit crops such as apple, grapes, *Citrus*, *Prunus etc.* but, the linkage map in commercial crops such as mango are still in preliminary phase (Kashkush *et al.,* 2001) while in many other minor crops such as sapota, pomegranate, *ber etc.* are yet to be developed. The map with increased marker density proves a useful resource for fine mapping of the quantitative trait loci (QTLs), identifying candidate genes and map based gene isolation (Troggio *et al.,* 2007), as such maps cover most of the genome.

The maximum advantage of a genetic linkage map is gained when it can be easily transferred to cultivars or populations other than those for which it was originally constructed (Liebhard *et al.,* 2003) An approach to increase the applicability of a map to diverse populations is to integrate maps from different population of the crop for example as in grapes (Vezzulli *et al.,* 2008) The map integration results in an increased loci density and provides a stronger framework for precise mapping of QTL and alignment with a genome sequence. Such maps serve as the reference map for the species. In these maps, the order of markers on the different maps is conserved and any discrepancy in the order is resolved with the help of available genome sequence. Due to conservation of the order of the markers, the integrated maps can prove helpful even in the cultivars whose linkage map has not been prepared.

Table 5 : Fruit crops with available molecular maps

Fruit crop	Parents/ populations used	Marker type and No.	Total Map length (cM) & linkage groups (LG)	Reference
Strawberry	*Fragaria vesca* x *F. nubicola* derived F_2 population	175 SSRs, 6 gene specific and 1 SCAR marker	424.3cM, 7 LG	Sargent *et al.*,2006
Sweet cherry	Microspore callus of Emperor Francis	2 allozymes, 90 RAPDs	503.3cM, 10 LG	Stokinjer *et al* ., 1996
Apricot	81 F_1 individuals from the	AFLP, RAPD, RFLP and SSR markers	Goldrich-(511 cM) Valenciano-(467.2cM)	Hurtado *et al.*, 2002
Apricot	F1 progeny of Stark Early Orange x Tyrinthos	180 AFLPs, 29 SSRs	602 cM	Vilanova *et al.*, 2003
Guava	Three mapping populations	AFLPs and SSRs	1,379 cM, 11 LG	Rodríguez *et al.*, 2007
European Hazelnut	OSU252.146 x OSU 414.062 F_1 progeny	249 RAPDs and 20 SSRs	OSU252.146-661cM OSU 414.062-812 cM and 11 LG	Mehlenbacher *et al.*,2006
Papaya	F_2 progeny of interspecific cross AU9 x Sun Up	712 SSR, 21 AFLP and 1 morphological marker	Major LG : 9, Minor LG : 5	Blas *et al.*, 2009
Grapes	Syrah x Pinot Noir	483 SNPs, 132 SSRs and 379 AFLP markers	1245 cM, 19 LG	Troggio *et al.*, 2007
Grapes[a]	Syrah (S) x Pinot Noir (PN), S x Grenache (G), Cabernet Sauvignon x Riesling	1,134 markers (350 AFLPs, 501 SNPs, 283 SSRs)	1,443 cM over 19 linkage groups	Vezzulli *et al.*, 2008
Apple[a]	Fiesta x Discovery progeny	475 AFLPs, 235 RAPDs, 129 SSRs and 1 SCAR	1,371 cM, 17 LG	Liebhard *et al.*, 2003,
Prunus[a]	Almond cv. Texas x Peach Early Gold	235 RFLPs, 11 isozymes and 60 SSRs	491 cM, LG	Joobeur *et al.*, 1998, Aranzana *et al.*, 2002
Prunus	*Prunus persica* X *P. ferganensis*) x *P. persica* BC1 progeny	78 RFLPs, 63 AFLPs, 57 SSRs, 16 RAPDs and two morphological Markers	665 cM LG : 8	Verde *et al.*, 2005
Olive	F_1 progeny of Leccino x Dolce Agogia	61 RAPDs, 21 AFLPs, 8SSRs & 4 RFLPs	Lecino: 2765 cM (22 major and17 minor LG) Dolce Agogia: 2,445 cM (27 major & 3 minor LG)	Rosa *et al.*,2003
Mango	Progeny of 'Keitt' x 'Tommy-Atkins'	34 AFLP markers	161.5 cM, 13 LG	Kashkush *et al.*, 2001
Litchi	Maguili × Jiaohesanyuehong F_1 population	312 (169 AFLPs and 143 RAPDs) markers	1040.43 cM, 16 LG	Liu *et al.*, 2008
Loquat	Algerie x Zaozhong-6 progeny	SSRs and AFLPs	Algerie -900cM, 17 LG, Zaozhong-6-870cM,	Gisbert *et al.*, 2009
Citrus	F_1 progeny of Clementine mandarin x Orlando tangelo	609 (385 SRAP, 97 RAPD, 95 SSR, 18 ISSR, peroxidase gene polymorphism (POGP) and 2 Resistance gene Analog (RGA) markers	Clementine- 858cM, 9 LG Orlando tangelo- 886 cM	Gulsen *et al.*, 2010

[a]These represent the reference maps in the respective fruit crop

The availability of reference maps can help in even determining the independent identity of the genes coding for the same traits. The *Prunus* reference map was of great help in defining the individual identity of plum and peach genes resistant to *Meliodogyne* complex. Based on the use of SSR and RFLP markers from the *Prunus* reference map, plum genes *Ma* (from *P. cerasifera*) and R_{jap} (from *P. saliciana*) were mapped on the same location on linkage group 7 with similar flanking markers while the peach genes were identified on linkage group 2 (Claverie *et al.,* 2004a).

9.2.6.1.2 Gene/QTL mapping and tagging

After the development of linkage map, the next step is to locate the gene/QTLs controlling traits of importance on various linkage groups of the map. To associate a gene or a QTL with the molecular markers, the correlation between the genotypic and phenotypic data is established using software packages like MapManager QTX (for single marker analysis, Manly *et al.,* 2001), Map Maker/QTL [or Simple Interval Mapping (SIM), Lincoln *et al.,* 1993 or QTL Cartographer [for Composite Interval Mapping (CIM), Basten *et al.,* 2001]. The mapping of the major genes/QTLs controlling traits of economic importance on the available linkage maps is indispensible to facilitate their use in markers assisted selection.

The genes, ranging from vegetative traits such as plant height, growth habit, disease resistance and fruit quality related traits have been mapped in fruit crops. The columnar growth habit gene '*co*' of apple was mapped on linkage group 10 of the map developed from Spur Fuji x Telamon cross, it was flanked by SSR markers COL at 3.9 cM on one side and CH03d11 on the other side (Tian *et al.,* 2005). In raspberry, the QTLs related with plant height and ripening were mapped on chromosome 3 (Graham *et al.,* 2009). The chilling requirement of the floral buds is controlled by the QTL and its location overlapped with already sequenced *Evg* region in peach (Fan *et al.,* 2010). The raspberry gene *H* is known to control the resistance to cane botrytis and spur blight was mapped on the linkage group 2 of the map developed from the progeny of European red raspberry cv. Glen Moy and the North American red raspberry cv. Latham (Graham *et al.,* 2006). In apple, a major QTL associated for resistance to fire blight (*Erwinia amylovora*) was identified on linkage group (LG) 7 of Fiesta (Calenge *et al.,* 2005). Similarly, a major QTL co-located with the deoxy-D : -xylulose synthase gene on linkage group 5 in Muscat grapes (Duchêne *et al.,* 2009). This gene codes for the first enzyme of the plastidial isoprenoid biosynthesis pathway. Recently, the QTL for resistance against *Phytophthora* were mapped on linkage group 1 & 5 of the *P. trifoliata* map (Siviero *et al.,* 2006).

The bioinformatic tools have also been used in determining the position of the candidate genes. In this approach, previously cloned and annotated genes from other species are used to find out homologous sequences on the parent genomes of interest, followed by its localization on the linkage groups in order to find the tightly linked markers to facilitate its map based gene cloning and use in the cultivar improvement programs at the later stages. In apple, the candidate genes specific for fruit softening (*Md-Exp 7*) was found out by comparing its native expansin specific sequences with fruit specific expansin genes of *tomato, peach, pear* and *strawberry.* The apple sequences showed highest similarity with pear expansin genes and with the help of available linkage map it was genetically mapped on linkage group 1 of Prima x Fiesta linkage map (Costa *et al.,* 2008). Similarly, late blooming is the sought after trait in almonds to avoid the frost damage (Silva *et al.,* 2005). Ten candidate genes specific for flowering, selected on the basis of their similarity with the annotated genes from model plants such as *Arabidopsis thaliana* were mapped on 'Texas' x 'Early Gold' *Prunus* reference map (Silva *et al.,* 2005).

When a gene is in close proximity of a DNA marker, so that the marker co-segregates with the gene of interest, it is termed as *'tagged'* and is suitable for markers assisted selection. The list of tagged genes in fruit crops is given in Table 6.

Table 6 : Traits of economic importance tagged with molecular markers in fruit crops

Trait	Fruit crop	Population used	Markers linked	Reference
Vegetative traits				
Evergrowing growth habbit (*evg*)	Peach	Empress op dwarf 3 X Evergrowing (P.I. 442380	EAT/MCAC (AFLP) from the gene at 1 CM	Wang *et al.,* 2002
Disease resistance				
Apple scab (*Vbj*)	Apple	A722-7 X Golden Delicious	gene mapped between SSR marker CH05e03 (0.6 cM) and T06-SCAR (3.9 cM)	Gygax *et al.,* 2004
Powdery mildew (*Pl-d*)	Apple	Fiesta X A871-14 F_1 progey	3 AFLP, 1 SCAR and 2 flanking SSR markers (CH03C02 and CH01D03) on linkage group 12	James *et al.,* 2004
Powdery mildew Pl-w	Apple	Fiesta X E295-4 (Gloster 69 x White Angel) progeny	EM M02 (SCAR) marker at 4.6 cM from *Pl-w* at LG 8	Evans and James, 2003
Citrus tristeza virus (*Ctv*)	Citrus	*Citrus* X *Poncirus trifoliata*	7 SCAR markers within 2.5 cM from the gene	Deng *et al.,* 1997
Eastern Filbert Blight (Anisogramma anomala)	European hazelnut	OSU252.146 X OSU 414.062	Two SSR markers linked to the gene present on chromosome 6	Mehlenbacher *et al.,* 2006
Downy mildew	Grapes	-	SCo06-1500 (SCAR marker)	Luo *et al.,* 2001
Pierce disease resistance *PdR1a*	Grapes	D8909-15 X F8909-17	VVCh14-56 A010. VVCh1 4-02 at 0.4 cM and UDV095 at 0.2 cM	Riaz *et al.,* 2008

Contd.

PdR1b	Grapes	F-27 X F8909-8	VVCh14-02 coseggregated and UDV095 VVCh 14-10 at 0.4 cM	Riaz *et al.*, 2008
Powdery mildew resistance (Unincula necator)	Grapes	Regent X Lemberger	9 RAPDs linked	Fischer *et al.*, 2004
Powdery mildew resistance (Unincula necator)	Grapes	Regent X Lemberger	ScORA7-760 (SCAR marker)	Akkurt *et al.*, 2007
Papaya ring spot virus (prsv1)	Papaya	F_2 progeny of V. parvilora X V. cundinamarcensis	Psilk4 (CAPS) marker at 2 cM from prsv locus	Dillon *et al.*, 2006
Black spot (Alternaria alternata)	Pear	Osa Nijisseiki X Okusankichi and Oushuu X Nansui derived F_1 plantlets	Flanking SSR markers Ch04h02 and Ch03d02 on linkage group 11	Terakami *et al.*, 2007
Anthracnose (Rca2)	Strawberry	Capitola X Pajaro	gene 0.6 cM from STS-Rca2_417 and 2.8 cM from STSRca2_240	Lerceteau-Kohler *et al.*, 2005
Insect and nematode resistance				
Aphid resistance (Sd1)	Apple	Prima X FiestaF1 progeny	Sd 1 between 2B1 2a (0.4 cM) and SdSSRa (0.9 cM) on LG7	Cevik and King, 2002
Dagger nematode resistance (Xiphinema index)	Grapes	D8909-15 (Vitis rupestris X V. arizonica)	VNg3a10 & VMC 5a10 (SSR), M4F3F (CAPS)	Xu *et al.*, 2008
Root knot Nematode (Melidogyne resistance (Ma)	Plum	Interspecific Prunus material	SCAL19 and SCAFL P2 SCAR markers	Lecouls *et al.*, 2004
Fruit related traits				
Anthocynin (Rf)	Apple		BC226 (RAPD) at < 2 cM on LG9	Cheng *et al.*, 1996
Anthocynin (Rni/ MdMYB10)	Apple	Sciros' X 91.136 B6-77	MdMYB marker co segregating	Chagne *et al.*, 2007
Fruit softening (Md-Exp7)	Apple	Progeny of Prima X Fiesta & 31 other genotypes	Md-$Exp7_{SSR}$ marker from the untrans lated region of the gene on LG1	Costa *et al.*, 2008
Seedlessness (S)	Grapes	Flame Seedless X Red Globe progeny	SCAR markers 39970524-5-564 and GSLP1 (569) at 0.6 cM and 1.2 cM respectively	Yang *et al.*, 2005
Sex determination	Papaya	F_2 progeny of Sun Up X Kapoho	SCAR T12 linked to Hermaphrodite and SCAR W11 to male type on LG 1	Deputy *et al.*, 2002

9.2.6.1.3 High resolution/fine mapping

High resolution/fine mapping involves further saturating the region with markers around the gene/QTL on the map. The genetic distance (cM) does not reflect the true physical distance in the genomes. But, it is quite difficult to delimit the gene containing region with the same population size. To obtain the new linked markers within the previously mapped region, the size of the population is increased to find out any possible recombinant between the gene and previously linked markers. Even different populations are also used to increase the density of the markers as in case of grapes (Hwang *et al.,* 2010). The fine mapping studies have also been undertaken in fruit crops around the previously mapped genes either to find the closely linked markers *eg.* for aphid resistance in *apple* (Cevik and King, 2002) and sex loci in papaya (Deputy *et al.,* 2002) or to use the fine map as a base for gene cloning *eg.,* grapes (Hwang *et al.,* 2010), citrus (Rai, 2006) and plum (Claverie *et al.,* 2004b).

For increasing the density of the markers in the vicinity of the gene containing region, most of the researchers have used AFLP markers, as these are easy to construct and no prior sequence information is required for their development. Once the tightly linked AFLPs are found, they are converted into more reliable SCAR (Sequence characterized amplified region) markers (Claverie *et al.,* 2004b). In Myrobalan plum, the *Ma* loci present on linkage group 7 with flanking SCAL and SCAFLP2 markers was enriched further with the tightly linked AFLP markers. The SCAR markers developed from these AFLPs helped in finding out the BAC clone of 280 KB size containing the resistance gene.

9.2.6.1.4 Association mapping

Association or linkage disequilibrium mapping has become very popular approach for dissecting the genetic basis of complex traits in plants. The benefits of association mapping, compared with traditional quantitative trait locus mapping, is, for example, a relatively detailed mapping resolution and that it is far less time consuming since no mapping populations need to be generated. This approach will become more useful in perennial fruit trees where generating mapping populations is difficult and time consuming. Association mapping shows great promise for helping us understand the genetic basis of complex traits of both economic and ecological importance (Jannink and Walsh 2001, Hall *et al.,* 2010). A search on ISI web of knowledge did not return any reference on association mapping in fruits, whereas it returned more than 200 references on association mapping in plants. Since dense linkage maps and whole genome sequence is now available for several fruit trees, it should be easy for mapping of QTL for economically important traits using association mapping.

9.2.6.1.5 Markers assisted selection

The use of molecular markers can greatly accelerate the pace of selection in fruit crops with inherent long juvenile period and when the trait under study is controlled by the recessive gene and incorporating more than one gene for disease resistance. However, the markers used for markers assisted selection should fulfill the following criteria :

a) Marker(s) should co-segregate or be closely linked (less than 1cM ideally) with the desired trait

b) PCR based

c) The screening technique should have high reproducibility across laboratories

Presently, the markers linked to resistance for dagger nematode (*Xiphinema index*) and pierce disease are being routinely used at University of California in the breeding programmes aimed at improving grapevine rootstock (Xu *et al.,* 2008).

The resistance to the most successful resistance gene (*Vf*) for scab has been overcome in apple by new races of the pathogen about one and half decade ago (Benauf and Parisi 2000). Thus to produce a scab resistant apple, the gene pyramiding is the suitable candidate by providing durable resistance. All major resistance genes to scab (*Vf, Vr, Vb, Va, Vm* and *Vbj*) have available markers associated with them (Gygax *et al.,* 2004). Thus, allow them to be used for gene pyramiding for scab resistance.

9.2.6.2 Map based gene cloning

It refers to isolation and cloning of the genes coding for a particular trait, based on the molecular marker information from the fine map. The tightly linked or co-segregating markers around the gene/QTL are used to identify the positive BAC clones. In fruit crops, two genes has been cloned on the basis of map based gene cloning approach (Table 7).

Table 7 : Map based gene cloning in fruit crops

Genes	Coding trait	Fruit crop	Flanking genetic markers	Reference
XiR1	Dagger nematode resistance (*Xiphinema index*)	Grapes	1N2R3B and M4F3R	Hwang *et al.,* 2010
Ctv	Resistance against *Tristeza* virus	Citrus	SSR markers	Yang *et al.,* 2003

The flanking markers to the locus encoding traits of importance can help in identifying the BAC clones containing the gene of interest.The cloned genes can be used in the fruit crop improvement through the transgenic technology and are also helpful in ascertaining the structure and function of the genes. With the use of SSR markers, the locus containing 10 different genes of resistance against citrus tristeza virus (*Ctv*) was delimited to 121 kb and the individual genes were later cloned (Rai, 2006). The cloned genes were transferred into susceptible citrus genotypes *viz.,* Ruby Red, Rio Red and Duncan grapefruit and the transgenic lines with R_1, R_2 and R_4 genes showed great resistance against the virus.

9.2.6.3 Whole genome sequencing

Whole genome sequencing is powerful tool in large scale gene discovery, their structural pattern (intron, exon and other regulatory elements), function of the non annotated genes, their organization on the chromosomes, relative abundance of a gene and their role in particular evolution of the fruit trees. The fruit crops, whose whole genomes have been sequenced, are listed in Table 8.

Table 8 : Sequenced fruit crops with genome size and number of genes

Fruit crop/cultivar	Genome size (Mb)	Predicted genes	Reference
Grapes cv. Pinot Noir	504.6	33,514	Jaillon *et al.*, 2007
Papaya cv. Sun Up	372	28,629	Ming *et al.*, 2008
Apple cv. Golden Delicious	742.3	57,386 putative genes + 31, 678 transposable related ORFs	Velasco *et., al* 2010

In apple, gene families regulating its typical fruit structure (*MADS-box* gene) and sorbitol metabolism (*A6PR* : rate limiting for sorbitol metabolism, *SDH* : converts sorbitol into fructose in fruit, $PcSOT_2$: sorbitol transporter) were discovered, which are specific to *Rosaceae* fruit (Velasco *et al.*, 2010). Similarly in grapes, the functions of the genes related with wine chracterstics were unveiled (Jaillon *et al.*, 2007) and the genes related to carbohydrate metabolism and volatile components were abundantly present in papaya genome (Ming *et al.*, 2008). The whole genome sequence of the grapes revealed that the present grapes (*Vitis vinifera*) originated from the contribution of three ancestor genomes in the past (Jaillon *et al.*, 2007). The unique chromosome number in apple is contributed due to autopolyploidization of *Gillenia* or related *taxon* followed by aneuploidization. Due to the absence of functional *Ph-1* like genes the genomes might have undergone rearrangements leading to the present state (Velasco *et al.*, 2010).

Most practical goal of genome sequencing is to accelerate the molecular breeding of the perennial fruit species. To accomplish this, the precise knowledge of the functional genes influencing desirable traits and disease resistance is a pre-requisite. A high number of disease resistance genes have been identified in apple (Velasco *et al.*, 2010) and grapes (Jaillon *et al.*, 2007) and many of them are even tagged with one or more SNPs to use them for introgression in cultivated varieties through markers assisted introgression.

9.3 CONCLUSIONS

The long juvenility and the problem of incompatibility and sterility are the biggest hindrance in the improvement of fruit crops through conventional fruit breeding. To mitigate the stresses of biotic and abiotic nature, improvement is desired in already cultivated cultivar for one or few traits. In that regard, the biotechnological tools will prove quite useful to expedite the rate of fruit crop improvement. Biotecnological tools such as somaclonal variation and *in vitro* mutagenesis are the choicest method for varietal development in banana. GCTCV 215-1 (Hwang and Ko 1992) and Novaria (Mac *et al.,* 1996) are the result of somaclonal variation and *in vitro* mutagenesis, respectively. Somatic hybridization has been explored for the rootstock and scion improvement of citrus. The somatic hybridization research of almost two decades in citrus is yielding fruit now (Guo *et al.,* 2004). The embryo rescue has helped in accelerating the pace of stenospermocarpic grape improvement (Ramming 1990) and the 'Sweet Scarlet' and 'Thomcord' are two of the seedless table grapevines, which were released using this technique. The transgenic varieties 'Rainbow' and 'SunUp' carrying resistance to papaya ring spot virus are being commercially cultivated in Hawaii of USA (Mendoza *et al.,* 2008). With the introduction of the concept of *cisgenics,* the genetically transformed fruits are likely to get public acceptance. Presently, the markers linked to resistance for dagger nematode (*Xiphinema index*) and pierce disease are being routinely used at University of California in the breeding programs aimed at improving grapevine rootstock (Xu *et al.,* 2008). Genome sequencing has unveiled the mystery about various genes, the phyllogenetic relationship and evolutionary history of the sequenced fruit crop (Velasco *et al.,* 2010). In the nutshell, biotechnolgical tools are the ideal mean for speedy fruit crop improvement.

9.4 REFERENCES

Aguero CB, Utasu SL, Greve C, Powell ALT and Dandekar AM (2005) Evaluation of tolerance to Pierce's disease and Botrytis in transgenic plants of *Vitis vinifera* L. expressing the pear PGIP gene. *Mol Plant Path,* 6 : 43-51.

Akkurt M, Welter L, Maul E, Topfer R and Zyprian E (2007) Development of SCAR markers linked to powdery mildew (*Uncinula necator*) resistance in grapevine (*Vitis vinifera* L. and *Vitis sp.*). *Molecular Breeding,* 19 : 103-11.

Anonymous (2004) Sweet Scarlet and Thomcord – two new table grape varieties released from ARS. *FPS Grape Program Newsletter*, October 2004. P. 4.

Anonymous (2010) *Indian Horticulture Database-2009.* National Horticulture Board, Gurgaon.

Aranzana MJ, Pineda A, Cosson P, Dirlewanger E, Ascasibar J, Cipriani G, Ryder CD, Testolin R, Abbott A, King GJ, Iezzoni AF, Arffls P (2002) A set of simple-sequence repeat (SSR) markers covering the *Prunus* genome. *Theor Appl Genet,* 106 : 819-25.

Basten C, Weir B, Zeng ZB (2001) QTL Cartographer. Department of Statistics, North Carolina State University, Raleigh, NC.

Bénauf G and Parisi L (2000) Genetics of host-pathogen relationships between *Venturia inaequalis* races 6 and 7 and *Malus* species. *Phytopathol,* 90 : 236-42.

Bhagwat B and Duncan EJ (1998) Mutation breeding of banana cv. Highgate (*Musa spp.* AAA group) for tolerance to *Fusarium oxysporum f. sp. cubense* using chemical mutagens. *Scientia Hort,* 73 : 11-22.

Blas AL, Yu Q, Chen C, Veatch O, Moore PH, Paull RE and Ming R, (2009) Enrichment of a papaya high-density genetic map with AFLP markers. *Genome,* 52 : 716-25.

Boscariol RL, Monteiro M, Takahashi EK, Chabreges SM, Vieira MLC, Vieiera LGE, Pereira LFP, Filho FAAM, Cardoso SC, Filho AB, Barbosa JM, Azevedo FA and Mendes BMJ (2006) *Attacin* A gene from *Tricloplusia ni* reduces susceptibility to *Xanthomonas axonopodis* pv. *citri* in transgenic *Citrus sinensis* Hamlin. *J Amer Soc Hort Sci,* 131 : 530-36.

Brito G, Costa A, Fonseca HMCA and Santos CV (2002) Response of *Olea europea* ssp. *maderensis in vitro* shoots exposed to osmotic stress. *Scientia Hort,* 97 : 411-17.

Caboni E, Anselmi S, Donato E and Manes F (2003) *In vitro* selection of *Actinidia deliciosa* clones tolerance to NaCl and their molecular and *in vivo* ecophysiological characterisation. *Acta Hort,* 618 : 77-83.

Calenge F, Drouet D, Denance C, Van de Weg WE, Brisset MN, Paulin JP and Durel CE (2005) Identification of a major QTL together with several minor additive for epistatic QTLs for resistance to fire blight in apple in two related progenies. *Theor Appl Genet,* 111 :128-35.

Cardoso SC, Mendes JMB, Camargo RLB , Christiano RSC, Filho AB, Vieira MLC, Mendes BMJ, Filho F de AM (2010) Transgenic Sweet Orange (*Citrus sinensis* L. Osbeck) Expressing the attacin A Gene for Resistance to *Xanthomonas citri* subsp. *citri*. *Plant Mol Biol Rep* 28 :185-92.

Cervera M, Navarro L and Pena L (2009) Gene stacking in 1-year-cycling *APETALA1* citrus plants for a rapid evaluation of transgenic traits in reproductive tissues. *J Biotech,* 140 : 278-82.

Cervera M, Ortega C, Navarro L and Pena L (2000) Generation of transgenic citrus plants with tolerance to salinity gene *Hal2* from yeast. *J Hort Sci Biotech,* 75 : 26-30.

Cevik B, Lee RF and Niblett CL (2006) Genetic Transformation of *Citrus paradisi* with Antisense and Untranslatable RNA-dependent RNA Polymerase Genes of *Citrus tristeza* closterovirus. *Turk J Agric For,* 30 : 173-82.

Cevik V and King GJ (2002) High-resolution genetic analysis of the Sd-1 aphid resistance locus in *Malus* spp. *Theor Appl Genet,* 105 : 346-54.

Chagné D, Carlisle CM, Blond C, Volz RK, Whitworth CJ, Oraguzie NC, Crowhurst RN, Allan AC, Espley RV, Hellens RP and Gardiner SE (2008) Mapping a candidate gene (MdMYB10) for red flesh and foliage colour in apple. *BMC Genomics,* 8 : 212.

Cheng FS, Weeden NF and Brown SK (1996) Identification of co-dominant RAPD markers tightly linked to fruit skin color in apple. *Theor Appl Genet* 93 : 222-27.

Chung KR, Shilts TLW and Timmer LW (2002) Engineering a genetic transformation system for *Collectotrichum acutatum* the causal fungus of lime anthracnose and postbloom fruit drop of citrus. *FEMS Microbiol Letters,* 213 : 33-39.

Claverie M, Bosselut N, Lecouls AC, Voisin R, Lafargue B, Poizat C, Kleinhentz M, Laigret F, Dirlewanger E and Esmenjaud D (2004a) Location of independent root-knot nematode resistance genes in plum and peach. *Theor Appl Genet,* 108 : 765-73.

Claverie M, Dirlewanger E, Cosson P, Bosselut N, Lecouls AC, Voisin R, Kleinhentz M, Lafargue B, Caboche M, Chalhoub B and Esmenjaud D (2004b) High-resolution mapping and chromosome landing at the root-knot nematode resistance locus Ma from Myrobalan plum using a large-insert BAC DNA library. *Theor Appl Genet,* 109 : 1318-27.

Collard BCY, Jahufer MZZ, Brouwer JB and Pang ECK (2005) An introduction to markers, quantitative trait loci (QTL) mapping and marker-assisted selection for crop improvement : The Basic Concepts. *Euphytica,* 142 : 169-96.

Costa F, Van de Weg WE, Stella S, Dondini L, Pratesi D, Musacchi S and Sansavini S (2008) Map position and functional allelic diversity of *Md-Exp7*,a new putative expansin gene associated with fruit softening in apple (*Malus × domestica* Borkh.) and pear (*Pyrus communis*). *Tree Genetics and Genomes,* 4 : 575–86.

Dantas AC de M, Fortes GRdeL, Silva JBda, Nezi AN, Rodrigues AC, de L-Fortes GR and da-Silva JB (2001) Tolerance to aluminum in apple rootstock somaclones in nutrient solution. *Persquisa Agropecuaria Brasileira,* 36 : 615-23.

Deng Z, Xiao S, Huang S and Gmitter FG Jr. (1997) Development and characterization of SCAR markers linked to the citrus tristeza virus resistance gene from *Poncirus trifoliata*. *Genome,* 40 : 697-704.

Deputy JC, Ming R, Ma H, Liu Z, Fitch MMM, Wang M, Manshardt R and Stiles JI (2002) Molecular markers for sex determination in papaya (*Carica papaya* L.). *Theor Appl Genet,* 106 : 107–11.

Dhillon S, Ramage C, Ashmore S and Drew RA (2006) Development of a codominant CAPS marker linked to PRSV-P resistance in highland papaya. *Theor Appl Genet,* 113 : 1159-69.

Donovan AM, Morgan R, Piagnani V, Ridout MS, James DJ and Garrett CME (1994) Assessment of somaclonal variation in apple. I. Resistance to fire blight pathogen *Erwinia amylora*. *J Hort Sci,* 69 : 105-13.

Drew RA, Siar SV, O'Brien CM, Magdalita PM and Sajise AGC (2006) Breeding papaya ringspot virus resistance in *Carica papaya* via hybridisation with *Vasconcellea quercifolia*. *Aus J Exp Agr* 46 : 413-18.

Duan YX, Fan J and Guo WW (2010) Regeneration and characterization of transgenic kumquat plants containing the *Arabidopsis APETALA1* gene. *Plant Cell Tiss Org Cult,* 100 : 273-81.

Duchêne E, Butterlin G, Claudel P, Dumas V, Jaegli N and Merdinoglu D (2009) A grapevine (*Vitis vinifera* L.) deoxy-D : -xylulose synthase gene colocates with a major quantitative trait loci for terpenol content. *Theor Appl Genet,* 118 : 541-52.

Echt C, Knapp S and Liu BH (1992) Genome mapping with non inbred crosses using Gmendel 2.0. *Maize Genet Coop Newsletter,* 66 : 27-29.

Evans KM and James CM (2003) Identification of SCAR markers linked to *Pl-w* mildew resistance in apple. *Theor Appl Genet,* 106 : 1178-83.

Fan S, Bielenberg DG, Zhebentyayeva TN, Reighard GL, Okie WR, Holland D and Abbott AG (2010) Mapping quantitative trait loci associated with chilling requirement, heat requirement and bloom date in peach (*Prunus persica*). *New Phytol,* 185 : 917-30.

Finn CE, Luby JJ, Rosen CJ and Ascher PD (1991) Evaluation *in vitro* of blueberry germplasm for higher pH tolerance. *J Amer Soc Hort Sci,* 116 : 312-16.

Fischer BM, Salakhutdinov I, Akkurt M, Eibach R, Edwards KJ, Tpfer R and Zyprian EM (2004) Quantitative trait locus analysis of fungal disease resistance factors on a molecular map of grapevine. *Theor Appl Genet,* 108 : 501-15.

Fuentes JL, Santiago L, Valdes Y, Guerra M, Ramrez IM, Prieto EF, Rodriguez NN and Velazquez B (2004) Mutation induction in zygotic embryos of avocado (*Persea americana* Mill.). *Biotecnologia Aplicada,* 21 : 82-84.

Gentile A, Deng Z N, Tribulato E, Albanese G, Grimaldi V, Catra A, Vardi A, Goren R and Goldschmidt E E (2000). *Acta Hort,* 535 : 259-63.

Gill MIS, Singh Z and Agrez V (2004) Factors affecting Agrobacterium mediated transformation in fruit and nut crops. An overview. *Food Agri Environ,* 2(2) : 327-47.

Giménez C, de García E and Haddad O (2008) Genetic and resistance stability to Black Sigatoka disease during micropopagation of *Musa* CIEN BTA-03 somaclonal variant. *Intl J Exp Bot* 77 : 65-79.

Gisbert AD, Martýnez-Calvo J, Llacer G, Badenes ML and Romero C (2009) Development of two loquat [*Eriobotrya japonica* (Thunb.) Lindl.] linkage maps based on AFLPs and SSR markers from different Rosaceae species. *Mol Breed,* 23 : 523-38.

Goldman MHS and Ando A (1990) Radiosensitivity of protoplasts of orange (*Citrus sinensis*). *Mut Breed Newsletter* 36 : 11.

Graham J, Hackett CA, Smith K, Woodhead M, Hein I, McCallum S (2009) Mapping QTLs for developmental traits in raspberry from bud break to ripe fruit. *Theor Appl Genet,* 118 : 1143-55.

Graham J, Smith K, Tierney I, MacKenzie K and Hackett CA (2006) Mapping gene *H* controlling cane pubescence in raspberry and its association with resistance to cane botrytis and spur blight, rust and cane spot. *Theor Appl Genet,* 112 : 818-31.

Grosser JW, Medina-Urrutia V, Ananthakrishnan G and Serrano P (2004) Building a replacement sour orange rootstock : Somatic hybridization of selected mandarin + pummelo combinations. *J Amer Soc Hort Sci,* 129 : 530-34.

Gulsen O, Uzun A, Canan I, Seday U and Canihos E (2010) A new citrus linkage map based on SRAP, SSR, ISSR, POGP, RGA and RAPD markers. *Euphytica,* 173 : 265–77.

Guo WW and Grosser JW (2004) Transfer of a potential canker resistance gene into citrus protoplasts using GFP as the selectable marker. *Acta Hort,* 632 : 255-58.

Guo WW, Prasad D, Serrano P, Gmitter FG Jr and Grosser JW (2004) Citrus somatic hybridization with potential for direct tetraploid scion cultivar development. *J Hort Sci Biotech,* 79 : 400-05.

Gupta P, Varshney R, Sharma P and Ramesh B (1999) Molecular markers and their applications in wheat breeding. *Plant Breed,* 118 : 369-90.

Gygax M, Gianfranceschi L, Liebhard R, Kellerhals M, Gessler C and Patocchi A (2004) Molecular markers linked to the apple scab resistance gene *Vbj* derived from *Malus baccata jackii. Theor Appl Genet,* 109 : 1702-09.

Hall DC, Tegstrom and Ingvarsson PK (2010) Using association mapping to dissect the genetic basis of complex traits in plants. *Briefings in Functional Genomics and Proteomics,* 9 : 157-165

Hammerschlag FA and Ognjanov (1990) Somaclonal variation in peach : screening for resistance to *Xanthomonas camperstris* pv. *pruni* and *Pseudomonas syringae* pv. *syringae*. *Acta Hort,* 280 : 403-08.

Hammerschlag FA, Warner DJ and Ritchie DF (1994) Stability of bacterial leaf spot resistance in peach regenerants under *in vitro*, greenhouse and field conditions. *Euphytica,* 76 : 101-06.

Hiramatsu M, Wakana A, Park –Sung M and Fukudome I (2003) Production of triploid plants from crosses between diploid and tetraploid grapes (*Vitis* complex) through immature seed culture and subsequent embryo culture. *J Faculty Agri Kyushu University,* 48 : 51-57.

Hurtado MA, Romero C, Vilanova S, Abbott AG, Llácer G and Badenes ML (2002) Genetic linkage maps of two apricot cultivars (*Prunus armeniaca* L.) and mapping of PPV (sharka) resistance. *Theor Appl Genet,* 105 : 182–91.

Hwang CF, Xu K, Hu R, Zhou R, Riaz S and Walker MA (2010) Cloning and characterization of *XiR1*, a locus responsible for dagger nematode resistance in grape. *Theor Appl Genet,* 121 : 789-99.

Hwang SC and Ko WH (1992) Breeding for resistance to *Fusarium* wilt of Cavendish banana by using tissue culture method. *Special Publication-Taichung District Agricultural Improvement Station,* 29 : 229-37.

Jaccoud D, Peng K, Feinstein D and Kilian A (2001) Diversity arrays : a solid state technology for sequence information independent genotyping. *Nucleic Acid Research,* 29 : e25.

Jacobsen, E and Schouten HJ (2007) Cisgenesis strongly improves introgression breeding and induced translocation breeding of plants. *Trends Biotechnol* doi :10.1016/j.tibtech. 2007.03.008.

Jaillon O, Aury, JA, Noel B et al., (2007) The grapevine genome sequence suggests ancestral hexaploidization in major angiosperm phyla. *Nature Letters,* 449 : 463-67.

Jain SM (2005) Major mutation-assisted plant breeding programs supported by FAO/ IAEA. *Plant Cell Tiss Org Cult,* 82 : 113-23.

Jain SM (2007) Recent advances in datepalm tissue culture and mutagenesis. *Acta Hort,* 736 : 205-11.

Jain SM (2010) Mutagenesis in crop improvement under the climate change. *Romanian Biotechnol Letters,* 15(2) **:** 88-106.

James CM, Clarke JB and Evans KM (2004) Identification of molecular markers linked to the mildew resistance gene *Pl-d* in apple. *Theor Appl Genet,* 110 : 175-81.

Joobeur T, Viruel MA, de Vicente MC, Jauregui B, Ballester J, Dettori MT, Verde I, Truco MJ, Messeguer R, Batlle I, Quarta R, Dirlewanger E and Arus P (1998) Construction of a saturated linkage map of *Prunus* using an almond x peach F2 progeny. *Theor Appl Genet,* 97 : 1034-41.

Kashkush K, Jinggui F, Tomer E, Hillel J and Lavi U (2001) Cultivar identification and genetic map of mango (*Mangifera indica*). *Euphytica,* 122 : 129-36.

Kayim M, Ceccardi TL, Berreta MJG, Barthe GA and Derrick KS (2004) introduction of a citrus blight associated gene into Carrizo citrange (*Citrus sinensis* x *P. trifoliata*) by *Agrobacterium* mediated transformation. *Plant Cell Rep,* 23 : 377-85.

Kung YJ, Yu TA, Huang CH, Wang HC, Wang SL and Yeh SD (2010) Generation of hermaphrodite transgenic papaya lines with virus resistance via transformation of somatic embryos derived from adventitious roots of *in vitro* shoots. *Transgenic Res,* 19 : 621-35.

Lander ES, Green P, Abrahamson J, Barlow A, Daly MJ, Lincoln SE and Newburg L (1987) Mapmaker an interactive computer package for constructing primary genetic linkage maps of experimental and natural populations. *Genomics,* 1 : 174-81.

Larkins PJ and Scowcroft WR (1981) Somaclonal variation - a novel source of variability from cell cultures for plant improvement. *Theor Appl Genet* 60, : 197-214.

Lecouls AC, Bergougnoux V, Rubio-Cabetas MJ, Bosselut N, Voisin R, Poessel JL, Faurobert M, Bonnet A, Salesses G, Dirlewanger E, Esmenjaud D (2004) Marker-assisted selection for the wide-spectrum resistance to root-knot nematodes conferred by the Ma gene from Myrobalan plum (*Prunus cerasifera*) in interspecific *Prunus* material. *Molecular Breeding,* 13 : 113-24.

Lerceteau-Kohler E, Guerin EG and Denoyes-Rothan EB (2005) Identification of SCAR markers linked to Rca2 anthracnose resistance gene and their assessment in strawberry germplasm. *Theor Appl Genet,* 111 : 862–70.

Li SC, Wen LZ, Zhang FQ, Luo FM, Yang MR, Zhang YZ and Huang DF (1983) A new cold and disease resistant grape cultivar – Beichun. *Chinese Bull Bot,* 2 : 28-30.

Li Y, Zhang Y, Feng F, Liang D, Cheng L, Ma F and Shi S (2010) Overexpression of a Malus vacuolar Na+/H+ antiporter gene (MdNHX1) in apple rootstock M.26 and its influence on salt tolerance. *Plant Cell Tiss Org Cult,* 102 : 337-45.

Liebhard R, Koller B, Gianfranceschi L and Gessler C (2003) Creating a saturated reference map for the apple (*Malus* x *domestica* Borkh.) genome. *Theor Appl Genet,* 106 : 1497-1508.

Lincoln S, Daly M and Lander E (1993) Constructing genetic linkage maps with MAPMAKER/EXP. Version 3.0. Whitehead Institute for Biomedical Research Technical Report, 3rd Edn.

Litz RE (2009) Recovery of mango plants with antrachnose resistance following mutation, induction and selection *in vitro* with the culture filtrate of *Colletotrichum gloesporoides* Penz. In : Induced Mutations in Tropical Fruit Trees, IAEA, Vienna, 2009. *IAEA -TECDOC*-1615 : 7-16.

Liu CM, Guo YS, Liu R, Zhang B, Xie JH, Hu YL, Ouyang R, Zhao YH and Fu JX (2008) Construction of a high density molecular linkage map for lychee based on AFLP and RAPD markers. *Acta Horticulturae,* 63 : 87-94.

Liu J, Xu X and Deng X (2005) Intergeneric somatic hybridization and its applications to crop improvement. *Plant Cell Tissue Org Cult,* 82 : 19-44.

Loko Y and Amoatey H (2001) Improvement of pineapple using *in vitro* and mutation breeding techniques. *IAEA TECH DOC* 1227 ISSN1011-4289 : 25-30.

Luo SL, He PC, Zhou P and Zheng XQ (2001) Identification of molecular genetic markers tightly linked to downy mildew resistant genes in grape. *Yi Chuan Xue Bao,* 28 : 76-82.

Mac C, Ho YW, Tan YP and Ibrahim R (1996) Novaria- a new banana mutantt induced by gamma irradiation. *Infomusa* 5 : 35-36.

Manly KF, Cudmore Jr RH and Meer JM (2001) Map Manager QTX, cross platform software for genetic mapping. *Genome,* 12 : 930-32.

Marino G, Beghelli S, Rombola AD and Carboni L (2000) *In vitro* performance at high culture pH and *in vivo* response to Fe deficiency of leaf derived quince BA 29 (*Cydonia oblonga*) somaclones regenerated at variable medium pH. *J Hort Sci Biotech,* 75 : 433-40.

Martinez-Gomez P and Dicentia F (2000) *In vitro* evaluation of apricot (*Prunus americana* L.) resistance to plum pox poty virus. *J Hort Sci Biotech,* 75 : 259-61.

Masuda T and Yoshioka T (1997) *In vitro* selection of a mutant resistant to *Alternaria* blotch disease in 'Indo' apple. *Tech News Institute Rad Breed,* 57 : 2.

Mehlenbacher SA, Brown RN, Nouhra ER, Gökirmak T, Bassil NV and Kubisiak TL (2006) A genetic linkage map for hazelnut (*Corylus avellana* L.) based on RAPD and SSR markers. *Genome,* 49 : 122-33.

Mendes JMB, Filho FdeAAM, Filho AB, Harakava R, Beer SV and Mendes BMJ (2009) Genetic transformation of *Citrus sinensis* cv. Hamlin with *hrpN* gene from *Erwinia amylovora* and evaluation of the transgenic lines for resistance to citrus canker. *Scientia Hort,* 122 : 109-15.

Mendoza EMT, Laurena AC and JR Botella (2008) Recent advances in the development of transgenic papaya technology. *Biotech Ann Rev* 14 : 423-62.

Midani AR, Sharma HC and Singh SK (2002) Effect of ovule age on ovulo-embryo culture in seeded and seedless grape genotypes. *Indian J Hort* 59 : 359-62.

Ming R, Hou S, Feng Y et al., (2008) The draft of the transgenic tropical fruit tree papaya (*Carica papaya* Linnaeus). *Nature Letters* 452 : 991-96.

Narayanaswami S and Norstog K (1964) Plant embryo culture. *Bot Rev* 30 : 587-628.

Nwauzoma AB, Tenkouano A, Crouch J H, Pillay M, Vuylsteke D and Kalio LAD (2002) Yield and disease resistance of plantain somaclones in Nigeria. *Euphytica* 123 : 323-31.

Ohgawara T, Kobayashi S, Ohgawara E, Uchimiya H and Ishii S (1985) Somatic hybrid plants obtained by protoplast fusion between *Citrus sinenesis* and *Poncirus trifoliata. Theor Appl Genet* 71 : 1-4.

Oliveira MLP de, Febres VJ, Costa MGC Moore GA and Otoni WC (2009) High-efficiency *Agrobacterium*-mediated transformation of citrus via sonication and vacuum infiltration. *Plant Cell Rep* 28 : 387-95.

Pavan A, Calixto MC, Cardoso SC, Mendes BMJ, Filho AB, Lopes JRS, Carvolho CR de and Filho F de AVM (2007) Evaluation of Hamlin sweet orange + Montenegrina mandarin somatic hybrid for tolerance to *Xanthomonas axonopodis* pv. *citri* and *Xylella fastidiosa*. *Scientia Hort* 113 : 278-85.

Pinet-Leblay C, Turpin FX and Chevreau E (1992) Effect of gamma and ultraviolet radiation on adventitious regeneration from *in vitro* cultured pear leaves. *Euphytica* 62 : 225-33.

Rai M (2006) Refinement of the Citrus tristeza virus resistance gene (*Ctv*) positional map in *Poncirus trifoliata* and generation of transgenic grapefruit (*Citrus paradisi*) plant lines with candidate resistance genes in this region. *Plant Mol Biol* 61 : 399-414.

Ram M and Majumdar PK (1981) Dwarf mutant of papaya induced by gamma rays. *J Nuclear Agri Biol* 10 : 72-74.

Ramming DW (1990) The use of embryo culture in fruit breeding. *Hort Science* 25(4) : 393-98.

Ramming DW and Tanner O (1987a) 'Gold Crest' peach. *Fruit Var J* 41: 52-53.

Ramming DW and Tanner O (1987b) 'May Fire' nectarine. *Fruit Var J* 41 : 80-81.

Rangaswami G and Mahadevan A (2002) *Diseases of fruit plants*. p 359 Prentice Hall of India. Pvt. Ltd., New Delhi, p.359.

Riaz S, Tenscher AC, Rubin J, Graziani R, Pao SS and Walker MA (2008) Fine-scale genetic mapping of two Pierce's disease resistance loci and a major segregation distortion region on chromosome 14 of grape. *Theor Appl Genet* 117 : 671-78.

Rodríguez N, Valdés-Infante J, Becker, D, Velázquez B, González G, Sourd D, Rodríguez J, Billotte N, Risterucci AM, Ritter E and Rohde W (2007) Characterization of guava accessions by SSR markers, extension of the molecular linkage map, and mapping of QTLs for vegetative and reproductive characters. *Acta Hort* 735 : 201-16.

Rosa RL, Angiolillo A, Guerrero C, Pellegrini M, Rallo L, Besnard G, Bervill A, Martin A, and Baldoni L (2003) A first linkage map of olive (*Olea europaea* L.) cultivars using RAPD, AFLP, RFLP and SSR markers. *Theor Appl Genet* 106 : 1273-82.

Rosati P and Predieri S (1990) *In vitro* selection of apple rootstock somaclones with *Phytophthora cactorum* culture filtrate. *Acta Hort* 280 : 409-13.

Sahijram L, Murthy BNS and Shikhamany SD (2005) Introgression of downy mildew resistance into grape cv. Thompson Seedless through hybrid embryo rescue. *Indian J Hort* 62 : 107-11.

Sargent DJ, Clarke J, Simpson DW, Tobutt KR, Arus P, Monfort A, Vilanova S, Denoyes-Rothan B,. Rousseau M, Folta KM, Bassil NV, Battey NH (2006) An enhanced microsatellite map of diploid Fragaria. *Theor Appl Genet* 112 : 1349-59.

Schaart JG (2004) Towards consumer-friendly cisgenic strawberry which are less susceptible to *Botrytis cinerae. Towards consumer-friendly cisgenic strawberry which are less susceptible to Botrytis cinerae* : 117.

Schouten HJ, Krens FA and Jacobsen E (2006) Do cisgenic plants warrant less stringent oversight? *Nature Biotechnol,* 24 : 753.

Schum A, Willers A and Schmidt H (1999) *In vitro* mutagenesis in *Malus* x *domestica* – induction of mildew resistance. *Erwerbsobstbau,* 41 : 131-37.

Semagn K, Bjornstad A and Ndjiondjop MN (2006) Principles, requirements and prospects of genetic mapping in plants. *African J Biotech,* 5 : 2569-87.

Sharma DR and Singh R (2006) Management of insect pest of temperate fruits. In : *Emerging Trends in Economic Entomology* (Chillar BS, Saini RK and Lal R eds.) pp. 111-22. Centre for Advanced Studies Department of Entomology, HAU, Hisar. pp. 111-22

Sharma DR, Kaur R and Kumar K (1996) Embryo rescue in plants – a review. *Euphytica,* 89 : 325-37.

Silva C, Garcia-Mas J, Sanchez AM, Arus P and Oliveira MM (2005) Looking into flowering time in almond (*Prunus dulcis* (Mill) D. A. Webb) : the candidate gene approach. *Theor Appl Genet,* 110 : 959–68.

Singh RP (2006) Desertification and its control. In : *Handbook of Agriculture pp.* 135. Indian Council of Agricultural Research, New Delhi.

Siviero A, Cristofani M, Furtado EL, Garcia AA, Coelho AS and Machado MA (2006) Identification of QTLs associated with citrus resistance to *Phytophthora* gummosis. *Journal of Applied Genetics,* 47 : 23-28.

Smith MK, Hamill SD, Langdon PW, Giles JE, Doogan VJ and Pegg KG (2006) Towards the development of a Cavendish banana resistant to race 4 of fusarium wilt : gamma irradiation of micropropagated Dwarf Parfitt (*Musa* spp., AAA group, Cavendish subgroup). *Aust J Expt Agric* 46 : 107-13.

Song JK, Guo WW, Yi HL, Liu JH, Chen CL and Deng XX (2005) Creation of triploid citrus plants by crossing ellite allotetraploid somatic hubrid pollen parents with diploid cultivars. *Acta Hort Sinica* 32 : 594-98.

Srivastva RK (1997) *In vitro induction of genetic variability for canker resistance in acid lime* (*Citrus aurantifolia* Swingle). Ph. D. dissertation Punjab Agricultural University, Ludhiana.

Stam P (1993) Construction of integrated genetic linkage maps by means of a new computer package : Join Map. *Plant Journal,* 3 : 739-44.

Stokinger EJ, Mulinix CA, Long CM, Brettin TS and Iezzoni AF (1996) A linkage map of sweet cherry based on RAPD analysis of a microspore-derived callus culture population. *J Hered,* 87 : 214-18.

Takahashi H (1993) Breeding of strawberry cultivars resistant to *Alternaria* black spot of strawberry. *Bulletin of the Akita Perfectural College of Agriculture,* 19 : 1-44.

Takahashi H, Matsomoto T and Takai T (1993) Somaclonal variants from strawberry cv. Morioka-16, M 16-AR 1,2 and 3 resistant to several isolates of Alternaria alternata strawberry phenotype occurring in Tohoku and Hokkaido. *J Jap Soc Hort Sci,* 61(4) : 821-26.

Terakami S, Adachi Y, Iketani H, Sato Y, Sawamura Y, Takada N, Nishitani C and Yamamoto T (2006) Genetic mapping of genes for susceptibility to black spot disease in Japanese pear. *Genome,* 50 : 735-41.

Tian L and Wang Y (2008) Seedless grape breeding for disease resistance by using embryo rescue. *Vitis* 47 : 15-19.

Tian YK, Wang CH and Dai HY (2005) Mapping of apple *co* gene using SSR markers. *Shi Yan Sheng Wu Xue Bao,* 38 : 272-75.

Troggio, M , Malacarne G, Coppola G, Segala C, Cartwright DA, Pindo M, Stefanini M, Mank R, Moroldo M, Morgante M, Grando MS and Velasco R (2007) A dense single-nucleotide polymorphism-based genetic linkage map of grapevine (*Vitis vinifera* l.) anchoring Pinot Noir bacterial artificial chromosome contigs. *Genetics,* 176 : 2637-50.

Tukey HB (1933) Artificial culture of sweet cherry embryos. *J Hered,* 24 : 7-12.

Unai E, Trujillo I and de García E (2004) Comparison of characteristics of bananas (*Musa* sp.) from the somaclone CIEN BTA-03 and its parental clone Williams. *Fruits,* 59 : 257-63.

Velasco R, Zhakikh A, Affourtit J *et al.,* (2010) The genome of the domesticated apple (*Mallus* x *domestica* Borkh.). *Nature Genetics* Doi :10.1038/ng.654.

Verde I, Lauria M, Dettori MT, Vendramin E, Balconi C, Micali S, Wang Y, Marrazzo MT, Cipriani G, Hartings H, Testolin R, Abbott AG Motto M and Quarta R (2005) Microsatellite and AFLP markers in the *Prunus persica* [L. (Batsch)] x *P. ferganensis* BC_1 linkage map : saturation and coverage improvement. *Theor Appl Genet,* 111 : 1013-21.

Vezzulli S, Troggio M, Coppola G, Jermakow A, Cartwright D, Zharkikh A, Stefanini M, Grando MS, Viola R, Adam-Blondon AF, Thomas M, This P and VelascoR (2008) A reference integrated map for cultivated grapevine (*Vitis vinifera* L.) from three crosses, based on 283 SSR and 501 SNP-based markers. *Theor Appl Genet,* 117: 499-511.

Vilanova S, Romero C, Abbott AG, Llcer G and Badenes ML (2003) An apricot (*Prunus armeniaca* L.) F2 progeny linkage map based on SSR and AFLP markers, mapping plum pox virus resistance and self-incompatibility traits. *Theor Appl Genet,* 107 : 239–47.

Viloria Z, Grosser JW and Bracho B (2005) Immature embryo rescue and seedling development of acid citrus fruit developed from interploid hybridization. *Plant cell Tiss Org Cult,* 82 : 159-67.

Wan SY, Deng ZA, Deng XX, Ye XR, Zhang WC and Kitto PH (1991) Advances made in *in vitro* mutation breeding in citrus. *Proc of Intl Symp on the role of Mutation Breeding on crop improvement,* 1 : 263-69.

Wang Y, Georgi LL, Reighard GL, Scorza R and Abbott AG (2002) Genetic mapping of the evergrowing gene in peach [*Prunus persica* (L.) Batsch]. *J Hered,* 93 : 352-58.

Wenzel G and Foroughi-Wehr B (1993) *In vitro* selection. In : *Plant Breeding : Principles and Practices* (Hayward MD, Bosemark NO and Romagosa I eds.), Chapman Hall, London, UK.pp. 353-71.

Xu HY, Yan AL and Zhang GJ (2005) Determination of the proper sampling period for embryo rescue from crosses between diploid and tetraploid grape cultivars. *Scientia Agri Sinica,* 38 : 629-33.

Xu K, Riaz S, Roncoroni NC, Jin Y, Hu R, Zhou R and Walker MA (2008) Genetic and QTL analysis of resistance to *Xiphinema index* in a grapevine cross. *Theor Appl Genet,* 116 : 305-11.

Yang KQ, Wang YJ, Zhang JJ, Wang XP, Wan YZ and Zhang JX (2005) Mapping of seedlessness gene in grapes using SCAR markers. *Yi Chuan Xue Bao,* 32 : 297-302.

Yoshokita T, Masauda T, Kotobuki K, Sanada T, Ito Y (1999) Gamma ray induced mutation breeding in fruit trees : breeding of mutant cultivars resistant to black spot in Japanese pear. *Japan Agri Res Quartely,* 33 : 227-34.

Zamir R, Ali N, Tariq SS, Mohammad T, and Ahmad J (2009) Guava (*Psidium guajava* L.) improvement using *in vivo* and *in vitro* induced mutagenesis. In : Induced Mutations in Tropical Fruit Trees, IAEA, Vienna, 2009. *IAEA -TECDOC*-1615 : 101-12.

Zanek MC, Reyes CA, Cervera M, Pena EJ, Velazquez K, Costa N, Plata MI, Grau O, Pena L, Garcýa ML (2008) Genetic transformation of sweet orange with the coat protein gene of Citrus psorosis virus and evaluation of resistance against the virus. *Plant Cell Rep,* 27 : 57-66.

Zhu YJ, Agbayani R, Jackson MC, Tang CS and Moore PH (2004) Expression of the grapevine stilbene synthase gene *VST* 1 in papaya provides increased resistance against diseases caused by *Phytophthora palmivora*. *Planata,* 220 : 241-50.

Chapter – 10

Exploitation of Genetic Resources of *Aonla*, *Bael* and *Noni* in Precision Horticulture

A. K. Singh, Sanjay Singh, S. K. Sharma, Raj Kumar and P. S. Sisodia

10.1 INTRODUCTION

Agriculture is a very important sector for the sustained growth of the Indian economy. About 70 per cent of the rural households and 8 per cent of urban households are still dependent on agriculture for their livelihood. Though, industrialization of the Indian economy has adversely affected the share of agriculture in the GDP, the fact can not be ignored that India has undergone a series of successful agricultural revolutions- starting with the 'green' revolution in wheat and rice in the 1960's and 1970's, the 'white' revolution in milk to the 'yellow' revolution in oilseeds in 1980's. As a result, India has achieved self-sufficiency in agriculture. Applications of agricultural inputs at uniform rates across the field without due regard to in-field variations in soil fertility and crop conditions does not yield desirable results in terms of crop yield. The management of in-field variability in soil fertility and crop conditions for improving the crop production and minimizing the environmental impact is the crux of precision farming. Geographically, India is widely distributed into several agro-climatic zones, and the information needs for the farming systems in these areas are entirely different. Integrating the application of available technologies to realize farmers' goals requires a systems approach to farming.

10.2 PRECISION FARMING

Precision farming is concerned with the management of variability in the dimensions of both space and time. Variability of resources, therefore, is a key factor of precision farming. Any component of production system ranging from natural resources to plants, production inputs, farm machinery and farm operations that is variable in some way is included in the realm of precision farming. Aspects of precision farming, therefore, encompass a broad array of topics, including variability of the soil resources base, weather, plant genetics, crop diversity, machinery performance, and most of the physical, chemical and biological inputs used in the production of crop. These are closely linked to the socio-economic aspects of production system, because precision farming should fit the needs and capabilities of the farmer and should be profitable. Success in precision farming is directly related to how well it can be applied to manage the space-time continuum in production system. The prospects of precision management increase as the degree of spatial dependence increases. However, the degree of difficulty in achieving precision management increase as the degree of spatial dependence and temporal variance increases. Within a given management parameters, the success to date of precision management is to a large extent determined by the degree to which the spatial variability is temporally stable. The precision farming seems to have efficient utilization of natural resources within the framework of time to enhance profitability of farm (Gary, 2000).

10.3 PRECISION FARMING-CONCEPT AND APPROACHES

The efforts for harnessing the natural resources for sustaining the human race have been started since the beginning of civilization. The basic needs of food, fuel and fodder are obtained from the land resources with the combination of water and sunlight. The approaches for harnessing these resources have been undergoing significant changes with the advancement of knowledge and availability of tools. The initial transformations in agriculture from hand operation to bullock/horse power in the late nineteenth century focused on the methods used to operate farm machinery. The revolution in information technology has now led to focus on the use of modern tools with high degree of accuracy and sophistication for maximizing returns. The combination of information technology and use of high-tech equipments paved the way for the introduction of precision farming. Precision farming (PF) could be defined as application of holistic management strategies that use information technology to bring data from multiple sources to bear decision associated with horticultural production, marketing, finance and personnel.

The concept of precision farming has developed over the years by way of addressing various components of traditional crop management systems, staring with soil fertility. Some aspects of precision farming are:

1) Precision farming is taking care of each and every plant in a farmer's field throughout its crop cycle.

2) Precision farming is technology enabled, information based and decision focused.

3) Information, technology and management are key elements of precision farming. Now-a-days individual components of precision agriculture are being adopted on the farm, but a distinctive precision farming system has not yet evolved. Some advancement has been made in precision farming application

in two ways i.e.variable rate fertilizer application and yield mapping. In a traditional system of fertilizer application the farmer obtains one or more soil samples, combines the samples and sends them off to a soil testing laboratory. The farmer receives a fertilizer mix recommendation and a suggested application rate based on the crop to be grown. The farmer applies the specified fertilizer mix at the recommended rate across the entire field. In this the key concept has been uniformity. However, it is presumed that most of the fields are anything but uniform. With modern methods the precision farmers apply fertilizer at variable rate with the help of a soil sampling programme by demarcating grids on his plot. It is possible to identify the exact location of the plot with the help of one of the main precision faming technologies of Global Positioning System (GPS). For applying the fertilizer at variable rates, advanced equipments are available which carry the fertilizer elements in separate tanks to the field and mixes them on the fly just before dispersal. To accomplish this, the farmer must mount a GPS receiver on the fertilizer tank so that the equipment knows its location in the field. An in-vehicle computer must contain the fertilizer need map, which compares to the field position data arriving from the GPS receiver (Ravi and Jagadeesha, 2002).

The another practice of precision farming, yield mapping has been quite popular all over the world. This is because information on crop yields is important to every farmer. Yield mapping has become possible with the use of GPS and sensors that are used to measure the flow rate of grain harvested through combines. As the combine works, an on board computer gather information from the GPS receiver and the sensors which are stored in a memory card and later download to a PC equipped with yield mapping software. The data generated through such a system revealed wide variation in the crop yield within a field (Shanwad *et al.,* 2002 and Iida *et al.,* 1998).

10.4 PRECISION HORTICULTURE

High-tech horticulture has been defined as "technology which is capital intensive, and less environment dependent having capacity to improve productivity and quality of produce. At present, emphasis is being given for precision farming which pertains with efficient management of resources through location specific high-tech interventions. The major component of precision farming are: micro-propagation, high-tech nursery, fertigation, greenhouse production, mulching, promotion of biofertilizers including organic/biodynamic production, IPM, IDM, INM mechanization in horticulture, integrated crop management, use of remote control including GIS, GPS etc (Gary, 2000 and Francis and Nowak,1999). Brief account of major one is enumerated below.

10.4.1 Micro-irrigation

Water is the critical input in crop production. Its efficient use in inevitable in context of achieving higher production from shrinking ground water resources. Micro irrigation, popularly known as drip irrigation, is an efficient way of irrigating the plant with high frequency application of water close to feeder roots through network of suitable emitting devices placed near root zone. It economizes water use by 40-60 per cent, enhances yield of quality produce and reduces weed population. Drip irrigation permits frequent application of water without soil moisture stress resulting into enhanced and continued growth of plants (Joan and Mar, 2000).

The minimal and optimal fraction of the soil volume to be wetted and management system in different horticultural crops under varying weather and soil conditions, application of nutrients at critical stages of growth and development. Creation of awareness, human resource development, pollution environment should also get due attention for popularization of drip irrigation (Pierre, 2000).

10.4.2 Fertigation

It pertains to supply of both water and nutrients to growing crop through drip irrigation. Through fertigation, many macronutrients like nitrogen, phosphorus, potash as well as micro nutrients e.g. Cu, Fe, Mn, Mg, Zn and B can be supplied directly to the active root zone. Besides ensuring efficient delivery of two important inputs, it also exploits the synergism of their simultaneous availability to plants. When the drip system is adopted, the root system proliferates only in the moist zone below the emitter making this as nutrient absorption zone. This system helps in improving productivity of quality produce and minimizes the losses of expensive nutrients. There is need to encourage fertigation through technological and policy support (Neilsen *et al.,* 2003).

10.4.3 Use of rootstock

While rootstocks for several fruit crops have already been standardized, some recent findings in use of rootstock for mango, grape and citrus are noteworthy. For Dashehari mango, Rumani as a rootstock has shown dwarfing effect. In West Bengal, Olour has been found to be a dwarfing rootstock for Himsagar and Langra. Use of Vellaikolamban as rootstock for Alphanso accounted for 45 per cent reduction in canopy size at Vengurla (Maharastra).

In citrus, Rangpur lime and Rough lemon have been found to be the most promising rootstock for different scions. Rangpur lime is reported to be the best for Mosambi and Kinnow in Punjab, Sathgudi in A.P. and some part of Gujarat, Coorg Orange in Karnataka and Nagpur mandarin in Maharastra. Rough lemon has shown better compatibility with Blood Red, Jaffa and Pineapple orange in Punjab (Rajput and Babu,1985).

A notable change in grape cultivation is the use of Dogridge rootstock by the grape growers of Maharashtra. Dogridge is difficult to root as such but tissue culture techniques have been standardized at IIHR for mass scale multiplication. St. George rootstock is recommended for its drought tolerance.

10.4.4 High density plantings in fruits

At IARI, New Delhi, Amrapali is reported to perform well when planted at a spacing of 2.5 X 2.5m accommodating 1600 plant/ha. At Pantnagar, a comparative study of high density orcharding with a planting distance of 3 X 2.5m aginst normal spacing of 12 X 12m in Dashehari showed higher final yield of 2.43 to 9.43 tonnes/ha during fifth-sixth year of plantation. Preliminary indications with Dashehari high density at CISH, Lucknow showed that medium density planting 5 X 5m, coupled with centre opening, trimming of side branches and removal of alternate tree in subsequent years may be appropriate proposition for mango planting. In an observation with NA-7 variety of aonla planted in double hedgerow system, accommodating 260 plant/ha, showed more than double yield over conventional square system under rainfed conditions of hot semi-arid ecosystem at CHES, Vejalpur, Panchmahals, Guarat (Bose *et. al.,* 1992., Singh and Singh, 2007, Singh *et al.,* 2011)

10.4.5 Soilless culture

In order to circumvent the heterozygosity of soil search, the alternative substitute resulted in usage of natural substrates like cocopeat, rock wool, gravel sand, saw dust, groundnut and paddy husk. Media constituent like cocopeat is successfully used for better management in both vegetables and flowers cultivation. Undoubtedly, the crop grown on cocopeat and rock wool has better growth and development as compared to soil grown plants. It has a special advantage due to high retention of water coupled with good aeration because of lesser bulk density and higher porosity. Because, the flowers and vegetables are lighter in weight when grown on these media which is of great significance in exports. Hydroponic techniques using deep flow and nutrient film technique is used to limited extent for commercial cultivation of vegetables and flowers.

10.4.6 Intercropping

Intercropping is intended to maximize land and space uses efficiency to generate supplemental income particularly during the initial unproductive phase of the orchard, to protect the interspaces from losses through, erosion, impact of radiation, temperature, wind and water and enriching the soil fertility by incorporating the biomass of intercrops. During the early phase of orchard establishment, interspaces left between tree rows can efficiently be utilized by raising suitable crops which not only enrich the soil but also generate additional income particularly during juvenile phase of plant. Compatible crop combination is necessary with regard to species, cultivars, planting method and sequence in location specific climatic conditions. Leguminaceous crops live cowpea, cluster bean and gram and okra can be taken as intercrops under rainfed conditions of semi-arid ecosystem.

10.4.7 Canopy management

Canopy management of the crop deals with the development and maintenance of their structure in relation to the size and shape for the maximum productivity and quality. Tree vigor, light, temperature and humidity play a vital role in production of quality fruits. The crux of canopy management lies in fact, as how best we manipulate the tree vigour favouring maximum use of available sunlight and temperature to increase the productivity and minimize the adverse effect of weather. Pruning is done to improve and regulate tree size and shape to achieve desired architecture of the canopy and also to reduce the foliage density by removing unproductive branches of the tree.

10.4.8 Biological control

It is use of organisms to regulate a pest or pathogen below its economic threshold level. It assumes importance in sustainable agriculture and organic farming.

The use of commercial nuclear polyhedrosis virus (NPV) is gaining importance all over the world. In India too, private industries are bringing out field compatible formulations. It has been found that using NPVs at early, stage bring an excellent control of *Helicoverpa armigera* on tomato. NPV is further, compatible with *Trichogramma* egg parasitoids, endosulfan and pheromone traps. These, in turn would constitute an ideal IPM. One of the advantages of NPV is its specificity. However, NPV of *Autographa californica* is known to infect several lapidopterous pests. It is necessary, therefore, to test specificity using restriction endo-nuclease analysis of viral DNA. The private and public sectors presently involved in mass production activities will not be in a position to meet the demand for supplying the biotic agents. There is a need to start more production units to meet the demand. The increasing demand for natural enemies combined with inadequate skill for production, release and maintenance of bio-agents has to be tackled. Limited financial resources are also coming in the way of mass production.

10.4.9 Post harvest technology

Indian fruit and vegetable sector is the largest in the world accounting for over 10 per cent of total world output. It will continue to expand rapidly, its growth being driven by rising income and increasing demand from current low levels of per capita consumption. Currently, the sector is dominated by fresh market. However, there is also an opportunity for players in the processed food sector to improve the quality and price performance of their products. Despite low per capita availability, India is already the second largest producer of fruits and vegetables in the world. Rise in income will ensure that this sector will continue to grow during the coming years. The bulk of the growth will continue to be in fresh produce market. The fruit and vegetable sector is marked by a very high degree of wastage and value destruction. The wastage is on the farm along the procurement chain and it is due to inadequate storage and transportation systems, and high trading margins of intermediaries. There is a need for more developed marketing mechanisms ensuring low wastage with improved storage and transport systems (Singh and Roy, 1984).

10.4.10 Cold storage/cool chain

The application of refrigeration techniques in the preservation and storage of perishables is important in the present context. The bumper production and limited seasons of harvest of fruits and vegetables have established the impressiveness of refrigerated storage. Cold storage is important vital organs of an efficient marketing system. Today, the cold storage industry is power starved and with greater running expenses. There is need for improving the design, operational cost and safety standards of cold storages. Introduction of pre-fabricated panel construction for cold storage has revolutionized the setting up of cold storage. Energy saving measures and use of alternative sources of energy are also to be thought off. Stationary cold storage are supplemented with cool chain technology especially for exports of fresh produce like mango, grapes, banana etc., wherein the produce is stored at low temperature from harvest to the consumer. The use of pre-cooling devices followed by cold storages has helped in the maintenance of low temperature. In addition to regulating the temperature of storage, retention of quality of the produce has been attempted by adoption of controlled/modified atmosphere packing of fruits and vegetables, as evidenced by the export trials of mango during the last few years. There are benefits and limitations in the use of this technology. A lot still needs to be done before this is successfully commercialized. Basic information on stationary storage under CA/MA on important fruits and vegetables has to be collected for the good success of this technology. A trial conducted by APEDA for export of mango from Maharastra and Andhra has indicated that those CA shipments have achieved a partial success. Problems with quality issues, like overcoming chilling injury, latex staining, uneven ripening of the stored fruit and disease control have to be solved.

10.4.11 Precision soil fertility management

Nutrient input to crop production is important because soils do not supply nutrients in sufficient quantities naturally to meet nutrient demands of commercial crops. Important inputs to crop production, fertilizers and manures are also identified as major sources of nutrient contamination of surface and ground water in agricultural areas. Soils vary in their ability to supply nutrients to plants and crops vary in their demand for nutrients. Since soil supply and plant demand vary in space and time and nutrient losses through leaching, erosion and runoff also vary temporally and spatially. These indicate that significant opportunities exist for precision management of soil fertility. The potential for improved precision in soil fertility management combined with increased precision in application control make precision soil fertility management an attractive, but largely unproven, alternative to uniform field management (Bhutani *et.al.,* 1994.)

10.4.12 Precision pest management

Insects and diseases are ever present and pure heavy problem in fruit production because significant infestations reduce fruit yield and quality and if severe, can limit fruit crop production options. Public concerns regarding the impacts of pesticide use include health risks related to food safety, water quality, worker safety and concerns over wildlife and ecosystem health. Therefore, horticultural management practices that reduce pesticide use, improve pest management or reduce risks of pesticides to human and ecosystem health are very desirable. The potential direct economic benefit of precision pest management is the reduction in chemical/non-chemical pest management costs, crop damage or both due to more efficacious or efficient application of pest control measures. A reduction in pesticide use, however, does not translate into profit if, the cost of obtaining information about pest populations and distribution exceeds the savings. Environmental benefits are presumed to result from a reduction in pesticide usage, particularly in sensitive environments. However, the results of the few studies on the potential of precision farming to provide environmental benefits have been inconclusive regarding its effect on pesticide use.

10.4.13 Precision soil moisture management

Soil moisture is one of the most important factors of fruit production. Its supply to crop in adequate quantity is essential. Depending upon the soil and land features, the availability of soil moisture vary. Besides, the moisture requirement of the tree would vary depending upon the stage of growth. The fact that crop water requirements vary over time and space indicate the scope of precision management of soil moisture. Irrigation scheduling using VRT combines knowledge of spatial variable soil water holding capacity, spatial and temporal variability of applied water, system delivery specifications and weather drove crop evpotranspiration (ET) calculation. On the other hand under conventional systems, a single value of crop ET is typically used for entire field for use in irrigation scheduling software.

10.4.14 Precision weed management

The application of precision farming to weed management is potentially beneficial to agriculture because (i) it offers an opportunity to reduce chemical/non-chemical inputs into crop production through site specific weed control and the use of precise application techniques and (ii) the acquisition of spatial and temporal information on weed occurrence and distribution made possible with precision agriculture technologies, will lead to an improved understanding of weed biology and ecology needed to develop more effective weed management strategies. Precision weed management is possible because weeds are spatially aggregated and not randomly distributed within most agronomic fields and because the efficacy, efficiency, and fate of weed control inputs varies with weed and crop conditions and with soil physical and chemical properties. All of which can vary spatially and to a varying extent, temporally. A common approach to precision weed management is site specific weed control achieved by (i) applying herbicides only where weeds are present or above economic threshold levels, termed intermittent herbicide application or patch spraying, (ii) Varying herbicide application according to soil physical and chemical properties or weed characteristics or (iii) some combination of the two approaches. For prevention or pre emergence weed control, site specific application requires prior knowledge of historical weed distributions since no weeds are visible at the time of application. This knowledge can be obtained by mapping weed aggregation in previous years. Weed control treatment or intensity can also be varied based on soil properties according to the label or other recommendations, if, knowledge of the spatial variation of these properties is adequately known. (Shirgure *et.al.,* 2003).

10.4.15 Biodiversity exploitation

India is bestowed with vast genetic resources of *aonla*, *bael* and *noni* having great diversity which can be exploited commercially in precision horticulture in various ways.

10.4.15.1 Exploitation of genetic resources of bael

The fruits are official in the Indian Pharmacopoeia. They are also highly valued in *Ayurvedic* medicines. The peripheral part just within the rind is fleshy and thick, and has a pleasant resinous odour. The walls separating the chambers have a light yellow tint which becomes yellowish brown on exposure and have slightly acrid bitter taste. The chambers are full of amber or honey coloured viscous very sticky or glutinous, translucent pulp, which is slightly sweet and feebly aromatic. The gummy substance surrounding the seeds serves as a good adhesive and is added to water-paints to improve strength and brilliancy. It is more abundant in young fruits. The gum has been used for the stabilization of drilling fluids. The stem also contains a gum similar to gum arabic.

The importance of bael fruit lies in its curative properties, which make the tree one of the most useful medicinal plants of India. Its medicinal properties have been dealt with in the '*Charaka Samhita*', an early medical treatise in *Sanskrit*. All the parts of the tree, whether it is stem, bark, root, leaf, flower, seed oil or fruits of any stage of maturity and ripening are used in various *Ayurvedic* medicines.

The demand of *bael* fruit is not much higher as other major fruits like mango, banana, guava, apple etc. The ripe fruit is eaten as a fresh and has a more demand amongst those, who wants it for therapeutic use. Sometimes the pulps are diluted with water and the requisite amount of sugar and tamarind is added to prepare a delicious cooling drink. The *bael* pulp is used as a base of various fruit products like squash, jam, slab, toffee, powder, nectar, RTS etc. Green *bael* fruits are used for preparing *murabba* (preserve), which is generally taken for stomach ailments. The green *bael* fruit slices often are dried and stored for future use. The unripe or half ripe fruit is regarded as astringent, digestive stomachic and good for heart and brain. The fruit is used in chronic diarrhoea and dysentery and to act as tonic for heart and brain. It is useful adjuvant, as it helps to remove constipation which hinders the healing ulcerated surfaces of intestine. Besides the fruits, the root is used as an ingredient of '*dasmula*' (ten roots) used as *Ayurvedic* medicine. The roots as well as bark are used in the form of a decoction as a remedy in melancholia, intermittent fever and palpitation of heart. Root has anti-amoebic and hypoglycemic properties. The young leaves and shoots are used as fodder for cattle, sheep and goats. The leaves are bitter and used as febrifuge, Poultice made of leaves is used for ophthalimia and ulcers. Fresh leaves are also used as remedy for 'dropsy, and *beri beri* associated with weakness of heart. The astringent rind of ripe fruits and bark are employed in dyeing and tanning. The timber is commonly used for making pestles of oil and sugar mills, for posts, shafts, axles and naves of cart. The shells of smaller fruits are used as snuff boxes. A yellow dye is extracted from the rind of unripe fruit. The diluted leaf juice is used for catarrh. The alkaloid aegelin present in the leaves are efficacious in asthma.

10.4.15.1.1 Varietal wealth

Previously, *bael* fruits were generally named after the locality where they are easily available. So far reports on the available cultivars are mainly from the states of Uttar Pradesh, Uttarakhand, Bihar and West Bengal. Singh (1961) described six varieties of *bael* varieties from U.P. ie *Mirzapuri, Darogaji, Ojha, Rampuri, Azamati and Khamaria.* It appears that the organoleptic quality of bael fruit depends upon three factors: mucilage, sugar and total phenolics. High sugars, particularly non-reducing sugars with low phenolics and mucilage make the bael fruit more palatable. It has been observed that the bigger size fruits have high pulp, often thick peel, less seeds, less sugar, low phenolics and less mucilage as compared to smaller ones. *Bael* trees with desirable fruit characteristics have been collected and evaluated for their growth, flowering, fruiting behaviour and quality attributes and many promising varieties have been developed through clonal propagation at ICAR Institutes / Regional Station and State Agricultural Universities.

Table 1 : Improved bael varieties developed in India

Varieties	Organizations from where developed
Narendra Bael-5, Narendra Bael-7 Narendra Bael-9, Narendra Bael-16 Narendra Bael-17	N. D. University of Agriculture and Technology, Kumarganj, Faizabad, U. P.
Pant Aparna, Pant Sujata, Pant Shivani, Pant Urvashi	G.B. Pant University of Agriculture and Technology, Pant Nagar, Uttarakhand
CISHB-1, CISHB-2	Central Institute for Sub-tropical Horticulture, Lucknow, Uttar Pradesh
Goma Yashi	Central Horticultural Experiment Station (CIAH-ICAR), Vejalpur, Panchmahals (Godhra), Gujarat

10.4.15.1.2 Description of varieties

The detailed description of varieties of bael is given below:

Narendra Bael-5: The plants are small and semi-spreading growth habit, precocious and prolific in bearing. The budded plants start fruiting in the 4th year. The average fruit yield of six-year old plant is 28.78 kg. The fruits are medium in size (12.50 cm × 11.50 cm), round, with smooth surface and very thin rind (0.16-0.17cm), straw yellow at maturity, low in mucilage, moderately fibrous and an attractive yellow pulp, with low seed content. Excellent in taste and flavour, the fruits have 33^0 Brix total soluble solids in pulp and 48^{0} Brix in mucilage and contain ascorbic acid 18.63 mg/ 100g of edible portion. The fruit weight range from 0.8-1.0 kg under rainfed conditions of semi-arid ecosystem during sixth years of orchard life. Its taste is good and can be used as fresh as well as processed into various value added products.

Narendra Bael-9: The plants are semi-vigorous and spreading having compact canopy. The variety is precocious and prolific bearer. The average fruit yield of a six-year old plant is 56 kg. Fruits are medium to large in size (16.00 cm x 13.50cm), roundish-oblong, with smooth surface and thick rind (0.31cm), light yellow at maturity, average in mucilage, moderately fibrous, slightly golden-yellow pulp with low seed content. The fruits are good to taste containing total soluble solids 38^{0} Brix in pulp and 41^{0} Brix in mucilage, slightly acidic having ascorbic acid 19.20 mg/100 g of edible portion.It can be used as fresh as well as processed into various value added products. Its keeping quality for storage is very good.

Narendra Bael-7: Plants are tall and semi-spreading. They are sparse in bearing with large size fruit. The budded plants start fruiting in the 4th year. The average fruit yield of six-year old plant i s 32.10kg (6th year). The fruits are medium in size (18.25 cm × 22.50cm), round and with smooth surface and very thick rind, yellow at maturity, low in mucilage and fibrous, an attractive yellow pulp, with low seed content. Fruits are good in taste and flavour, having total soluble solids 30^{0} Brix in pulp and 42^{0}Brix in mucilage, with slight acidity and 19.78mg/100g ascorbic acid content.

Pant Aparna: Its trees are dwarf with drooping foliage, almost thornless, precocious and heavy-bearer. The leaves are large, dark green and pear shaped. Fruit is globose in shape with average size of 13.00 cm × 12.00cm and weight of 0.8-1.25kg. Fruit pulp is yellow and rind is thin. TSS 34^{0}Brix in pulp and 47^{0}Brix in mucilage, titratable acidity 0.67% and ascorbic acid 17.15mg/100g of pulp. Mucilage, seed and fibres are low. Mucilage and seeds are enclosed in separate segments. Flavour and taste are very good. The plant yields 40.25 kg/plant at the age of 6 years.

Pant Shivani: It is an early mid-season variety. Trees are tall, vigorous, dense, upright growing, precocious and heavy-bearer. Fruit shape is ovoid, oblong and the size being 18.50cm × 15.00cm. Fruit weight ranges from 2 to 2.4kg. Colour of fruit is lemon-yellow and its storage quality is good. Rind is medium-thin, pulp is lemon-yellow with pleasant flavour and mucilage, seeds and fibre are low to medium. Taste is very good. It has 69 per cent pulp, TSS 36^{0} Brix in pulp, mucilage TSS 48^{0}Brix, total titratable acidity 0.47 per cent and ascorbic acid 19.55mg/100g of flesh.

Pant Sujata: It is an early mid-season variety but has problem of fruit splitting. So far it has not been reported under rainfed conditions of hot semi-arid ecosystem of western India. Trees are medium-dwarf with drooping and spreading foliage, dense, precocious and heavy bearer. Thorns are stout and bigger. Fruit is globose shaped, depressed at both ends with average size of 14.50cm × 13.50cm and weight varied from 1.12 to1.40kg under rainfed condition of hot semi arid ecosystem of western India. Fruit and pulp are light yellow. Storage life is better, rind is thin, and the seeds, mucilage and fibre are low. Its flavour is pleasant and taste is very good. Flesh is 72 per cent, TSS

32^0Brix in pulp and 42^0Brix in mucilge, acidity 0.44 per cent and ascorbic acid 17.10mg/100 g of flesh.

Pant Urvashi: It is a mid-season variety. Trees are tall, vigorous, dense, upright growing, precocious and heavy bearer. Fruit is ovoid-oblong with average size of 14.50cm x 17.20cm and fruit weight ranges 1.5-2.50kg. Fruit is yellow, rind is medium to thin and pulp is light yellow. Fruit has 62.35 per cent pulp with pleasant flavour. Seeds and mucilage are medium, fibre content low, TSS 33^0Brix in pulp and 41^0Brix in mucilage, titratable acidity 0.49 per cent and ascorbic acid 17.15mg/100g pulp.

CISHB-1: It is early maturing variety. The plants are semi-tall and having spreading growth habit. The budded plants start fruiting in the 4^{th}year. The average fruit yield of a six-year old plant is 42.64 kg. The fruits are medium in size (16.50cm × 12.00cm), oval-oblong, with smooth surface, yellow at maturity, low in mucilage and fibrous, an attractive yellow pulp, with high seed content. Excellent in taste and flavour, the fruits have 32^0Brix total soluble solids in pulp and 43^0Brix in mucilage. The fruit weight varies from 0.8 to1.40kg under rainfed condition of western India.

CISHB-2 The plants are dwarf and spreading. The average fruit yield of six-year old plant is 38.45kg. The fruits are medium in size (16.00cm x 14.00cm), rounded, with smooth surface, yellow at maturity and the rind is thick, low in mucilage and fibrous, an attractive yellow pulp, with low seed content. Good in taste and flavour, the fruits have total soluble solids 31^0Brix in pulp and 38^0Brix in mucilage, titratable acidity (0.41%). The fruit weight ranges from 1.7 to 2.6kg/fruit.

Goma Yashi: It was released from CHES, Godhra in the year, 2010. It is early maturing variety with semi-spreading, drooping growth habit, less lateral branches growth (20-35cm annually), The tree Stative relatively small, spineless, central leaf size (13.20 × 6.70cm), lateral leaf size (9.20 × 5.30cm), ovate having acute apex. Initiation of flowering takes place during last week of April and end of flowering is observed during last week of June, bud sizes 12mm × 8mm, anthesis is seen between 6 - 8 AM and the dehiscence between 6.30 - 8.30 AM. Flower sizes 15mm × 29mm, petals 4 - 5 in number.

The plant yield 51.00kg at 6^{th}year are, The fruit size is 13.00cm × 12.50cm, fruit girth 41 - 45cm, shell thickness 0.17cm, seed weight 25-30g, fibre weight 40.24-51.20g, shell weight 180-210g, fruit weight 1.00-1.62 kg, locules in cross section 13-15, TSS of pulp 35-39^0B, TSS of mucilage 41-43^0B, titratable acidity 0.26-0.32% and vitamin C 22.00 mg/100 g pulp. It is highly suitable for high density planting.

10.4.15.2 Exploitation of genetic resources of aonla

Aonla (*Emblica officinalis* Gaertn), also known as Indian gooseberry, is an important fruit crop of 21st century. It has been identified as an ideal plant for various kinds of wastelands viz, eroded, ravines, upland, riverbed and the areas with undulated topography as it possesses some specific characters like intensive and deeper root system, summer dormancy of zygote, reduced leaf area, synchronization of fruit growth and development with moisture availability period and selective absorption of cations and anions which enable it to grow in fragile agro-climatic condition particularly in saline and alkaline soils of arid and semi-arid regions. India is principal *aonla* growing country and Uttar Pradesh is the leading state and Pratapgarh district is well known for *aonla* cultivation throughout the country. At present, in India *aonla* is grown in over 50,000 ha area with annual production of 1.75 lakh tonnes. Outside India, *aonla* trees are also reported to be found in the natural forest of Cuba, USA, Pakistan, Sri Lanka, Malaysia, China, Java and West Indies. The area under *aonla* cultivation is increasing rapidly in arid and semi-arid regions of Gujarat, Maharastra and Rajasthan, Haryana, Andhra Pradesh, Karnataka, Tamil Nadu, Punjab and Himachal Pradesh (Pathak, *et al.,* 1993). However, the productivity per unit area is relatively low owing to moisture stress and lack of proper management strategies, which can be enhanced by adopting suitable varieties, improving soil moisture conservation practices, suitable crop geometry (high density planting), and appropriate package of practices including plant nutrition, protection, etc.

The fruit of *aonla* is the richest source of vitamin C next only to *Barbados cherry*. The fruit contains chemical substances known as leucoanthocyanins (polyphenols) that retard the oxidation of vitamin C making it rich source of vitamin C. The fruit is also rich in pectin and minerals such as iron, calcium and phosphorus. *Aonla* fruit is very well known for its medicinal properties as mentioned in literature of both *Ayurvedic* and *Unani* system of medicines. It is acidic, cooling, refrigerant, diuretic and laxative. Although the fresh *aonla* has no table value, juice of the fresh fruit is popularly consumed by masses. After storing in brine for few days, it is fondly consumed. Dried fruit is useful in chronic dysentery, diarrhoea, diabetes, dyspepsia, cough, anaemia and jaundice. The excellent nutritive value of the fruits has great potentiality for processing into various products like *Murabba*, candy, shred, jam, pickles, herbal jam, *chawanprash* and *trifala*. Fruits are also used in shampoos and hair dyes.

10.4.15.2.1 Varietal wealth

The important cultivars of *aonla* which are being commercially exploited under arid and semi-arid ecosystem of the country are Banarasi, Chakaiya, Francis, NA-4, NA-5, NA-6, NA-7, NA-9, NA-10 and Anand-1, Anand-2, Anand-3 and BS-1.

The salient features of commercial cultivars are as under

NA-7: It is seedling selection from open pollinated strains of Francis, semi spreading growth habit, about 10.05 female flower per branchlet. Fruits are medium to large and slightly conical in shape, smooth skinned, segment six and easily separated. It is precocious, profuse bearer and mid season variety. It is highly suitable for processing having T.S.S. 8.35 per cent, acidity2.15 per cent and vitamin C 527mg/100g.

Kanchan: It is a chance seedling of Chakaiya. It tall and spreading growth habit. Fruits are small to medium, flattened oblong, segment six and difficult to separate, flesh fibrous and hard and highly suitable for pickle making. It matures late and remains free from necrosis, T.S.S. 9.5 per cent, acidity 2.11 per cent and vitamin C 422.20mg/100g of fruit pulp.

Krishna: It is a chance seedling of Banarasi from Pratapgarh district of Uttar Pradesh Tree is semi tall and spreading growth habit, fruit large, triangular, red spots on exposed surface; segment six, easily separated and highly astringent. It is shy bearer and matures early, T.S.S. 11.00 per cent, acidity 2.32 per cent and vitamin C 549.20 mg/100g.

Banarasi: A semi-tall variety with spreading growth habit with three branchlets per node. Fruits are large, triangular and slightly conical at apex, segment six and easily separated fibreless and soft. It is shy bearing owing to less number of female flowers and self-incompatibility. It is yearly maturing variety with poor keeping quality.

Chakaiya: A variety with tall and upright growth habit, profuse bearing and has 4.03 female flowers per branchlets, fruits are small to medium, flattened round, segment six, difficult to separate, flesh fibrous and whitish green in colour. It is late in maturity with good keeping quality and highly suitable for pickle and processed products preparations.

Francis: Erect and tall with drooping branching habit, moderate in bearing, fruits are medium, flattened oblong thick at upper side and thin at basin. Segments are six and easily separable, soft and nearly fibrous. Fruit are highly susceptible to necrosis hence not suitable for preserve making. It is mid season variety.

NA-6: It is selection from open pollinated chance seedling of Chakaiya. It is prolific bearer having 10 female flowers per branchlet. Fruits are small to medium and oval in shape, skin smooth semi-traslucent and light green in colour. Flesh is fibrous and semi-hard. Seeds are small light brown in colour. Keeping quality of fruit is suitable for processing.

NA-10: It is chance seedling of Banarasi locally known as Agra Bold. Tree is semi tall and having spreading growth habit. It is moderate bearer with 4.6 female flowers per branchlet. Skin is rough having yellowish green colour, flesh slightly fibrous, whitish green, soft, juicy and highly astringent. Keeping quality good and moderately susceptible to fruit necrosis.

Goma Aishwarya (CHES selection-1): Goma Aishwarya is a selection from the tree identified from the large population. It is an early drought tolerant variety; average yield potential is 102.9kg/tree. It has low fibre content and is suitable for processing and export. Semi-spreading plants at 5 year age are 3.63m in height and 3.78 x 3.49m in spread. Medium sized fruits (45.17g / fruit) fruit contains 47.00 per cent juice, pulp: stone ratio is 26.65. TSS is 10.34^0 Brix ascorbic acid content is 554.78 mg/100g.

Anand-2: Trees are tall with right growth habit. It is moderate bearer having 2-2.5 female flowers per branchlet. Fruits are small to medium, skin rough with medium keeping quality. Besides this, Anand-1 and Anand-2 has been selected as promising strains in Gujarat.

10.4.15.3 Exploitation of genetic resources of noni

Noni (*Morinda citrifolia*) belongs to family Rubiaceae. It is considered a sacred plant in *Ayurveda* and is mentioned in ancient text as *Ashyaki*, which is *Sanskrit* of longevity. It is underutilized medicinal fruit plant which is highly useful from therapeutic and nutraceutical points of view. *Noni* has distinct chemical constituents which are responsible for healthful characteristics. *Noni* and its products are used as health panacea against high blood pressure, respiratory problems and immune deficiencies. Its roots are applied as paste as antidote for venom of poisonous creatures. Its decoction is used to cure chronic skin infections like boils. Ash of its leaves is also made into paste in oil base and applied on boils. Its fruits, after discarding seeds are used as poultice. Its bark and wood is used to make yellow pigment. Its wood is used as timber for making stakes and pillers and handles of farm implements. Its fragrant flowers are offered to Lord *Ganesha*. A researcher in Hawaii claims to have identified the main active ingredients in *noni* fruit called xeronines and proxeronine. The seeds are triangular radish to brown in colour have air sac attached to one end which makes the seeds buoyant. *Noni* grows in tropical and deciduous forest such as those found in Western *Ghat*. It is sturdy plant and can tolerate to wide range of agro-climatic conditions and grows well in wet, dry, acidic, alkaline and even saline soil. In Gujarat, *Morinda* spp. are found scattered in patches in almost all over the state mainly in forest areas (*Chhota Udipur, Jambugoda, Dharampur, Gir and Ratan Mahal*) having diversity in their growth, flowering, fruiting behaviour and quality attributes (Singh *et al.*, 2009 and 2010). Looking into the availability of variability in genotypes, an attempt was made for collection, introduction and evaluation of *Morinda* genotypes under semi-arid ecosystem of Western India. A toatal of 31 genotypes have been collected and they are being evaluated.

10.4.15.3.1 Qualitative attributes

The seed weight per ranges fruit from 24.76-40.00g. The TSS and pH of fruit juice varied from 8.00-12.50 and 4.25-5.00 and colour of fruit became brownish black upon ripening.

10.5 SCOPE FOR PRECISION FARMING IN HORTICULTURE

Precision farming technology has been deployed predominately for agricultural crops like corn, soybean, rice, cotton, peanut, sugarbeet and sugarcane. Among horticultural crops this technology has been applied for crops like potato, banana and oil palm. Studies conducted at the Irrigated Agricultural Research in Extension Centres, Wash, USA on using site specific crop management in potato demonstrated promising results. The tools available commercially for the potato growers are yield monitoring and VRT (Variable rate application technology) fertilizer application equipment. All aspects of precision farming involving VRT water application, VRT fertilizer application, yield monitoring, VRT insect and pest control can be applied to horticultural crops. The critical factor would be cost effectiveness (Pierre, 2000 and Joan, 2000).

10.6 CONCLUSION

Precision horticulture is a holistic management strategy that uses information technology to bring data from multiple sources to bear decision associated with horticulture production, marketing, finance and personnel. Basically, it integrates information technology with application systems to enhance crop productivity with saving resources including time.The enabling technology of precision farming includ geographic information system, global position system, sensors and application control. Variability management centre and moisture, soil fertility, insect and pest management, weed control, etc has been deployed mostly for agricultural crops in the developed countries. In India, the small size of farms and fields limits economic gain from currently available precision horticulture technologies while the population density and public concern for the environment, food safety and animal welfare means that those potential benefits of precision agriculture are being given more attention. For instance ISRO Centre, Ahmedabad, Gujarat has started experiment to study the role of remote sensing, GIS and GPS in mapping the variability. Horticultural crops like potato and banana have been addressed. A beginning has been made in India for promoting precision farming concepts by establishing 17 PFDCs. A scheme is also been contemplated for delivering the technology to the farmers in an affordable manner.

10.7 REFERENCES

Bhutani, V. R., Khokhar U. U. and Raina V. P. (1994) Effect of herbicides, mulching and clean cultivation on weed population growth and cropping of apple trees. *Horticulture Journal,* 7 :7-13.

Bose T K, Mitra S K, and Chattopadhyay P K. 1992. Optimum planting density for some tropical fruits, *Acta Horticulturae,* 296 : 171-76.

Francis, J. P. and Nowak, P. (1999). Aspects of Precision Agriculture. *Advances in Agronomy*, 17:1-85.

Gary, T. R. (2000) Precision technology of Horticulture Cropm production, *Horticulture Technology*, July September, pp. 448-51.

Iida,M., Umeda, M., Kaho, T., Lee, C. K. And Suguri, M. (1998).Measurement of annual crop, *International conference on Precision Agriculture, St. Paul, MN.*19-22, ASA, CSSA and SSSA, Madison WI.

Joan, R. D.and Mar, J. H.(2000).Using site specific approaches to advance potato management in irrigated system, pp. 452-57.

Neilsen, G. H., Hogue, E. J., Forge, T. and Neilsen, D. (2003) Mulches and biosolids affect vigour, yield and leaf nutrition of fertigated high density apple. *Hort. Science,* 38 (11): 41-5.

Pathak, R. K., Srivastava, A. K., Dwivedi, R. and Singh, H. K. (1993). *Aonla descriptor*. N. D. Uni. of Agri. & Tech., Faizabad

Pierre, C. R. (2000). Site specific management for the twenty first century, *Horticulture Technology*, July-September, 444-47.

Rajput, C. B. S. and Babu, S. H. (1985). *Citriculture.* Kalyani Publishers, New Delhi.

Ravi, N. and Jagdeesha, C, J.(2002).*Precision Agriculture, training courses on remote sensing and GIS applications in Agriculture*, RRSSC- Bangalore, May27- 7th June, pp.225-28.

Shanwad, U. K. Patil , V. C.Dasog, G. S.Masur, C. P.and Shashidhar, K. C. (2002).*Global positioning system in precision Agriculture*. The Asian GPS Conference, 24-25 October, New Delhi, India

Shirgure, P. S., Sonkar, R. K., Singh, S. and Panighrahi, P. (2003) Effect of different mulches on soil moisture, weed reduction, growth and yield of drip irrigated Nagpur mandarin (*Citrus reticulata. Indian Journal of Agricultural Science* 73 : 148-152.

Singh, L. B. (1961).Some promising selection of bael *A. R. Hort. Res. Inst.,* Saharanpur, 111-19.

Singh, R. N. and Roy, S. K.(1984). The bael (Cultivation and Processing), DIPA, ICAR, New Delhi.

Singh, A. K., Singh, Sanjay, and Joshi, H. K., Bagle, B. G.More, T. A. and Sisodia, P. S. (2009).Survey, identification, collection and evaluation of *Morinda* spp. under semi-arid ecosystem of Gujarat.Procedings. *Third National Symposium on Noni for Nutrition and Health held at Chennai,* Oct. 18-19, 2008, pp-133-39.

Singh, A. K., Singh, S., and Joshi, H. K., Bagle, B. G. More, T. A. and Sisodia, P. S._(2010). Genetic variability of *Morinda* spp. for morphological features of Western India. First World Noni Congress, *Noni for Global Wellness*, 1st - 3rd October, 2010 at Chennai. pp-72-80.

Singh, A. K, Singh, S., Appa Rao, V. V., Bagle, B. G. and More, T. A. (2011). Effect of high density planting systems on the productivity of NA-7 *aonla* (*Emblica officinalis* Gaertn) under rainfed conditions of semi-arid ecosystem. *Indian Journal of Horticulture.*

Singh, A.K and Singh, S. (2007). Preliminary studies on high density planting system in NA-7 *Aonla* (*Emblica officinalis* Gaertn). *Orissa J. Hort.*, 35 (1): 81-85.

Chapter – 11

Redressal of Spatial Variation in Soil Fertility and Site Specific Nutrient Management for Rationale of Fertilizer Use in Citrus

A. K. Srivastava

11.1 INTRODUCTION

Citrus as a perennial crop occupies globally a place of prime importance amongst major fruit crops including India. The cultivation of citrus in India is dominated by three major cultivars viz., mandarin, sweet orange, limes and lemons occupying collectively an area of 9.23 lakh ha with a production of 86.08 lakh tonnes and productivity of 9.3 tonnes/ha. Nagpur mandarin in India is grown in 0.97 lakh ha area with total production of 8.81 lakh tonnes and productivity of 9.1 tonnes/ha, respectively. The productivity of mandarins within the country is highly variable in space and time, the major constraint is inadequate and unbalanced fertilizer use. The commercial cultivation of Nagpur mandarin is extensively confined along the Satpuda foothills of central India comprising Chindwara region of Madhya Pradesh and Vidarbha region of Maharashtra. The nutrient demand is highly variable in space (due to heterogenous soils) and time (due to varying weather conditions). Therefore, determination of crop nutrient requirement is a subject of great complexity. Experiments have shown that soils initially rich in fertility developed a variety of nutrient constraints following the non-synchronisation in demand and supply of nutrients with the orchard age (Srivastava *et al.*, 1998). For example when exchangeable K is not rapidly replenished, crops start drawing on the non-exchangeable K, resulting in soil mining and depletion in soil K reserve. A nutrient level sufficient for one particular productivity level may not hold sufficient for other higher productivity levels. An extensive survey of 18,000 ha area of Nagpur mandarin orchards of central India showed large scale deficiency of three nutrients viz., N, P and Zn, with most of the sites expressing multiple nutrient deficiencies (Srivastava and Singh 2002; 2005). Reduced longevity of commercial citrus orchards due to varying nutritional constraints is one of the major production related constraints highly responsible for poor orchard efficiency (Srivastava and Singh, 2008a; 2008b).

Knowing the required nutrients for all stages of growth, and understanding the soil's ability to supply those needed nutrients are critical to profitable crop production. The recommendations on fertilizer application may not, however, produce the same magnitude of results when practised on an orchard of large area, because of its inability to accommodate variation in soil fertility status (Srivastava *et al.*, 2010). Slight changes in the nature of soil, local climate, and agronomic practices etc. may seriously affect the nutrient utilisation capacity of the plant. However, application of a single rate of nutrients may result in over-application of nutrients at some sites and under-application at others often lead to reduced fertilizer use efficiency. Under such circumstances, site specific nutrient management is adopted in big orchards requiring different treatment in patches as per the nature of surface and sub-surface soil properties so as to improve the orchard efficiency (average yield of specified trees in relation to average orchard yield) in ultimate terms (Srivastava *et al.*, 2007).

Site specific nutrient management (SSNM) is a dynamic concept. It should not mean that every time, a crop is grown, all the nutrients should be applied in a particular proportion. Rather fertilizer application should be tailored according to the crop's need keeping in view, the capacity of these soils to fulfil various demands (Srivastava *et al.*, 2006; 2009). To achieve this, it is necessary to keep an overall nutrient balance in relation to total crop load. This may indicate the need for the application of different nutrients at specific times, in a particular order to derive the maximum benefit from the application of a given quantity of nutrients. Conventionally designed long term fertilizer trials revealed that: i) omission of limiting macro- or micronutrients led to their progressive deficiencies due to heavy removals; ii) sites initially well supplied with P, K or S become deficient when continuously cropped using N alone and iii) fertilizer rates considered optimum still resulted in nutrient depletion at higher productivity levels, if continued, become sub-optimum rates. There is a strong necessity to keep overall nutrient balance in relation to total crop load. Application of a single rate of nutrients may result in over-application of nutrients at some sites and under-application at other sites, often lead to reduced fertilizer use efficiency. Under such circumstances, SSNM is adopted in big orchards requiring variable precision application as per soil variability so as to improve the orchard efficiency (average yield of specified trees in relation to average orchard yield) in ultimate terms.

11.2 SITE SPECIFIC NUTRIENT MANAGEMENT (SSNM) OBJECTIVES

The site specific nutrient management is usually carried out with the following objectives:

i. Identification and quantification of the variability of soil physical and chemical properties

ii. Understanding the impact of soil variability on crop growth, yield profitability and

iii. Management of soil variability to improve production, profits, and reduce environmental impact.

11.3 SPATIAL VARIABILITY IN SOIL FERTILITY

An exhaustive grid-based (36 x 36 m grid size) soil sampling within an orchard (Fig.1) showed a large variation in all the macro-(NPK) as well as micronutrients (Fe Mn Cu Zn). These values within the orchard varied extensively viz., alkaline $KMNO_4$ distilled N from 90-188 mg/kg, Olson-P 6-14 mg/kg, NH_4OAc-K 94-134 mg/kg, DTPF-Fe 6-18 mg/kg, DTPA-Mn 4-14 mg/kg, DTPA-Cu 6-14 mg/kg and DTPA-Zn 0.40-0.68 mg/kg compared to conventional soil test values of 102.00 mg/kg alkaline $KMNO_4$-N, 7.1 mg/kg Olson-P, 106.2 mg/kg NH_4OAc-K, 13.3 mg/kg DTPA-Fe, 11.0 mg/kg DTPA-Mn, 8.1 mg/kg DTPA-Cu and 0.58 mg/kg DTPA-Zn. Using these variations, spatial variaograms on GIS (Geographical Information System) version Arc 3.5 and accordingly the spatial variograms were developed (Fig. 2). Based on these variations, two soil types viz., Typic Ustorthent (Site 1) and Typic Haplustert (Site 2) of varying nutrient supplying capacity was identified for carrying out SSNM experiment in order to rationalize the fertilizer use for precision citriculture.

11.4 SSNM IN NAGPUR MANDARIN

The study included two distinct yet representative soil types. Site 1 had a relatively shallow soil profile classified as a Typic Ustorthent (Entisol), while Site 2 was a Vertisol with a deeper soil profile classified as Typic Haplustert (Table 1) at Narkhed teshil, Nagpur. These soil types are both derived form basaltic parent material with typical soil profiles predominantly rich in expanding-type, 2:1 montmorillonitic clay minerals characteristics of the sub-humid typical climate of central India. The Vertisol at Site 2 has interesting slickensides strongly expressed within 52cm to 1.48m depth, an indication of significant shrink and swell activity. Established orchards were 12-years old at Site 1 and 8-years old at Site 2. Plant to plant and row to row distances were 6m. Both orchards used a scion of Nagpur mandarin (*Citrus reticulata* Blanco) budded on Rough lemon rootstock (*Citrus jambhiri* Lush). A total of 16 treatments were applied based on soil analysis and the principles of SSNM (Table 2).

Table 1 : Soil physicochemical characteristics and fertility for soil surface horizons

	Site 1, Entisol	Site 2, Vertisol
pH	7.3	7.6
EC d/Sm	0.21	0.18
$CaCO_3$, g/kg	21.2	20.2
Texture, g/kg		
Sand	384.0	296.6
Silt	203.8	222.4
Clay	412.2	482.0
Available nutrients, mg/kg		
N	88.2	96.2
P	7.6	11.4
K	132.6	162.8
Fe	6.1	8.2
Mn	8.0	7.6
Cu	0.9	1.2
Zn	0.7	0.8

Table 2 : Range of fertilizer treatments applied to both orchards

Current Rec.	N	P_2O_5	K_2O	M_1[1]	S_1[2]
		Rate, g/tree			
	600	200	100	X	X
Low N					
T_1	600	400	600	√	√
T_2	600	200	600	√	√
T_3	600	0	600	√	√
T_4	600	600	600	√	√
T_5	600	400	0	√	√
T_6	600	400	300	√	√
T_7	600	400	600	X	X
T_8	600	400	900	√	√
High N					
T_9	1,200	600	600	√	√
T_{10}	1,200	600	900	√	√
T_{11}	1,200	600	1,500	√	√
T_{12}	1,200	600	1,500	√	√
T_{13}	1,200	600	0	√	√
T_{14}	1,200	600	1,200	X	√
T_{15}	1,200	600	1,200	X	X

[1]M_1 = 300 g each of $ZnSO_4$, $FeSO_4$, $MnSO_4$ and 100g borax/tree

[2]S_1 = 400 g $MgSO_4$/tree and 100g elemental S/tree

Two levels of input intensity were incorporated in the design based on high and low nitrogen (N) rate. These treatments were replicated four times in a randomized block design. Timing of fertilizer applications were kept the same at both sites. Nitrogen was applied in the months of April, August and October; phosphorus (P) and potassium (K) were applied in August and October. Two seasons of data collection included measurements of tree canopy growth, fruit yield and quality, leaf nutrient concentrations, and a cost:benefit analysis. Only the effective treatments and current recommendations (CR) are discussed in the article.

Significant changes in leaf nutrient concentrations expressed as parts per million (ppm), occurred in response to fertilization. Micronutrients inputs particularly affected zinc (Zn) concentrations of leaves in some cases elevated leaf N, P and K concentrations. Application of K increased leaf Zn concentrations irrespective of soil-type or whether any micronutrient was included in the treatment. However, the effect of K was greatest when co-applied with the micronutrients and the effect increased as K supply increased. Hence, K application improved the efficacy of soil Zn and applied Zn, a result of similar metabolic pathways during the course of Zn absorption.

Canopy of fruit growth response differed between sites. The more mature trees at Site 1 responded more favourably to the more input intensive regimes. Differences between high and low N regimes were much less apparent at Site 2. Thus, at Site 1, T_9 and T_{10} registered the highest increases in canopy volume over initial measurements and T_{11} produced a comparable result (Table 3). These treatments provided the highest levels of N, P, micronutrients, and secondary nutrient fertility plus either 600, 900 or 1,200 g K_2O/ha. Significant yield responses to fertilization followed responses observed in leaf nutrient concentrations. Fruit yield response to micronutrients was highly evident at both sites under either the high or low input regimes. Yield failed to respond K application beyond 600 g K_2O/tree under the high input at both sites. However, a differential response to K was noted between sites under set of low N input treatments, as Site 1 responded up to 900 g K_2O/tree while yield at Site-2 reached plateau at 300g K_2O/tree while yield at Site-2 reached a plateau at 300 g K_2O/tree. Highest fruit yields of 14.7 t/ha (52.7 kg/tree) and 19.0 t/ha (68.3 kg/tree) were obtained with T_9 (Site 1) and T_6(Site 2), respectively.

Table 3 : Canopy volume and fruit yield response to fertilization (mean of 2 years)

	Site 1			Site 2		
Treatments	**Canopy[1] volume (m^3)**	**Fruit yield (kg/tree)**	**Fruit yield, (t/ha)**	**Canopy[1] volume (m^3)**	**Fruit yield (kg/tree)**	**Fruit yield (t/ha)**
Current Rec.	3.5	31.5	8.7	3.0	53.75	14.9
Low N						
T_1	3.9	37.4	10.4	3.5	58.90	16.4
T_2	3.7	30.6	8.5	2.7	57.25	15.9
T_3	3.4	27.9	7.7	3.1	57.15	15.9
T_4	4.6	39.2	10.9	2.9	58.00	16.2
T_5	4.2	33.4	9.3	2.4	55.30	15.4
T_6	4.7	33.9	9.7	5.4	68.30	19.0
T_7	3.8	25.1	7.0	2.6	39.25	10.9
T_8	5.7	49.9	13.9	4.3	48.70	13.5
High N						
T_9	6.6	52.7	14.7	3.7	60.95	16.9
T_{10}	6.6	41.8	11.6	3.3	50.40	14.0
T_{11}	5.8	39.3	10.9	3.9	56.10	15.6
T_{12}	4.6	36.3	10.1	4.3	56.36	15.7
T_{13}	3.8	33.3	9.3	3.3	46.55	12.9
T_{14}	4.5	33.9	9.4	2.9	46.35	12.9
T_{15}	3.9	30.0	8.3	2.9	45.50	12.6
LSD (p=0.05)	1.2	8.0	2.2	1.0	8.10	2.2

[1]Expressed as increase over initial values

A cost/benefit analysis of T9 at Site 1 produced a net return of Rs.58,569/ha (US$1,325/ha) or Rs.2.12 per rupee invested in fertilizers and other inputs. At Site 2, T6 produced a net return of Rs.46,260/ha (US$1,045/ha) or Rs.1.68 per rupee invested. Across sites, micronutrient and secondary nutrient application had little impact on juice content total soluble solids (TSS), or fruit acidity (Table 4). However, both sites and input regimes demonstrated significant quality responses to K. Maximum fruit juice contents corresponded with conditions of high K fertility, as did fruit acidity. This latter observation suggests that K fertilization will play a role in influencing the time to fruit maturity since fruits with higher juice acidity take more time to attain the colour break stage. Total soluble solids showed a negative response to increased K application. Significant response to improved fertilization strategies over currently recommended doses of fertilizers warrants addressing nutrient requirements on a site-specific basis.

Table 4 : Fruit quality response to fertilization (mean of 2 years)

	Site 1			Site 2		
Treatments	**Juice (%)**	**TSS (%)**	**Acidity (%)**	**Juice (%)**	**TSS (%)**	**Acidity (%)**
Current Rec.	44.0	8.6	0.57	43.1	8.5	0.68
Low N						
T_1	45.7	8.2	0.56	45.5	8.1	0.77
T_2	44.5	8.5	0.64	41.6	7.6	0.68
T_3	44.1	9.1	0.60	42.4	8.4	0.75
T_4	44.7	8.8	0.63	43.7	7.9	0.68
T_5	41.9	9.6	0.56	42.4	8.7	0.64
T_6	44.9	9.3	0.58	49.8	8.6	0.67
T_7	45.2	8.6	0.62	46.5	7.8	0.81
T_8	48.3	8.2	0.75	48.2	7.9	0.82
High N						
T_9	45.4	8.9	0.55	42.7	8.8	0.62
T_{10}	42.6	8.6	0.59	43.6	8.2	0.71
T_{11}	44.9	8.5	0.63	44.8	8.1	0.76
T_{12}	48.4	8.2	0.80	46.3	7.6	0.86
T_{13}	41.6	8.2	0.51	42.5	9.1	0.66
T_{14}	43.2	9.5	0.64	43.7	8.5	0.77
T_{15}	44.6	9.6	0.63	43.8	8.2	0.74
LSD (p=0.05)	3.1	0.5	0.09	3.2	0.6	0.08

11.5 SSNM IN SWEET ORANGE ON VEITISOL

A field experiment was conducted for 3 years (2006-07 to 2008-09) at Narkhed Tehsil in Nagpur, Maharashtra, to evaluate whether soil test-based SSNM improves 'Mosambi' productivity, fruit quality, and economics of production. An 8-year-old 'Mosambi' sweet orange orchard was used which was having plant with scion of sweet orange (*Citrus sinensis* Osbeck) budded on Rough lemon rootstock (*Citrus jambhiri* Lush). The plant-to-plant and row-to-row distance was 6 m each, which results in a plant population of 278 trees/ha. The site had an alkaline, calcareous soil (Typic Haplustert) with available N, P, K, Fe, Mn, and Zn contents of 231, 25, 417, 25, 18 and 2.20 kg/ha, respectively. The climate of 'Mosambi' growing belts in the Marathwada region is characterised by hot and dry pre-monsoon summer months (March to May), followed by well expressed monsoon months (June to September). The mean summer (April, May, and June) to mean winter (December, January, and February) temperatures vary from 42° to 38°C. The average annual rainfall of the region is 800 mm, of which 80 to 90% is received during monsoon months. For the experiment, 17 different fertiliser treatments was designed as outlined in Table 1. These fertiliser treatments were designed based on: a) the standard analysis of soil macronutrient, secondary nutrient, and micronutrient status of the experimental soil prior to the start of the experiment, and b) fertiliser recommendations designed to evaluate if up to 300% of the recommended doses can improve yield and/or fruit quality. In each of the experimental years, fertiliser application was split into three equal doses coinciding with the emergence of new flush in the months of April, August, and October. Different fruit quality parameters viz., TSS was determined using hand refractometer, juice content volumetrically, and acidity tritrimetrically as per commonly followed procedures.

Table 5 : Response of different treatments on growth and yield of 'Mosambi' sweet orange (Pooled data of 3 years)

Treatments[1]	Fruit yield		Quality	
	(kg/tree)	Juice (%)	TSS (%)	Acidity (%)
$T_1 = N_0\text{-}P_{200}\text{-}K_{300}\text{-}M_1$	37.9	47.2	8.5	0.46
$T\text{-}_2 = N_{400}\text{-}P_0\text{-}K_0\text{-}M_1$	37.7	45.1	8.3	0.41
$T_3 = N_0\text{-}P_0\text{-}K_{300}\text{-}M_1$	36.2	45.8	8.3	0.46
$T_4 = N_{400}\text{-}P_{200}\text{-}K_0\text{-}M_1$	42.0	46.5	8.3	0.40
$T_5 = N_{400}\text{-}P_{200}\text{-}K_{300}\text{-}M_1$(RDF)	44.4	48.3	8.9	0.44
$T_6 = N_{400}\text{-}P_{200}\text{-}K_{300}\text{-}M_2$	46.4	47.7	8.6	0.46
$T_7 = N_{400}\text{-}P_{200}\text{-}K_{300}\text{-}M_0$ (FFP)	40.2	46.9	8.3	0.48
$T_8 = N_{800}\text{-}P_{400}\text{-}K_{600}\text{-}M_1$ (SSNM)	61.4	50.9	9.5	0.44
$T_9 = N_{800}\text{-}P_{400}\text{-}K_{900}\text{-}M_1$	58.8	49.6	9.3	0.51
$T_{10} = N_{800}\text{-}P_{400}\text{-}K_{1200}\text{-}M_1$	57.9	49.9	9.3	0.61
$T_{11} = N_{800}\text{-}P_{400}\text{-}K_{1200}\text{-}M_2$	56.7	49.8	9.2	0.57
$T_{12} = N_{1200}\text{-}P_{400}\text{-}K_{300}\text{-}M_1$	53.6	47.9	8.7	0.47
$T_{13} = N_{1200}\text{-}P_{400}\text{-}K_{600}\text{-}M_1$	54.2	48.9	8.8	0.49
$T_{14} = N_{1200}\text{-}P_{400}\text{-}K_{900}\text{-}M_1$	51.2	49.7	8.9	0.58
$T_{15} = N_{1200}\text{-}P_{400}\text{-}K_{1200}\text{-}M_1$	50.8	50.5	8.8	0.63
$T_{16} = N_{1200}\text{-}P_{400}\text{-}K_{300}\text{-}M_0$	48.3	46.7	8.4	0.53
$T_{17} = N_{1200}\text{-}P_{400}\text{-}K_{300}\text{-}M_1S_1$	48.7	48.3	8.8	0.47
CD(*P=0.05*)	**1.98**	**1.2**	**0.27**	**0.031**

[1]Subscripts after N, P and K indicate rates applied, kg/ha

M_0 = no micronutrients

M_1 = micronutrients consisting of 250g each of $FeSO_4$, $MnSO_4$ and $ZnSO_4$/ tree

M_2 = micronutrients consisting of 500g each of $FeSO_4$, $MnSO_4$ and $ZnSO_4$/ tree

S_1 = $CaSO_4$ and $MgSO_4$ each @ 250g/tree

11.5.1 Yield response

Fruit yield is a good index of orchard productivity. A signifycantly higher 'Mosambi' yield was obtained with SSNM as compared to RDF and FFP (Table 5). This indicates the potential of SSNM to reduce the gap between actual and potential productivity of 'Mosambi' sweet orange orchards. The RDF, FFP, and SSNM treatments all had similar N:P:K ratios, and the only change was in the levels of N, P, and K applied with SSNM using double the amounts of macronutrients (Table 5). This indicates that in crops where the traditional macronutrient ratio approach to guiding fertiliser application is well established, SSNM does not try to change the approach. Instead, it tries to include the effect of other related factors (like levels, etc.) to better nutrient management decisions. Also, as the established macronutrient ratio was altered (T_8 vs T_9 and T_{10}), a significant decline in 'Mosambi' yield was observed Micronutrient application had a significantly positive effect on 'Mosambi' yield (FFP vs RDF treatments) under the experimental conditions (Table 5). Their effect was more pronounced at lower levels of N, P, and K (T_5 vs T_6 vs T_7) than at higher levels (T_{10} vs T_{11}). However, the application of secondary nutrients, i.e., Ca and Mg, caused yield declines (T_{12} vs T_{17}). This was probably due to the increased competition for plant uptake between these nutrients and K. A variety of nutrient deficiencies was observed in nutrient omission plots as a cumulative effect of 3 years of experimentation. For example, N deficiency was observed where no N was applied for three successive years (T_1), K deficiency was observed in the form of small fruit size where no K was applied (T_2), and Fe, Mn, and Zn deficiencies were observed where no micronutrient application was done in the 3 years of experimentation (T_7). These deficiencies were confirmed using leaf analysis, which indicated that there was a continuous mining of nutrients in 'Mosambi' sweet orange orchards. Unless the nutrients using a SSNM strategy is supplemented attaining potential productivity of these orchards is almost impossible.

11.5.2 Fruit quality response

Juice content, TSS, and juice acidity are the three most important parameters used to determine orange quality. Just like the yield response, SSNM had a significant positive impact on these parameters as compared to RDF and FFP treatments (Table 5). Omission of K (T_4) from the RDF (T_5) significantly reduced juice percentage and TSS, suggesting a strong influence of K on quality parameters of sweet orange (Table 5). However, even when more K was applied of disturbed the balanced ratio of macronutrients (T_8 vs T_9 and T_{10}), the juice content also declined significantly. Additionally, higher K rates increased juice acidity, regardless of the level of NP input. Inclusion of micronutrients produced a significantly favourable response on juice and TSS (T_7 vs T_5 and T_{16} vs T_{12}). Inclusion of secondary nutrients increased juice yields significantly, but did not have a significant effect on TSS (Table 5).

11.6 ECONOMICS OF NUTRIENT MANAGEMENT APPROACHES

Just like its favorable response on 'Mosambi' yield and quality, SSNM provided a comparatively higher net return than either RDF or FFP (Table 6). The cost of cultivation increased marginally with SSNM compared with RDF and FFP, but this increase was offset by a remarkable increase in net benefit, realised mainly through increased 'Mosambi' yields. The results of this study clearly show the need to a) maintain a balance between macronutrients and micronutrients in deciding need-based, optimum fertiliser doses and b) revise the current fertiliser recommendation system to realise full productivity potential on a given soil type.

Table 6 : Analysis of economic returns from SSNM versus RDF and FFP

Treatment	Cost[1] (000' Rs/ha)	Benefit[2] (000' Rs/ha)	Net returns (000' Rs/ha)
T_7 (FFP)	16.5	110.8	94.3
T_5(RDF)	21.7	121.9	100.2
T_8(SSNM)	32.5	169.0	136.5

[1]Includes operational charges consisting of two weedings, basin cleaning, and labour charges for fertilizer application (Rs. 10,000/ha) plus cost of fertilizers including urea (Rs. 8 /kg), SSP (Rs. 7/kg), KCl (Rs. 8/kg), gypsum (Rs. 2/kg), $FeSO_4$ (Rs. 15/kg), $MnSO_4$ (Rs. 30/kg), and $ZnSO_4$ (Rs 30/kg)

[2]As per existing farm rate (Rs. 10,000/t)

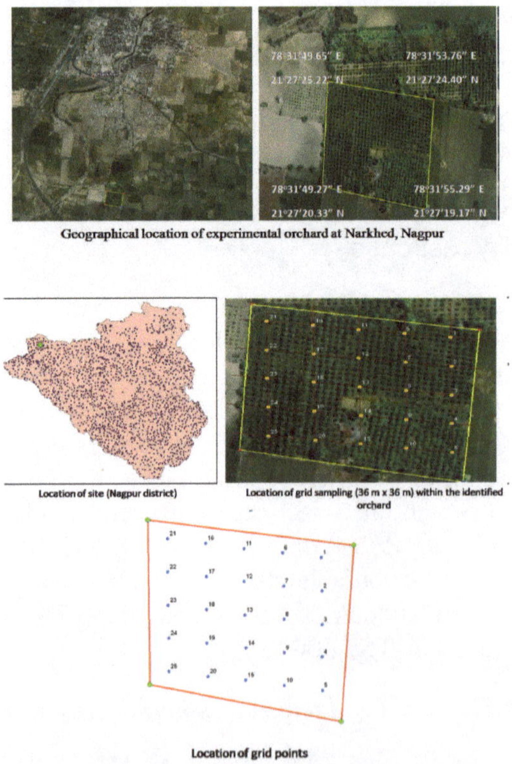

Fig. 1 : Geographical location of experimental orchard and distribution of grid sampling

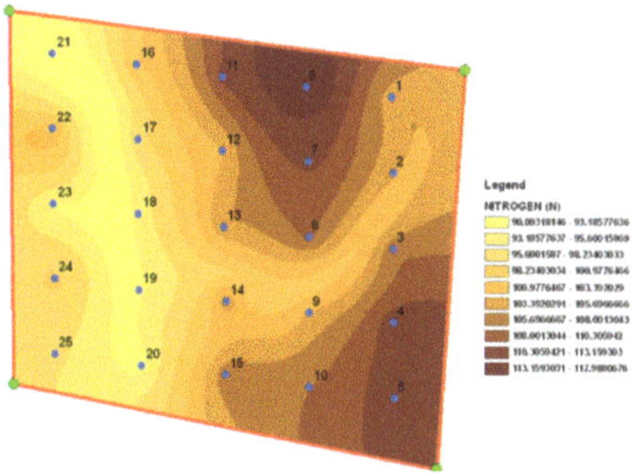

Spatial variability in alkaline $KMNO_4$ extractable N in soil

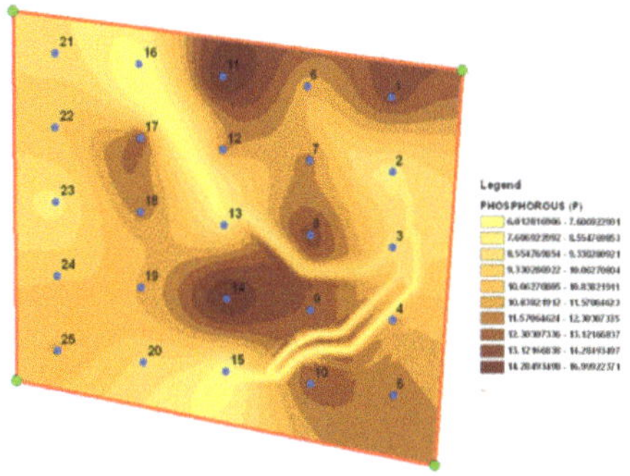

Spatial variability in Olson – P in soil

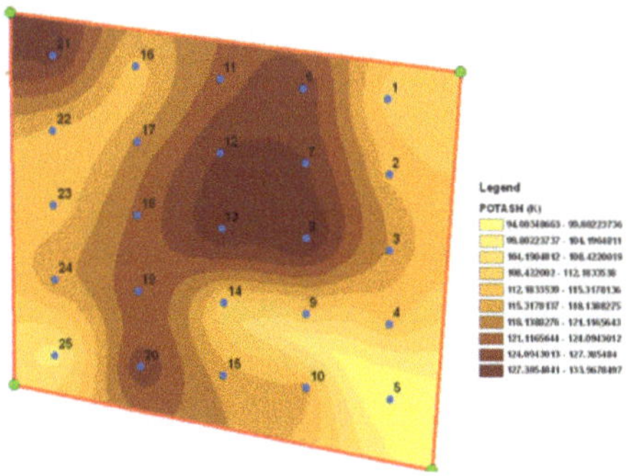

Spatial variability in neutral NH_4OAc – K in soil

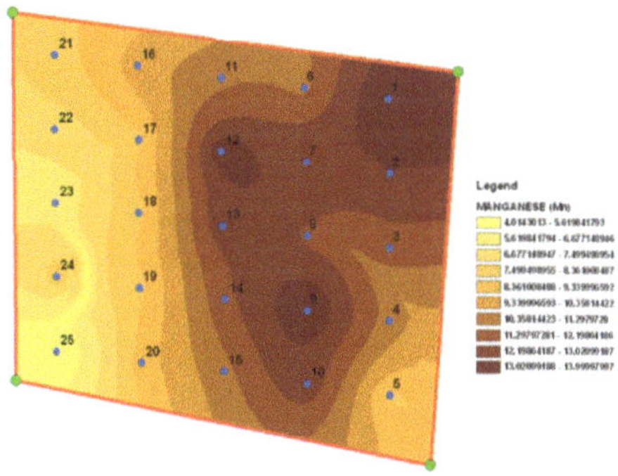

Spatial variability in soil DTPA – Mn in soil

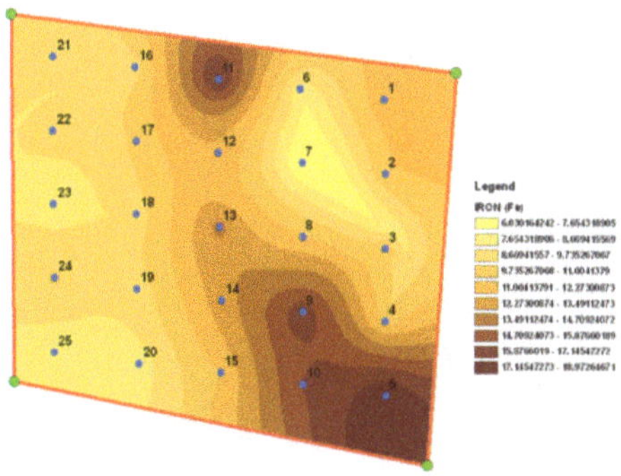

Spatial variability in soil DTPA – Fe in soil

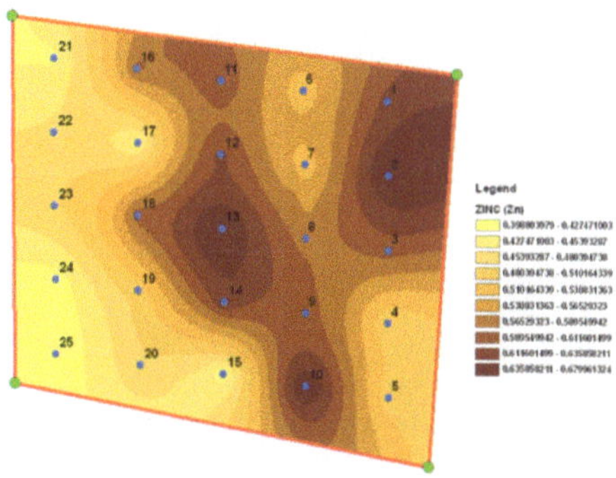

Spatial variability in soil DTPA – Zn in soil

Fig. 2 : Spatial variability in soil fertility across the grid sampling within the identified orchard

Hence, these studies in two commercial cultivars on soils of similar smectite rich mineralogy showing a much remunerative multifarious response than conventional fertilizer recommendation. In years to come, SSNM will find a greater application in precision citriculture using variable rate application technology via canopy sensors and integrating further with programmable fertigation so that nutrients are applied synchronizing crop physiological nutrient demand and supply from soil.

11.7 REFERENCES

Srivastava, A.K., Kohli, R.R. and A.D.Huchche (1998). Relationship of leaf K and forms of soil K at critical growth stages of Nagpur mandarin (*Citrus reticulata* Blanco). *J. Indian Soc. Soil Sci.* 46(2): 245-248.

Srivastava, A.K. and Shyam Singh (2002). Soil analysis based diagnostic norms for Indian citrus cultivar. *Commun. Soil Sci. Pl. Anal.*.33(11&12):1689-1706.

Srivastava, A.K. and Shyam Singh (2005). Diagnosis of nutrient constraints in citrus orchards of humid tropical India.. *J. Pl. Nutri.* 29(6) : 1061-1076.

Srivastava, A.K. and Shyam Singh (2008a). DRIS norms and their field validation in Nagpur mandarin (*Citrus reticulata* Blanco). *J. Pl. Nutri.* 31(6):1091-1107.

Srivastava, A.K. and Shyam Singh (2008b). Analysis of citrus orchard efficiency in relation to soil properties. *J. Pl. Nutri.* 30 : 2077-2090.

Srivastava,A.K. Shyam Singh, Das, S.N. and Harmandeep Singh (2010). Delineation of productivity zones in mandarin orchards using DRIS and GIS. *Better Crops* – South Asia 92 : 9-13.

Srivastava, A.K. Shyam Singh, Diware,V.S. and Haramandeep Singh (2009). Site-specific nutrient management in 'Mosambi' sweet orange. *Better Crops – India* 3(1):10-11

Srivastava, A.K., Shyam Singh and K.N. Tiwari (2006). Site specific nutrient management for Nagpur mandarin (*Citrus reticulata* Blanco) *Better Crops* 88(2): 22-25.

Srivastava, A.K., Shyam Singh and K.N. Tiwari (2007). Diagnostic tools for citrus: Their use and implications in India. *Better Crops – India* 1(1):26-29.

Chapter – 12

Nutrient Use Efficiency in Oil Palm Plantations Under Indian Conditions

B. Narsimha Rao, K. Suresh and S. Sunitha

12.1 INTRODUCTION

Oil Palm (*Elaeis guineensis* Jacq.) is a heavy feeder and needs a balanced and adequate supply of nutrients for growth and yield. Fertilizer is one of the most costly inputs, which accounts for greater part of production costs in the cultivation of oil palm. Large amount of nutrients are removed by the fruit bunches harvested from the mature palms. It is estimated that 25 tonnes fresh fruit bunches per hectare per year removes 93.5 Kg Nitrogen, 11.0 Kg Phosphorous, 92.7 Kg Potassium, 19.3 Kg Magnesium and 20.3 Kg Calcium.

Economic yields of oil palm are achieved when the supply of nutrients from the soil, palm residues and fertilizers are properly integrated in required quantities. Leaf analysis is an analytical tool for detecting the nutrient requirements in oil palm and other perennial crops, which are slow growing and can provide easily, defined standard leaf material for analysis due to leaf phyllotaxis. Nutrient deficiency symptoms are also an important means for determining the nutritional problems in oil palm, since the crop presents easily identifiable N, K and Mg deficiency symptoms. However, by the time symptoms are evident, nutrient deficiency might have already affected growth and yield.

Also the nutrient supply may be insufficient before the appearance of leaf symptoms. So there is a need for a thorough inspection of the leaf canopy to corroborate the results of leaf analysis and identify the areas with severe nutrient deficiencies. The research on oil palm nutrition was carried out under rainfed conditions at Palode, Kerala and irrigated conditions at Pedavegi, Andhra Pradesh.

In India, a great variety of soils ranging from red sandy loams/sandy clay loams (Alfisols), alluvial silt loams (Inceptisols / Entisols), heavy black soils (Vertisols) and acidic red laterite soils (Oxisols) are put to oil palm cultivation. In Andhra Pradesh, it is grown in red sandy, red loamy, red loam with clay base, deltaic alluvium pH - 6.5 to 9.0. In Karnataka, it is grown in red and brown sandy loams and black clayey soils, with pH - 6.5 to 9.0. Red soils, red loamy soils and clay loam soils are used for growing oil palm in Tamil Nadu. Laterite soils are used for oil palm cultivation in Kerala.

12.2 FERTILIZER MANAGEMENT IN NURSERY

A fertilizer experiment was conducted using nursery seedlings consisting of four levels - 0-0-0, 10-5-10, 20-10-20 and 30-15-30 g N, P and K. in order to find out an optimum fertilizer schedule. Based on the growth performance and biomass production of seedlings at planting 10-5-10g per seedling, was found as the best treatment. Frequency of fertilizer application was tried in another experiment with four treatments namely – monthly, bimonthly, trimonthly and quarterly. Application of fertilizers in equal splits at three monthly intervals was found sufficient and economic for nursery.

An ideal potting mixture was also arrived for raising nursery by studying the performance of seedlings in nine potting mixtures involving different combinations of soil, sand and FYM. The mixtures studied were soil and sand in 1:1,2:1 and 3:1 proportions, soil, sand and FYM in 1:1:1, 2:1:1 and 3:1:1 proportions and soil, sand and oil palm waste compost in 1:1:1, 2:1:1 and 3:1:1 proportions. The standard pot mixture 1:1:1 soil – sand – FYM was found ideal for oil palm, but 1:1:1 soil - sand - Palm waste compost was also good for optimum growth of seedlings. The nutrient uptake of seedlings based on this study confirmed that the fertilizer dose arrived is sufficient to meet the nutrient requirement of seedlings. (Sunitha, 2003; Sunitha *et al.*2005,).

In order to find out the fertilizer requirement of young palms during pre-bearing phase, six doses were studied namely 0-0-0, 200-100-200, 300-150-300, 400-200-400, 500-250-500 and 600-300-600g NPK during first year of planting and 0-0-0, 400-200-400, 600-300-600, 800-400-800, 1000-500-1000 and 1200-600-1200g NPK during second year of planting. Based on the growth performance, initial flowering and yield characters, 300-150-300 was found optimum for first year and 600-300- 600 dose for second year of planting. From third year onwards, the present fertilizer recommendation of 1200-600-1200 holds good for maximum yield.

12.3 FERTILIZER MANAGEMENT IN MATURE PALMS

The first fertilizer trial was conducted with three levels each of nitrogen, phosphorus and potassium and two levels each of calcium and magnesium. The results revealed that NPK at 800, 400 and 1800 g per palm per year are required for maximum production of fresh fruit bunches and growth of plantation. Application of calcium and magnesium did not show any influence on growth and yield of the crop. Nutrient contents in tissues significantly influenced the yield of palms through increasing the concentration of the respective nutrients in tissues. Even though Ca and Mg concentrations in tissues also increased, the same did not reflect on yield. The study also revealed that nitrogen increased leaf concentration of all the elements (N, P, K, Ca, and Mg) and reduced the chlorine concentration. A two way antagonism between the leaf concentration of K and Ca + Mg was observed. Similarly, an antagonism of chlorine on the uptake of potassium was observed. Studies on seasonal variation in nutrient concentrations of foliar tissues of oil palm indicated that there was a reduction in nitrogen concentration in foliar tissues during summer months and increased during rainy season. But a reverse trend was noticed in phosphorus, potassium, calcium and magnesium concentration in leaf did not show any definite pattern (Nair, 1981).

Research on Oil palm nutrition under the AICRP centers was concentrated on the irrigation cum fertilizer experiments at different agro-climatic locations namely Gangavathy, Mulde, Vijayarai and Aduthurai. Fertilizer schedules were recommended for oil palm cultivation for the different regions namely - Coastal Tamil Nadu region, Konkan coastal region of Maharashtra and Coastal Andhra Pradesh (1200-600-2700 g NPK/palm/year) and 1200-600-1200 g NPK/palm/year for the Tungabadra command area of Karnataka.

Leaf nutrient analysis of 1086 samples received from mature plantations of Nellore, Krishna, East and West Godavari District (Suresh *et al.,* 2001) revealed that majority of plantations had optimum leaf nitrogen content except in Krishna, where 87.5 per cent of the plantations were deficient in nitrogen. Leaf phosphorous content was deficient in all plantations. Only fifty per cent of plantations in West Godavari and Krishna districts possessed optimum levels of potassium. Majority of the plantations in East Godavari and Nellore recorded deficient K levels. About 42.7 per cent of the plantations of West Godavari had excess leaf potassium content. Leaf magnesium levels were optimum in majority of the plantations in West Godavari and Krishna districts but it was excess in East Godavari and Nellore plantations.

12.4 NUTRIENT UPTAKE BY OIL PALM AND ASSOCIATED CROPS

To work out the nutrient budgeting in oil palm based cropping system, the nutrient uptakes by oil palm and intercrops were estimated (Reddy and Suresh, 2008). The native soil fertility of the soil was estimated and the soil analysis indicated that the soils had acidic pH with normal EC. The organic carbon content was normal and the contents of P and K were medium. The nutrient uptakes by oil palm during the first three years of juvenile phase were calculated. The uptake of nitrogen was more followed by potassium and phosphorus during the initial three years. The intercrops were able to add as much as 0.53 to 13.79 t/ha of organic matter which in turn added nutrients in the magnitude of 7-182 kg N, 2.77-210.5 kg P and 12-271 kg of K per ha per year. All the nutrients would be subsequently available to the main crop i.e., oil palm. In oil palm, the uptake of nitrogen was more followed by potassium and phosphorus during the initial three years. Leaves were the best sinks for nutrients followed by trunk and roots. Among the different intercrops, the uptake of nitrogen by guinea grass was maximum followed by maize and chillies. While the uptake of K was more in banana followed by chillies and guinea grass.

Leaf samples received from oil palm gardens were categorized as high yielding (>15 tons FFB) and low yielding (<15 tons) in each of the age group (10, 7, 6 and 5 year old). There was not much difference in nutrient content of high and low yield gardens and leaf nutrients levels did not follow any trend. High yielding gardens had both higher and lower nutrients levels whereas low yielding gardens had both low and high nutrient levels. Poor yields in plantations with low leaf nutrient levels may be attributed to poor nutrient, but plantations with high leaf nutrient levels may indicate poor water management. A similar phenomenon could be seen in the data reported by Uday Kumar (1994), wherein large differences in yield were observed between high and low yielding plantations with small or no differences in leaf nutrient levels. Low nutrient levels in high yielding gardens may indicate large mobilization of leaf nutrients into fresh fruit bunches as the samples were taken before the application of fertilizer, which necessities the immediate replenishment of nutrients through fertilizer application (Suresh *et al* 2001).

Studies undertaken in ten high and low yielding oil palm plantations revealed that the soils were acidic to neutral in reaction, EC was normal, Organic carbon was low in all gardens (Suresh *et al.,* 2005). Soil Phosphorus was medium to high, Soil Potassium was very high and Ca and Mg less than adequate. Leaf N was optimum except in two gardens. Leaf P was less than optimum, Leaf K was optimum except in one case and Ca and Mg were optimum. In most of the gardens, N and K application was more than recommended, while applied P, Ca and Mg fertilizers were less than recommended. K fertilizers were applied in excess of recommended dose. Soil P was positively linked with yield of the just concluded year and soil Mg was negatively correlated with yield of just concluded year. All the major nutrients applied in the form of commercial fertilizers were positively correlated with current year yields. None of the other nutrients in soil, leaf or fertilizers were significantly correlated with yield of just concluded year.

To establish the index leaf for nutrients in oil palm, Suresh *et al* (2002) studied the nutrient contents of all the leaves in oil palm. The concentrations of N, P, K and Mg were highest in the younger fronds and decreased with the frond age. In contrast, the concentration of Ca increased with increasing leaf age. With regard to micronutrients, no trend was observed between nutrient content and age of the frond except for that of Zinc where the nutrient content decreased with age of the frond. A synergism between N, P, K and Mg contents and their antagonistic effect with Ca content is observed. Positive interaction between P and K is also seen.

12.5 BIO-RECYCLING OF OIL PALM WASTES

The palm components were analyzed for nutrient content and quantification was made as regard to the nutrient contribution expected from a mature plantation. Five different techniques were tried viz., aerobic digestion, anaerobic digestion, chemical digestion, microbial digestion and vermi composting for composting palm wastes. Aerobic digestion was found to be economic and efficient technique for large scale production of compost. Based on the nutrient value of compost, it was found that almost 90 per cent of N, 50 per cent of P and 75 per cent of K of nutrient requirement of palms could be met through palm waste compost (Sunitha and Varghese, 1999; Swayamprabha *et al*, 2001).

A field experiment was also conducted to evaluate the compost in mature palms involving different combinations of inorganic and organic fertilizers. The results revealed that a combination of one-third nutrient requirement as inorganic fertilizers and two-third requirement as palm waste compost was ideal (Sunitha and Varghese, 2002). This combination was found economic when economics was worked out for producing compost and applying to palms (Sunitha and Varghese, 2004). The biomass production of palms at the time of felling of an oil palm plantation was quantified. The palm parts were analyzed and nutrient contribution expected from a hectare of plantation was estimated. It was estimated that from an expected stand of 143 palms, about 1503 kg N,129 kg P, 2345 kg K, 513 kg Ca and 438 Kg Mg could be expected from a plantation with an estimated biomass of 163 tonnes per ha (Sunitha and Varghese, 2003; Sunitha and Krishnakumar, 2006). In a study conducted at Palode, two thirds of nutrient requirement of palms could be met through compost made from wastes available from plantation without affecting its yield. Organic substitution up to this level was found to be economical.

12.6 DIAGNOSTIC NORMS DEVELOPED USING SOIL TESTING PLANT ANALYSIS

A factorial experiment was conducted in mature palms with four fertilizer levels and three levels of irrigation. The fertilizer doses tried were 0-0-0, 600-300-600, 1200-600-1200 and 1800-900-1800g NPK and irrigation treatments were 0, 45 and 90L per palm per day during summer months. A fertilizer dose of 1200-600-1200g and an irrigation level of 90L using drip system were found optimum for getting maximum yield. From this experiment nutrient uptake of mature oil palm was determined and optimum leaf nutrient ratios were determined and Diagnosis and Recommendation Integrated System was developed for oil palm. (Varghese, 1994)

12.7 REFERENCES

Reddy, V.M. and K. Suresh. (2008) Nutrient uptakes in oil palm based cropping system during juvenile phase. Souvenir and Abstracts, National Conference on Oil Palm, 2-4 Feb., 2008, Vijayawada, Andhra Pradesh.

Nair, A., (1981) Nutritional status of Oil Palm (*Elaeis guineensis*, Jacq.). Ph.D thesis submitted to College of Agriculture, Vellayani, Trivandrum, 1981.

Sunitha,.S. (2003) Studies on nutrient management in oil palm nursery. Ph.D Thesis, Annamaai University, Tamil Nadu

Sunitha, S and Thomas Varghese, P. (1999) Composting of oil palm wastes for efficient recycling of nutrients in palm plantations. *The Planter* 75 (885): 677-681.

Sunitha, S and Thomas Varghese, P. (2002) Effect of integrated nutrient management on FFB yield of oil palm, PLACROSYM XV, 13-15 Dec, 2002, Mysore.

Sunitha, S and Thomas Varghese, P. (2003) Quantification of oil palm biomass at felling. *Int. J. Oil Palm* 3&4: 53-56.

Sunitha, S and Varghese P,T. (2004) Yield of oil palm as influenced by integrated nutrient management. *J. Plant. Crops (suppl.)* 2004: 263-265.

Sunitha, S, Thomas Varghese, P,T, Panneerselavm, P. (2005) Evaluation of different potting mixtures on growth and vigor of oil palm seedlings in nursery. *Int. J. Oil Palm* 5(1&2):33-36.

Sunitha, S and Krishnakumar,T, (2006) Biomass production and potential nutrient contribution from oil palm at felling. *J. Plant. Crops* 34(3): 309-311.

Suresh, K. Reddy, V,M, Sivasankarakumar, K,M, Ravinder Rao, (2005) Nutritional status and yield performance of high and low yielding oil palm gardens in West Godavari district of Andhra Pradesh. Poster presented during Hindi Seminar –"Thel thad anusandhan avam vikas, chunuithiya aur sambhavanai" on 26th October, 2005 at NRCOP, Pedavegi, A.P.

Suresh, K, Reddy, V,M, Bhanusri, A, Sivasankara Kumar, K.M. (2002) Ontogenetic changes in the leaf nutrient contents of oil palm under irrigated conditions. Proceedings PLACROSYM_XV,Edited by K.Sreedharan, P.K. Vinod Kumar, Jayarama and Basavaraj M. Chulaki.P. 362-364.

Suresh, K, Reddy, V,M, Rethinam, P, Sugunamani, M, Bhanusri, A, Suja Nair, Avasthe, R,K, (2001) Leaf nutrient status of oil palm plantations in Andhra Pradesh. *Int. J. Oil Palm* 2(1):25-32.

Swayamprabha, H.L., Thomas Varghese, P. and Sunitha, S, (2001) Changes in certain chemical parameters during composting of oil palm wastes. *Int. J. Oil Palm* 2(1): 33-35.

Uday Kumar, M, (1994) Report of the DBT Project entitled "Mineral nutrition of oil palm – Assessment of nutrient status and nutrient response".

Varghese, P.T. (1994) Effect of nutrition as influenced by irrigation on growth and yield of oil palm (*Elaeis guineensis* Jacq.). Ph.D.Thesis,. Kerala Agricultural University, Kerala.

Chapter – 13

Precision Farming in Arid Horticultural Crops : Status and Prospects

S. K. Sharma and B. D. Sharma

13.1 INTRODUCTION

The hot arid regions are spread over about 31.7 million ha area mainly in the States of Rajasthan, Gujarat, Andhra Pradesh, Punjab and Haryana, which inhabit on an average 61 persons per square km making up a home for population of nearly 20 million people. The Indian arid zone is characterized by high temperature and low and variable precipitation which limit the scope for high horticultural productivity. However, these conditions greatly favour development of high quality production of fruits such as date palm, *ber*, pomegranate, citrus, *aonla*, *bael*, grapes, guava and vegetables such as cucurbitaceous crops, spices and some medicinal plants. The optimized technologies and inputs could increase the existing low productivity. It is now realized that there is a limited scope for quantum jump in fruit and vegetable production in the traditional production areas. The amelioration of the extreme conditions is also considered vital for life support to the inhabitants of this area. The recent awareness regarding the potential of these ecologically fragile lands for production of quality horticultural produce has not only opened up scope for providing economic sustenance for the people of this region, but also for bringing new areas to increase production through horticulture.

The arid horticultural crops have developed and/or modified to perform certain vital physiological functions even under stress. Strong deep root system (*ber*, *bael*, *aonla*, wood apple, *jamun*, etc.), moisture synchronous flowering and fruit development (*ker*, *lasora*, *aonla*, *pilu*, etc.) and other xerophytic characters i.e. leaf shedding in summer (*ber*), scanty foliage (*ker*), spiny cladode (cactus pear), mucilaginous sap in plant part (*ker*, *gonda*, *pilu*, *bael*, etc.), sunken stomata and fur/ hairiness and waxy coating on the leaf surface (*phalsa*, *ber*, *gonda*, fig, etc.) thorny nature, and selective or reduced absorption of cation (Na^+) and anions (Cl^-, $SO4^{--}$) for survival under adverse arid conditions are peculiar adaptations. Therefore, cultivation of suitable species of fruit and vegetables in the arid areas will increase the sustenance of the inhabitants and provide the alternate sources of income through development of nurseries, small scale industries and exports.

13.2 AREA AND PRODUCTION

India is the second largest producer of fruits in the world with an annual production of 49.30 million tonnes from an area of 4.96 million hectares contributing 10 per cent production of fruits in the world. In arid region, considerable area has come up under fruits like *aonla*, *ber*, pomegranate, *annona*, fig and *phalsa* in different parts of the country, which had negligible spread in the recent past. *Ber* has spread from northern states to the western and southern India from a mere 12000 ha in 1978 to nearly 86,000 ha in 2004-2005 with a production of about 0.90 million tonnes. Similarly, the area under pomegranate has also leaped to over 1.25 lakh ha. Likewise *aonla* is presently cultivated over 50,000 ha with the production of 2, 50, 000 tonnes. This has become possible as a result of the research and developmental efforts. The All India Coordinated Research Project on Arid Zone Fruits initiated during the IVth Five Year Plan led to the development of new varieties and agro-techniques for fruit crops.

Majority of Indian vegetable growers are small scale cultivators. Post WTO scenario is enforcing stringent demands on vegetable growers and processors in India. India continues to be the second largest producer (next to China) of vegetables in the world with 101.43 million tonnes from an area of 67.55 million hectares. The country contributes about 13 per cent of the world's vegetable production and occupies first position in the production of cauliflower, second in onion and third in cabbage in the world. In the past three decade, India made a quantum jump in vegetable production, but it is still much less than the total requirement of vegetables @ 300 g/day/person for the balanced diet. There is challenge to achieve the target of 250 million tonnes of vegetable production to meet the requirement of the country by 2020. The vegetable seed industry much developed in the past decade, but the focus has been primarily on a few vegetables like onion, tomato, chilli, bell pepper and okra. Public sector should come forward for diversification to specialized indigenous vegetables in each state for domestic market as well as for export.

Under-utilized horticultural crops are another area which needs attention. There is need to develop large orchards of under-utilized fruits like *jamun*, *mahua*, *chironji*, *rambutan*, fig, *kokam* and the like for commercial exploitation both for internal consumption and export. In arid zone fruits like *aonla*, *ber* and the like, the possibility of expanding cultivation and export of indigenous fruits like *bael*, *phalsa*, wood apple, mulberry, etc. need to be explored. Some of the fruits are of high nutritive value and are rich in anti-oxidants which are important for health. Varieties were also identified in flowers and medicinal and aromatic plants through researches conducted in different parts of the country.

The vast land resources offer great opportunity for horticultural development in the Indian arid zone which is spread over nearly 31.7 million ha area of which 41.5 per cent is arable and 19 per cent is culturable wasteland (Table 1).

Table1 : Area under arid region in India

States	Total area (m ha)	Arable area (m ha)	Culturable wasteland (m ha)
Rajasthan	19.60	9.20	4.20
Gujarat	6.20	0.02	1.66
Punjab	1.50	1.05	0.03
Haryana	1.30	1.15	0.02
Peninsular India	3.10	1.63	9.11
Total	**31.70**	**13.15**	**6.03**

Varied agro-climatic condition of India offer the best opportunity to export produces in the off-season to importing countries. Fruits like mango, grapes and pomegranate can also be harvested in India at such a time when they are not harvested anywhere else in the world. Production zone could be identified for production of these novel crops.

13.3 PROSPECTS OF ARID HORTICULTURE

The region receives high solar radiation. Multistorey cropping system models based on fruit trees can be developed to utilize this valuable resource for production of quality fruits. Vegetables and medicinal and aromatic plants can be a component of such models. In high radiation zone, which also has low humidity, the attack of insect pests and diseases are very low. However, conditions of high radiation and temperature coupled with low humidity cause water stress in plants owing great water losses from soil and plant surface but by following proper soil and water management techniques, optimum productivity can be maintained. The high summation of heat units especially during the long summer is a valuable resource for development of high total soluble solids in fruit crops such as date palm, grape, etc. Besides, the conditions of high temperature and low humidity help in solar drying of fruits and vegetables. The practice is already common for drying pods of *khejri (sangri)* and fruits of *kachri*, round gourd, and fruits of *ber* and date palm. The sharp fluctuations in day and night temperatures during autumn, spring and summer help in development of sweetness in sweet orange, *kinnow*, *ber* and date palm, flesh colour and sweetness in pomegranate arils and *mateera* pulp.

13.4 SELECTION OF HORTICULTURAL CROPS AND THEIR VARIETIES

While selection of fruit crops for dryland horticulture, one of the basic requirements is that those crops which complete their vegetative growth and reproductive phase during the period of maximum moisture availability. The fruit such as *ber,* guava, pomegranate, custard apple, *aonla* and sour lime, conform to this prerequisite. Among vegetable crops, the hardiest types belong to the cucurbitaceous and solanaceous groups. Cowpea, cluster bean and *okra* are also very drought hardy. In these crops also, there is need to select the most drought hardy cultivars for rain fed production. Varieties have also been identified in fruits, vegetables, flowers and medicinal and aromatic plants through researches conducted in different parts of the country. The important crops and cultivars which can be grown successfully in hot arid region are presented in Table 2.

15.5 PLANT PROPAGATION

Techniques for propagation of *ber*, pomegranate, fig, custard apple, date palm and other fruit crops have also been standardized; details are presented in Table 3.

Table 2 : Promising crops and their cultivars of fruits and vegetables

Crops	Cultivar
Fruits	
Aonla	Kanchan, Krishna, Balwant, NA-6, NA -, NA-10
Bael	NB-5, NB-9, Pant Aparna, Pant Suwarna
Ber	Gola, Mundia, Kaithali, Banarasi Karaka, Early Umran
Custard apple	Balanagar, Mammoth, Island, Gem, Red Sitaphal, Arka Sahan
Date palm	Halawy, Barhee, Medzool, Khuneji
Fig	Poona Fig, Conadria, Excel, Black Ischia
Guava	Sardar, Allahabad Safeda, Kohir Safed, Safed Jam, Chittidar
Pomegranate	G-137, GKVK-1, Mridula, Bhagawa, Phule Arakta, Ruby, Jalore Seedless
Sapota	Kalipatti, Cricket Ball
Tamarind	PKM -1, Pratisthan, Yogeshwari
Sweet orange	Blood Red Malta, Mosambi, Valencia
Vegetables	
Cluster bean	Pusa Navbahar, Durga Bahar, HBG 4, ARG 80
Cowpea	Pusa Dofasali, Pusa Phalguni, Pusa Rituraj, Pusa Barsati
Chilli	Pusa Jwala, Mathania
Bottle gourd	Pusa Summer Prolific Round, Pusa Summer Prolific Long, Pusa Meghdoot, Pusa Manjari
Muskmelon	Tonk melon, Durgapura Madhu,

Table 3 : Standardized methods of propagation

Name of fruits	Methods of propagation	
	Standardization	**Commercial**
Aonla	Budding (patch, shield)	Patch budding
Bael	Root cutting, budding, (patch and shield), air layering	Patch budding
Ber	Cutting, air layering, budding, (T, I, ring and forkert)	Ring and T-budding
Custard apple	Soft wood grafting	T-budding, inarching,
Date palm	Seed, offshoots	Offshoots
Fig	Cutting, budding and air layering	Hard wood cuttings
Grape	Hard wood stem cutting, chip budding, layering and micro propagation	Hard wood stem cuttings
Guava	Cutting, air layering and budding	Stooling, Inarching, air layering
Jamun	Seed and budding	Shield and patch budding
Karonda	Seeds, soft wood stem cuttings in mist	Seeds, air layering
Phalsa	Seeds and cutting	Seeds
Pomegranate	Budding (chip, patch and forkert) air layering, inarching	Hardwood stem cuttings and air layering

The suitable rootstock in different fruit crops has also been standardized to achieve the optimum productivity of the crops. The details for each crop are given in Table 4.

Table 4 : Rootstocks for arid fruit crops

Fruit crops	Rootstocks	Distinguishing features
Ber	*Zizyphus nummularia,Z. rotundifolia*	DwarfingDwarfing
Custard apple	*Annona glabra* (pond apple)	-
Fig	*Ficus glomerata*	Nematode resistant
Bael	Seedling	-
Guava	*Psidium friedrichsthalianum, Psidium friedrichsthalianum* var. *lucidum P. pumilum*	Dwarfing and wilt resistant Wilt resistant, Most dwarfing respectively
Jamun	*Syzigium fruticosum*	Termite resistant

13.6 AGRO-TECHNIQUES

Agro-techniques related to water and nutritional requirement, pruning, training and for management of pests and diseases have been developed. The technologies have also opened up scope for the development of the vast wastelands in different parts of the country. However, there is need to further refine these technologies with emphasis on optimum utilization of water and nutrients and on reducing dependence on chemicals for the control of diseases and insect pests. There is also need to commercially exploit the existing stress hardy genetic resources for development of horticultural industry in the region.

13.6.1 Water management

Studies have shown that following strategies can be adopted to enhance the water use efficiently in horticultural crops :

1. Use of water conservation techniques such as mulching, pitcher irrigation, double walled pot, use of soil amendments, water absorbing polymer, etc.
2. Use of pressurized irrigation, tank irrigation, *ex-situ* and *in-situ* water harvesting methods.
3. Application of water at critical stages in relation to growth and development of fruit and vegetable crops.

13.6.1.1 Rainfed region

Most part of the vast arid region depends on little rainfall. Horticulture production under these conditions is only possible through proper management of rain water, utilization of the natural biophysical resources including the hardy plant species and management of the biotic and abiotic stresses. Watershed management has been used by mankind since antiquity. For this, special techniques have been used for growing trees and crops in desert areas. The *khadins* or *kharins* in Jaisalmer which are low laying basins collecting runoff from surrounding rocky catchments for cultivation are believed to be about 300-500 years old. In the year 1908, 377 *khadins* have been reported to exist around Jaisalmer. These, however, remained abandoned for about 100 years, but are now being revived again. Even under the extremely arid agro climate of Jaisalmer, fruit trees like mango, Citrus, *Cordia*, etc. can be seen growing in the *khadins*. The *in-situ* runoff concentration technique is the most suited particularly for the fruit crops because: i) being deep rooted they are able to utilize the moisture stored in lower soil layers and can produce optimum crops even in low rainfall year, and ii) once established they continue to produce for many years. In India, some work has been done at different AICRP centres which led to recommendations for micro-catchments for rainfed horticulture. Even in the indigenous hardy plant species such as *ker (Capparis decidua), lasora (Cordia myxa),* custard apple (*Annona squamosa),* etc., productivity will increase if due consideration is given to watershed management. The watersheds are divided into micro catchments considered optimum either for a tree or a group of trees, arranged in a row or other patterns suited to the locality. In nearly flat areas, catchments could be provided on two sides of a tree row or around each tree. The optimum catchments are worked out considering water requirement and root architecture of the fruit species, on one hand, and the expected water input through rainfall and runoff on the other hand (Table 5). Study led to estimate the catchments area required per tree and tree per hectare of different fruits for growing in sandy catchments. Using the formula, A=TS/RC where

A is catchments area per tree in m^2, T is canopy are of the tree in m^2, S is runoff supplement required in mm per tree to meet the rainfall deficit, R is average annual rainfall in mm and C is run-off coefficient of the soil.

Table 5 : Catchments area per tree and plant population per hectare with different slopes of sandy catchments

Fruit	Canopy area (m^2)	Runoff supplement (mm)	Catchment's slope (%)	Catchment's area (m^2)	Area per tree (m2)	Population per ha
Ber	36	300	0.5	100	136	74
			5.0	75	111	90
			10.0	60	96	104
Pomegranate	20	800	0.5	148	168	60
			5.0	111	131	76
			10.0	89	109	92
Guava	56	800	0.5	415	471	21
			5.0	311	367	27
			10.0	249	305	33
Fig	56	800	0.5	415	471	21
			5.0	311	367	27
			10.0	249	305	33
Aonla	56	500	0.5	259	375	32
			5.0	194	250	40
			10.0	156	212	47
Sour Lime	36	1000	0.5	333	369	27
			5.0	250	286	35
			10.0	200	236	42
Custard apple	20	300	0.5	56	76	132
			5.0	42	62	161
			10.0	33	53	189

Ber plantation has been raised with *in-situ* runoff concentration system under hot arid climatic conditions of North West India. The fruit yield in Gola and Seb cultivars of *ber* increased with increase in degree of slope and decreased with length of run as a result of increase in runoff and moisture storage in a 3m soil profile. The highest fruit yields were obtained when 0.5 and 5 per cent respectively had 8.5 and 7 m length of run and 72 and 54 m^2 catchments area per tree giving 2 and 1.5 contributing per planted area ratios.

To conserve the moisture losses from the surface of plant and soil, use of mulches, cover crops and anti transpirants have been used. Black polythene (400 gauge) has been found effective mulch material in *ber*, pomegranate, date palm and *aonla* orchards. Some local weeds like *bui (Aerva tomentosa), sarkanda (Saccharum munja),* etc. are effective live mulch materials. Anti transpirants like 10 per cent Kaolin and 10 ^{3}M PMA have been found effective to check moisture losses from plant surface resulting in increased fruit production in pomegranate.

13.6.1.2 Irrigated arid region

The main sources of irrigation in arid region are wells and tubewells as in Jodhpur, Barmer, Churu, Sikar, Jhunjhunu, Pali, Jaipur, Nagaur and Jaisalmer districts of Rajasthan and in a considerable part of Gujarat and Haryana; canals as in Ganganagar, Hanumangarh, Bikaner and Jaisalmer districts of Rajasthan, northern Gujarat and in a considerable part of Punjab and Haryana. About 36 per cent area of the arid region is irrigated by wells or tubewells having brackish water and 60 per cent area is under canal command (Pareek, 1982). At present most of the horticultural production in arid areas comes from the small pockets having assured irrigation covering less than 4 per cent of these lands.

Efficient water use technique such as mulching for soil moisture conservation, use of barriers in the soil to reduce percolation, irrigation by drip and sprinklers have been developed for fruit and vegetable crops.

Drip irrigation or more broadly known as micro-irrigation is mainly suited for orchard and plantation crops where it saves 30-70 per cent irrigation water and increases yield by 25-80 per cent (Table 6). Evidences from drip irrigation trials have clearly indicated the advantages like water saving, higher productivity, limited weed growth, better management of assets, off-season maturity, better fruit quality and reduced incidence of insects and pests and disease. The salient findings of micro-irrigation research at various institutions in India are summarized by Sivanappan, (1998) as detailed in Table 6.

Table 6 : Relative performance of crops with drip irrigation in comparison with that of traditional irrigation methods

Crop	Location	Yield(q ha^{-1})		Irrigation water (cm)		WUE (Q ha^{-1}cm^{-1})		Advantages of drip irrigation	
		Surface	Drip	Surface	Drip	Surface	Drip	Saving of water (%)	Increase in yield (%)
Ash gourd	Jodhpur	108	120	84	74	1.3	1.6	12	10
Beet	Coimbatore	5.7	8.9	86	18	0.07	0.5	79.1	36
Bottle gourd	Jodhpur	380	558	84	74	4.5	7.5	12	31.9
Bitter gourd	Chalakudy	32	43	76	33	0.42	1.3	56.6	25.6
Brinjal	Akola	91	148.0	168.0	64.0	0.55	2.3	62.0	38.5
	Delhi	280	33.8	45.0	35.0	6.2	9.7	22.2	17.2
	NCPA	280	320.0	90.0	42.0	3.11	7.6	53.3	12.5
	Pune	225	245.0	78.0	51.0	2.9	4.8	34.6	8.2
	Rahuri	280	280.0	90.0	42.0	3.11	6.7	53.3	0.0
Broccoli	Delhi	140	195.0	70.0	60.0	2.0	3.25	14.3	28.2
Cauliflower	Akola	83	116.0	39.0	26.0	2.1.0	4.5	33.3	28.4
Chilli	Pantnagar	171	274.0	27.0	18.0	6.3.0	15.2	33.3	37.6
Cucumber	NCPA	42.3	60.9	109.0	41.7	0.30	1.5	61.7	30.5
Lady' finger	Pune	155	225.0	54.0	24.0	2.9	9.4	55.6	31.1
	Coimbatore	100	113.1	53.5	8.6	1.87	13.2	84.0	11.6
	Delhi	360	480.0	42.0	26.0	8.6	18.5	38.1	25.0
	Rahuri	189	203.0	219.0	113.0	0.86	1.8	48.4	7.0
Onion	Delhi	284	342.0	52.0	26.0	5.5	13.2	50.0	17.0
	Hisar	93	112.0	50.0	45.0	1.6	2.5	25.0	17.0
Potato	Delhi	172	291.0	60.0	27.5	2.9	10.6	54.2	41.0
	Hisar	235.7	344.2	20.0	20.0	11.8	17.2	0.0	31.5
	Parbhani	334	480.0	30.0	22.0	11.1	21.8	26.7	30.4
Raddish	Coimbatore	10.5	11.9	46.0	11.0	0.23	1.1	76.1	11.8
Sugarbeet	Hisar	418	489	50.0	37.0	8.4	13.2	26.0	14.5
Sweet potato	Coimbatore	42.4	58.9	63.0	23.0	0.67	2.4	60.3	28.0
Tomato	Akola	45	58.0	102.0	77.0	0.44	0.75	24.5	22.4
	Coimbatore	6108	88.7	49.8	10.7	1.24	8.28	78.5	30.3
	Delhi	257	396.0	47.0	25.0	5.5	15.8	46.8	35.1
	Ncpa	320	480.0	30.0	19.0	10.7	25.3	36.7	33.3
	Pantnagar	104.0	137.0	22.0	14.0	4.7	9.8	36.4	24.1
	Parbhani	320/0	480.0	32.4	22.2	9.9	21.6	31.5	33.3
	Pune	292.0	413.o	31.0	20.0	9.4	20.7	35.5	20.3
	Rahuri	16.4	17.2	29.7	20.8	0.6	0.82	30.0	4.7
	Udaipur	144.0	175.0	41.0	28.0	3.5	6.3	31.7	17.7
Banana	Hawanisagar	277.0	329.0	186.0	172.0	1.5	1.9	7.5	15.8
	Kharagpur	290.0	400.0	106.0	106.0	2.74	3.8	0.0	27.5
Ber	Belvatgi	13.7	18.0	15.4	12.5	0.9	1.4	18.8	23.9
Grapes	Dharwad	101.0	101.0	53.0	28.0	1.91	3.6	47.2	0.0
Guava	Allahabad	0.16/plant	0.22/plant	6.4	5.21	0.03	0.04	18.6	27.3
Kinnow	Delhi	68.0	98.0	22.1	17.3	3.1	5.7	21.7	30.6
Lemon	Delhi	15.0	27.0	23.0	17.5	0.65	1.54	23.9	44.4
Pomegranate	Belvatgi	7.4	14.4	10.7	8.7	0.7	1.7	18.7	48.6
	Hyderabad	15.0	37.0	183.0	178.0	0.08	0.21	2.7	59.5
Watermelon	Jodhpur	294.6	882.0	80.0	80.0	3.7	11.0	0.0	66.6
	Pune	82.1	504.0	72.0	25.0	5.9	20.2	65.3	16.3

Magar and Firke (1994) reported the results in respect to the quantity of water applied in different fruit crops on the basis of age of the plant and open pan evaporation (Table 7).

Table 7 : Average water requirement of fruit crop by drip method

Name of fruit crops	Age (Y)	Water requirement (litres/tree/day)											
		Jan.	Feb.	Mar.	Apr.	May	Jun.	Jul.	Aug.	Sep.	Oct.	Nov.	Dec.
Sapota,	1	-	-	-	-	-	22	23	27	24	22	25	21
Mango,	2	27	38	56	77	82	60	41	34	34	41	31	26
Tamarind	3	45	63	93	128	137	94	68	56	56	62	51	43
	4	65	92	135	186	199	137	98	81	81	97	74	62
	5	76	107	158	217	232	159	195	95	101	114	86	73
Coconut, *Anola*	1	-	-	-	-	-	-	-	-	-	-	-	-
	2	18	25	36	50	54	37	27	22	22	26	20	17
	3	22	31	46	66	67	46	33	27	27	33	25	21
	4	32	45	66	91	97	67	48	40	40	48	36	31
	5	37	53	77	106	114	78	56	47	47	56	42	36
Orange,	1	-	-	-	-	-	-	-	-	-	-	-	-
Lime,	2	13	18	28	37	40	27	20	16	16	19	15	12
Guava	3	16	23	34	46	49	34	24	20	20	24	18	16
	4	23	33	49	67	72	49	35	29	29	35	27	23
	5	29	41	61	84	89	61	41	37	47	47	32	28
Ber,	1	-	-	-	-	-	-	-	-	-	-	-	-
Pomegranate,	2	12	17	25	34	36	25	18	15	15	18	13	11
Fig	3	16	23	34	46	50	34	18	15	15	18	13	11

The data in Table 8 revealed that although tree growth was not affected, the fruit yield and water use efficiency were considerably increased by drip irrigation in *cv.* Ganesh of pomegranate at Rahuri in Maharashtra.

Table 8: Effect of methods of irrigation on growth and yield of Ganesh pomegranate

Method	Water used litres	Tree height (m)	Stem diameter (m)	Fruit yield (q/ha)	WUEQ/ha/cm)
Check basin					
0.6 IW/CPE	4800	2.35	3.78	40.6	0.82
0.8 IW/CPE	6600	2.56	4.02	47.5	0.69
Trickle Daily					
20% WA	3580	2.16	3.79	52.5	1.25
30% WA	5322	2.44	4.02	55.9	0.72
Alternate day					
20% WA	3580	2.30	3.84	48.6	1.16
30% WA	5322	2.24	4.01	57.3	0.92

Source : Pareek, O. P. (1993)

IW - Irrigation water; CPE - Cumulative pan evaporation WA - Water applied.

Nutritional requirement for several fruits, vegetables and species have been worked out. Techniques for training and pruning particularly in *ber* and grape have now been standardized. Chemical as well as cultural measures for the control of diseases and pests in different crops have been standardized. Growth regulators to improve quality and productivity of produce are now commercially used in grape, onion and cucurbits. Thus, there are possibilities of their production under arid agro-climate. Crops like tomato, brinjal, chillies, cauliflower and onion can be produced in dry lands by using runoff concentration technique by transplanting seedlings raised in a nursery at the onset of monsoon in early July.

13.7 CHALLENGES

Crop farming, in horticulture too, is the risky proposition and production cost is also rising than productivity gain, eroding profit margins. It calls for the development and deployment of new technologies for lowering production costs, raising crop yields, enhancing the quality of produce and reducing the post harvest losses. Market driven diversification in a global perspective has become the new paradigm driving future horticultural growth of India. Specialty commodities such as off-season varieties and production systems, new crops and novel varieties should be identified to capture new opportunities. Hort-business centres equipped with facilities like cold storage, processing unit, packaging units, quality control facilities, financial linkage points, insurance facilities, etc. should act as a price determination centres at the producer level. Value addition of horticultural produce offers great scope in domestic as well as export markets. Strict adherence of international standards of quality is essential in this area through GMP and HACCP. Farmer's organizations, prospective entrepreneurs, cooperatives self help groups need to encourage develop/strengthen value additive type of marketing infrastructure at the farm level.

13.8 REFERENCES

Pareek, O. P. (1993) Water management in fruit crops. In: *Advances in Horticulture*, Vol. 2. MPH, New Delhi (Eds: Chadha, K. L. and Pareek, O. P.) pp. 726-761,

Magar, S. S and Firke, D. (1994) Water management of fruit crops under limited water supply. Shetkari. pp. 17-21.

Pareek, O. P. (1982) *Proceedings of workshop on Problems of deserts in India*, Jaipur, 16-18 September 1975. Geological Society of India, Calcutta, Misc. Publ. No. 49: 313-320.

Sivanappan, R. K. (1998) Status and perspective of micro-irrigation research in India. In: *Proceeding of the National Seminar on Micro-Irrigation Research in India: Status and Perspectives for the 21st century*, Bhubaneswar, July 27-28, 1998.

Chapter – 14

Exploitation of Underexploited Crops in Precision Horticulture

Sanjay Singh, A. K. Singh and K. Lata

14.1 INTRODUCTION

The efforts for harnessing the natural resources for sustainining the human race have been 'on' since the beginning of the civilization. The basic needs of food, fuel and fodder are obtained from the land resources with the combination of water and sunlight. The combination of information technology and use of hi-tech equipments paved the way for the introduction of precision horticulture. Precision horticulture may be defined as application of a holistic management strategy that uses information technology to bring data from multiple sources to bear decision associated with agriculture production, marketing, finance and personnel. Enhanced input efficiency is an important component of horticultural production under precision horticulture site specific management. Enhanced input efficiency in crop management can be obtained only when suitable varieties, desired amount of nutrients and water based on calculations, at right time and by proper method are used.

With recent advances in geographic positioning system, remote sensing, data processing, harvesting equipment, fertilizer and pesticide applicators, the approach can be used from a more practical stand point to identify variability and to use the variable rate of inputs over an entire field.

In recent years, there is considerable awareness about the nutritional security and food safety. More emphasis is given to underutilized fruits due to their high nutritive and medicinal value in addition to being resistant/tolerant to many biotic / abiotic stresses. The demand of these fruits are gradually increasing owing to tremendous potential for commercial exploitation aimed at improving the economic status of the poor and marginal farmers. An area, which needs immediate attention, is the collection, documentation, conservation and utilization for their sustained production and popularization.

The properties of the jamun has been extensively studied, particularly with reference to the pulp and seed. It has many valuable properties and virtually every part of the tree has been utilized by both urban and rural dwellers. The fruit is good source of iron, sugars, minerals, protein and carbohydrate etc. Fully ripe fruits are eaten fresh and can be processed into products like jelly, jam, squash, wine and vinegar. The *jamun* fruit has sub- acid spicy flavour and squash is very refreshing drink for quenching the thinust in the summer season. A little quantity of fruit syrup is much useful for curing the diarrhoea. The vinegar prepared from juice extracted from slightly unripe fruit is stomachic, carminative and diuretic apart from having cooling and digestive properties. Fruits are used as an effective medicine against diabetes, heart and liver trouble. Glucose and fructose are the principal sugars in the ripe fruit, not even a trace of sucrose was detected. Malic acid is the major acid (0.59 per cent of the weight of fruit). In addition, a small quantity of oxalic acid is also present. Gallic acid and tannins account for the astringency of the fruit. The purple colour of the fruit is due to anthocyanin. The waxy component of the fleshy pericarp contains sterol and small quantity of essential oil. The blossoms are important source of honey from *Apis dorsata* in North India. The powder of seeds has high value being useful in the treatment of diabetes. Seeds contain alkaloid jambosin and glycoside, jambolin or antimellin, which control the diastatic conversion of starch in to sugars. Therefore, the *jamun* fruits are having high therapeutic and nutritional values. Fruits and seeds of tamarind has tremendous potential for industrial use. *Mahua* is a boon for tribal farmers. The corolla commonly called as *mahua* flowers is a rich source of sugar containing appreciable amount of vitamins and minerals. Flowers (fresh and dried) and spent flowers (after fermentation) are also used as feed for livestock. Fruits are eaten as raw or cooked. Seeds are good source of oil. The oil obtained from *Mahua* kernel is used for edible purpose and permitted for preparation of vegetable oil. *Mahua* oil is used in manufacture of soap, lubricating grease, fatty alcohols and candles. *Chironji* is excellent tree of agro forestry and social forestry. At present, it is growing under forest condition as an under exploited fruit and gives monitory reward to the tribal community of the country and seems to be boon for them. The flesh of ripe fruit is very palatable and the oily kernels are good source of protein (28.00-30.00 per cent) and are used in garnishing of sweets. Kernel oil can be used to substitute olive and almond oils. *Karonda* is also very hardy plant, it is planted as protective hedge on the fringes of the orchard. *Khirni* thrives well on rocky, gravelly, saline and sodic soils and is good source of sugars and carotenoids. It is being used commercially as rootstock for sapota. These fruits have great potential under different agro climatic conditions of India.

14.2 BIODIVERSITY UTILIZATION

14.2.1 *Jamun* (*syzygium cuminii* Skeels)

It is a member of family Myrtaceae. It is one of the important under utilized fruits. A survey was carried out in North Goa to identify the elite genotypes of *jamun*. The individual fruit weight ranged from 3.42 to 13.67g, length from 3.31 to 5.26cm, girth from 5.21 to 9.82cm, pulp percentage from 58.57 to 84.55, TSS from 12.00 to 26.80 per cent, titratable acidity from 0.59 to 1.63 per cent, total sugars from 6.87 to 25.31 per cent and sugar: acid ratio from 15.39 to 27.92 (Devi *et al.*, 2002). An intensive survey was made to find out the existing natural variability among the *jamun* seedling trees and to identify superior genotypes with good fruit quality from Uttar Pradesh (Lucknow, Varanasi and adjoining areas) and Jharkhand (Ranchi and adjoining areas). Out of surveywork, the genotypes, RNC-26, RNC-11, V-6, were reported V-7 and V-8 promising.

Singh and Singh, (2005a) also made a survey in Gujarat to identify the elite genotypes of *jamun* among its population. Flowering, fruiting and physico-chemical attributes of ripened fruits from seventeen selected genotypes were studied. The study revealed that there was a wide variation among the genotypes. The earliest flowering (Mid February) took place in GJ-24, and the latest (Last March) in GJ-23, GJ-30, GJ-31, GJ-32 and GJ-33. Maximum panicle length (15.24cm) and number of fruits per panicle (26.00) were found in GJ-19. Collection numbers GJ-19, GJ-21, GJ-22 and GJ-25 were found earliest (First Week May) in ripening, while GJ-17, GJ-28 and GJ-32 ripened at the last (Last June). Maximum yield per plant was recorded in GJ-19 (175.00kg) followed by GJ-23 (170.00kg), GJ-22 (160.00kg) and GJ-18 (150.00kg). Individual fruit weight ranged from 10.10 to 22.50g, length from 1.99 to 3.24cm and pulp percentage from 73.66 to 85.68. There was a wide variation in chemical characters also. Total soluble solids varied from 10.30 to 12.34 per cent, total sugar 8.58 to 9.13 per cent and vitamin C 32.12 to 46.37 mg/100g. On the basis of overall performance GJ-18, GJ-19, GJ-23, GJ-24 and GJ-25 were found to be promising among the genotypes studied based on its yield and physico - chemical characters.

At CHES, Godhra, total 26 genotypes were evaluated (2001-2009) for their growth, flowering and fruiting attributes and GJ-2 (Goma Priyanka) was released as variety. Salient characters of Goma Priyanka are as under:

Goma Priyanka

Goma Priyanka was released by CHES, Godhra during the year 2009. It was collected from *Ode* village of Anand district of Gujarat. The selected genotype was propagated through patch budding and tested under field conditions for 8 years (2001-2009). The selection performed better in respect to flowering pattern, fruiting and fruit quality attributes. The height and girth of tree was recorded 5.25m and 64.66cm respectively. North- South and East- West spread was found to be 5.00m and 4.76m respectively. It is semi-dwarf, with spreading growth habit, dense foliage and drooping branches, precocious bearer (starts flowering in 4th year) and suitable for high density planting. It starts flowering in the month of March, ripens in the fourth week of May and had19.86g average fruit weight, 85.06 per cent pulp, 16.80°Brix TSS, 0.38 per cent titratable acidity, 12.10 per cent total sugar, 6.11 per cent reducing sugar, 45.44 mg/100 g vitamin C, 14.38 mg/100g calcium, 34.78mg/100 g magnesium, 15.83mg/100g phosphorus and 56.31mg/100g potassium. It recorded 43.80 kg fruit yield during 8th year of orchard life under rainfed conditions of hot semi-arid ecosystem of western India.

Other promising genotypes of *jamun*

GJ-8:

It was collected from *Ode* village of Anand district of Gujarat. Peak period of flowering for this cultivar is the month of March. It ripens during the second week of June and recorded 17.00g average fruit weight, 81.82 per cent pulp, 14.20 per cent, TSS, 0.39 per cent, acidity, 11.35 per cent total sugar and 45.10mg/100g vitamin C.

GJ-40:

It was also collected from *Ode* village of Anand district of Gujarat. Its peak period of flowering is the month of March. It ripens during the fourth week of June and recorded average fruit weight 16.00g, 82.00 per cent pulp, 14.00 per cent TSS, 0.39 per cent acidity, 11.00 per cent total sugar and 46.10mg/100g vitamin C.

Jamun : Goma Priyanka

Tamarind : Goma Prateek

Mahua : MH 63

Cape gooseberry : CHESC 2

14.2.2 Tamarind (*Tamarindus indica* L.)

Tamarind, a member of sub family caesalpiniaceae of the family leguminosae, is an under-utilized fruit and widely distributed throughout tropic and sub-tropics as stray plantation or avenue. Owing to its varied use in household and export market, it has emerged as a potential fruit crop. Singh *et al.*, 2008 reported variability in tamarind. Singh and Singh (2005b) reported maximum panicle length of 15.12 cm and number of fruits per panicle 15.00 in tamarind. They reported early ripening during 4th week of March and also late ripening during the 2nd week of May. Among the genotypes, fruit yield ranged from 80.00 to 175.00 kg/ plant. The highest pulp per cent of 53.87 per cent was recorded in a genotype.

At CHES, Godhra, total 24 genotypes were evaluated (2001-2009) for their growth, flowering and fruiting attributes and Tamarind-13 (Goma Prateek) was released as variety. Salient characters of Goma Prateek are as under:

Goma Prateek

Goma Prateek was released by CHES, Godhra during the year 2009. It is early, spreading in growth habit, regular in bearing, semi-dwarf in stature and starts flowering during 4th year. Peak period of flowering for this cultivar is first week of June and fruits ripen in the first week of March. The fruit recorded 26.70g average pod weight, 16.70 cm pod length, 50.50 per cent pulp, 14.06 per cent acidity, 71.00 °Brix TSS, 55.81 per cent total sugar, 27.27 per cent reducing sugar, 17.53mg/100g vitamin C, 3.29g/100g protein, 177.70mg/100g calcium, 44.17mg/100g magnesium, 70.83mg/100g phosphorus. Fruit shape was found slightly curved with reddish pulp. Fruit yield per plant was 58.50kg during 9th year of orchard life under rainfed conditions of hot semi-arid ecosystem. It is suitable for processing. The fruits have great potential under semi-arid conditions of western India.

Varietal wealth: Some promising tamarind selections as a result of evaluation at different institutions are listed below:

PKM 1: It is a selection made at Horticulture College and Research Institute, Periyakulam. Grafts come to flowering three years after planting. The pods have 35 per cent pulp, 17.1 per cent acidity and 3.90mg/100g ascorbic acid.

Urigam: It is progeny of more than 200-year–old tree identified near Urigam by Department of Horticulture, Tamil Nadu. The average length of fruits is 30 cm.

Pratisthan : It was released from Fruit Research Station, Aurangabad. It has 61 per cent pulp, 12 per cent seed, 27 per cent shell and 7-9 per cent acidity. It is sour – sweet tamarind. Its pulp can be stored for longer period.

Tamarind - 263: It was also released by Fruit Research Station, Aurangabad (MS). The fruits have pinkish and light yellow pulp and 18-19 per cent acidity.

Yogeshwari: This variety has been released from Taluk Seed farm, Ambajogai (Beed) in Maharashtra .The fruits are large and have red and sour – sweet pulp with 6-7 per cent acidity.

DTS-1: It is a selection identified at University of Agricultural Sciences, Dharwad. The pods are straight having semi-curved shape, 23.6cm length, 3cm width, 19.5g weight, 51.00 per cent pulp and 13.60 per cent acidity. It is a late variety and takes 310 days from fruit set to maturity.

DTS-2: This is a selection made at University of Agricultural Sciences, Dharwad. The pods are straight having semi- curved shape, 17.60cm length, 2.60cm width, 18.00g weight, 53.00 per cent pulp and 12.20 per cent acidity. It is an early variety and the pods mature in 280 days after fruit set.

14.2.3 Chiraunji (*Buchanania lanzan* Spreng.)

It is excellent tree of agroforestry and social forestry. In the wasteland development and dryland horticulture, it assumes great significance due to its multifarious uses and capacity to withstand adverse climatic conditions. At present, it is growing under forest condition as an under exploited fruit and gives monetary reward to the tribal community of the country and seems to be boon for them. The thickset tissue of ripe fruit is very palatable and is largely eaten raw or roasted and the oily kernels are the most important part and are used in preparation of puddings. Very good juice may be prepared from the pulp of *chiraunji* fruits. The kernel is highly nutritious and rich in protein (25.0-30.0 per cent) and yields sweet oil, which can be used to substitute olive and almond oil. Kernel contains 33.50 per cent oil, 1.90 per cent of which is unsaponifiable. The saponifiable part contained 20.00 per cent of linoleic acid. An ointment made out of the kernels is used to cure itch of the skin and to remove blemishes from the face. *Chironji* provides quality timber wood for various uses.

Fruits and Kernels of Chiraunji CHESC: 7

Khirni: GK-1 and GK-11

Singh and Singh, (2006) observed earliest flowering (first week of February) in 'CPT 1', 'CPT 5', 'CPT 10', 'CPT 15', 'CPT 17', 'CPT 19' and 'CPT 22', while 'CPT 13' was observed as the latest in which flowering was observed during 4th week of February. Maximum panicle length (35.11cm) was found in 'CPT 1', while 'CPT 18' recorded the least panicle length (14.20cm). Highest percentage of perfect flowers (13.10) and fruit per panicle (37.50) was recorded in 'CPT 12' while, it was found at least in 'CPT 18'. Ripening time varied from third week of April to second week of May in different genotypes. The highest fruit yield per plant was found in 'CPT 7' (28.00kg/plant).

At CHES, Godhra, fifteen genotypes of *chiraunji* were evaluated for flowering and fruit quality attributes. Based on the horticultural traits studied, the genotype, CHESC-7 was found promising. Salient characters of CHESC-7 is as under :

Chiraunji CHESC-7

It was collected from *Kada* Dam of Panchmahal district, Gujarat. Peak period of flowering and fruit set was during the first and 3rd week of February, respectively. It recorded maximum panicle length (26.00cm). Peak period of ripening time was third week of April. It recorded 1.20g fruit weight, 22.00 per cent TSS, 13.00 per cent total sugar and 50.00mg/100g vitamin C, 0.12g kernel weight and 30.50 per cent kernel protein.

14.2.4 Mahua (*Madhuca indica* J. F. Gmel.)

Survey of *Mahua* was made in Gujarat. It is one of the important crops for tribal farmers of Gujarat. Singh and Singh, (2005c) recorded early flowering i.e 1st week of March in MH-1, MH-4, MH-5, while, it was recorded late (2nd week of April) in MH-6. The number of flowers (45.00) and fruits (8.00) per fascicle were noted highest in MH-2. MH-8 recorded maximum leaf area (128.00cm^2). Range of ripening span in different genotypes was observed from 2nd week of May to first week of June. Dry flower yield varied from 30.00 to 46.00kg/plant being highest in MH-2. Variation in fruit yield among the genotypes ranged from 35.00kg to 82.00kg/plant and it was maximum in MH-2. Detailed description of 4 four promising genotypes, selected at CHES, Godhra are as under:

Mahua-10:

It was collected from Vejalpur village of Panchmahal district, Gujarat. Its peak period of flowering is the second week of March. Flowers recorded 2.29g weight, 65.00 per cent juice and 26.37 per cent TSS. It ripens in second week of May and recorded 30.50g fruit weight, 13.50g seed weight and 11.00g kernel.

Mahua-14:

It was collected from *Otala* village (*Baria* Road) of Panchmahal district, Gujarat. Peak period of flowering in it was the second week of March. Flowers recorded 2.24g weight, 66.00 per cent juice and 25.00 per cent TSS. It ripens in the second week of May and recorded 29.00g fruit weight, 12.70g seed weight and 9.53g kernel.

Mahua-35:

It was collected from *Reechawani* village (*Baria* Road) of Dahod district. It had peak flowering in the second week of March. Flowers recorded 2.13g weight, 64.12 per cent juice and 24.90 per cent TSS. It ripens in third week of June and recorded 28.50g fruit weight, 12.23g seed weight and 10.08g kernel weight.

Mahua-63:

It was collected from *Reechawani* village (*Baria* Road) of Panchmahal district, Gujarat. Flowering reaches its peak during the 4th week of March. Flowers recorded 2.50g weight, 68.00 per cent juice and 24.37 per cent TSS. It ripens in 4th week of May and recorded 23.50g fruit weight.

14.2.5 Khirni (*Manilkara hexandra* (Roxb.) Dubard)

Khirni or Rayan is an economically multipurpose tree of the family sapotaceae. Thirty genotypes were selected and evaluated for their fruiting and fruit quality attributes. Detailed description of promising genotypes are as under :

GK-1: It was collected from *Kharsalia* village of Panchamahal district, Gujarat. The peak period of flowering was recorded in the month of November. It ripens during the first week of March and recorded 5.66g fruit weight, 4.93g pulp weight, 21.00 per cent TSS, 16.20mg/100g vitamin C and 4.30g/100g carotene.

GK-10: It was collected from *Parwadi* village of Panchamahal district, Gujarat. The peak period of flowering was recorded in the month of December. It ripens in the third week of May and recorded 6.64g fruit weight, 5.71g pulp weight, 26.40 per cent TSS, 24.25mg/100g vitamin C and 6.6.06g/100g carotene.

14.2.6 Detailed description of promising genotype of *Karonda* (*Carissa carandas* Linn.) :

CHESK-1: It was collected from CHES campus, Vejalpur, Panchamahal district, Gujarat. In this type the peak period of flowering was recorded in the month of March. It ripens in the month of June and recorded 6.00g fruit weight, 4.64g pulp weight, 10.63 per cent TSS, 0.97 per cent acidity and 26.39mg/100g vitamin C.

14.2.7 Cape Gooseberry (*Physalis peruviana* L.)

Survey work was made to select the promising genotypes of cape gooseberry. Detailed description of two promising genotypes are as under:

CHESC-1: It was collected from Bihar. It ripens in the month of January- February and recorded 11.50g fruit weight, 12.00 per cent TSS, 1.64 per cent acidity, 7.50 per cent total sugar and 220.10mg/100g vitamin C.

CHESC-2: It was also collected from Bihar. It ripens in the month of January- February and recorded 10.75g fruit weight, 11.20 per cent TSS, 1.70 per cent acidity, 7.25 per cent total sugar and 190.22mg/100g vitamin C.

14.3 PRODUCTION OF QUALITY PLANTING MATERIALS OF UNDERUTILIZED FRUITS

14.3.1 Mother block

The planting stock that is maintained as a source for commercial propagation is referred to as a mother block. Therefore, for preservation of pathogen free planting stock, proper isolation, sanitation, periodical inspection, testing and maintenance are very important. The mother plant should be healthy, precocious, prolific bearer, disease free, genetically true-to-type and free from insect-pests attack. As far as possible, mother block should be in the close proximity to the nursery site. The scion shoots should be taken only from bearing plants. There should be proper record and certification of planting stock. The plants should have known pedigree records regarding bearing potential, fruit quality and problems, if any.

Commercial planting stock can be referred as certified stock if it is grown under supervision of a legally designated agency with prescribed regulations designed to maintain minimum standards of cleanliness and clonal identity.

14.3.2 Raising of rootstocks in the polythene bags

Polythene bags having the size of 25cm × 10cm, 25cm × 15cm and 30cm × 15cm as per requirement are being used for raising the rootstocks in almost all parts of the country. Small holes are made in the bottom and sides of polythene bags for drainage and aeration and filled with porous rooting medium or pot mixture for raising rootstocks. The locally available seeds (preferably one tree seedling) should be sown in the polythene bags, filled with soil and FYM (3:1) for raising the rootstocks. Seeds can also be sown in the raised nursery beds or polyhouses and transplanting of very young seedlings in the polythene bags at the 2-4 leaf stage has been found successful. About 9-12 months old seedlings of uniform size having stem of pencil thickness should be used as rootstock. Plants raised in the polythene bags can easily be transported to distant places with higher transplanting success.

14.3.3 Propagation techniques of underutilized fruits

In *jamun* (*Syzygium Cuminii* Skeels), highest success through patch budding was noted in March (70-80 per cent), whereas, soft wood grafting recorded maximum success in the month of July and August (75-80 per cent). Irrespective of scion and rootstock, maximum accumulation of nitrogen and carbohydrate content was recorded in March, while it was found in lower concentration during September, October, November, December and January. Patch budding in March and soft wood grafting in July-August may be followed for multiplication of elite *jamun* plants. Chovatia and Singh (2000) recorded 41.67 per cent success through soft wood grafting during the month of June under Gujarat conditions. In *jamun*, soft wood grafting was found successful in the month of July- August under semi- arid environment of western India (Singh and Singh, 2006 and Singh *et al.*, 2007).

In tamarind (*Tamarindus indica* L.), the highest percentage of budding success was noted when performed during the month of July-August (75-80.00 %). While, soft wood grafting recorded maximum success during the month of May, June, July and August (60-75%). It was concluded that patch budding during the month of July-August and soft wood grafting during the month of April-May may be adopted in the region for multiplication of elite tamarind genotypes (Singh *et al.*, 2007).

Soft wood grafting in *Khirni* (*Manilkara hexandra*), *Chiraunji* (*Buchanania lanzan* Spreng) and *Mahua* (*Bassia latifolia*) has been studied under Gujarat conditions. In Khirni highest percentage of graft success was noted when done during the month of March (76.66 %) followed by June, July and August, whereas, it was recorded the least when performed during the month of February (10.00 %). In *chiraunji*, highest percentage of graft success was noted in July (66.66 %) followed by August and June, whereas, it was recorded the least when grafted during the month of February (3.33 %). The highest percentage in *Mahua* was there when grafting was done during the month of March (70.00) closely followed by April, July and August. Naturally defoliated scion sticks of *Mahua* were found suitable to achieve good success through softwood grafting during the month of March-April. Soft wood grafting during the month of June, July and August may be practised with 65.00-80.00 per cent success for multiplication of wood apple and custard apple genotypes. Singh *et al.* (2003) reported that *lasoda* (*Cordia myxa*) may be propagated through patch budding with 70-80 per cent success under different agro-climatic conditions. Nursery raised plants should be shifted at least 2 times during rainy season to discourage the growth of tap root. Hardening of grafted plants should also be followed to reduce mortality of the grafts in the field. Name of the cultivars with age should be properly tagged. Care should be taken for proper nutrition, weeding, irrigation and plant protection of planting materials. It may be concluded that most of the under-utilized fruits may be multiplied through soft wood grafting and patch budding during the month of July, August, March and April. Hi-tech glass house may also be used to multiply quality planting materials throughout the year.

14.4 SITE SPECIFIC NUTRIENT MANAGEMENT

There are several aspects of nutrient management for precision farming that challenge the applicability of using single element soil test values as the basis of predicting variable rate nutrient needs across the large fields. The availability of sophisticated information handling systems, i.e. computers with spreadsheets and geographic information system programmes, now make it relatively easy to use more than a single value to predict nutrient needs.

14.5 TREE CANOPY MANAGEMENT

Among the production packages, tree canopy management, especially size control, has become a priority for the orchadist due to the demands imposed by modern markets in terms of production costs, yield and fruit quality. Early height control and tree management are practised in many countries but these strategies being relatively recent, are generally not practised in India. Similarly, the problem of large tree size in fruit crops has been ameliorated by using topping and hedging because large and crowded trees pose many disadvantages. This has resulted in widespread efforts to reduce tree size and increase tree density. In order to increase efficiency utilization of improved technologies need to be accelerated in India. Fruit quality of *jamun* fruits improved in terms of fruit weight and pulp content after following 50 per cent pruning of annual extension growth during the month of September. To maintain the dwarf framework of the *jamun*, main stem may be headed back at desired height (4-6 meters) to facilitate easy harvesting of the fruits.

Canopy management has its implication in light regulation for the plant. The total dry matter production of an orchard is significantly influenced by the amount of light energy. The tree intercepts light energy through which it produces chemical energy and in turn dry matter. Both, total dry matter production and total yield of fruit are related to light interception. Light interception is mainly influenced by the amount and arrangement of the leaves, fruits and branches within the tree crown, the tree shape and size, spacing, row orientation and the angular distribution of light from the sun and sky. Light interception can be increased by reducing the spacing, increasing the height and spread of the trees and increasing the leaf area index. Increase in yield of large well-coloured fruit in orchards will only occur, if, increase in tree density are combined with an improvement in the light distribution in the trees. Thus, proper understanding of the relationships between light interception and tree density is one of the important components of planting system as shading lays serious set back on growth, production and quality aspects. Canopy management of proper light distribution is the key factor in accommodating higher number of productive trees per unit area.

14.5.1 Height control

If, vegetative growth is not controlled properly in high density planting, there is over crowding and shading, which reduces flower bud formation, fruitlet retention, fruit size and fruit colour. Control of apical growth must begin within the first year of planting and continue each year in high density planting system. Various, methods suggested for economical control of tree vigour and size in high density planting (HDP) of fruit trees are :

14.5.2 Genetic control

Genetic control of growth is achieved through selection of dwarf scion varieties and slow growing selections of a particular cultivar. There is an urgent need to develop rapid screening technique for identifying dwarfing rootstocks.

14.5.2 Mechanical control

An interest in mechanical tree size control (hedging, topping, pruning and tree thinning) has increased at the farmers field. This may be the only practical methods of tree size control of inherently large trees in HDP. Tree height control is accomplished by removal of the multiple central leaders within the canopy of the young tree. The dominant, upward shoot or shoot from each growth umbel (whorl of new shoots) is removed and when there are parallel branches within the canopy, the more dominant upright shoots are removed. Today, many workers accomplish early height control pruning. As most pruning in advanced countries is done mechanically, these size control measures are rarely used. Mechanical pruning has not yet been adopted to serve this purpose in India. Experimental evidence indicates that topping and hedging are eficient approach to control tree size in several fruit crops. However, height management should begin in the first year. Single stemmed trees are headed between 40 and 70 cm depending on the desired height of the lowest scaffold limbs. Multiple shoots develop from the buds below the heading cut. However, the response to hedging depends on the nature and growth pattern of fruit crops. Planting densities are decided considering vigour of tree, climatic conditions, soil fertility, water availability and market requirements.

14.6 CONCLUSION

Site-specific management forms cornerstone of future for horticulture system of the 21st century because it offers a variety of potential benefits in profitability, productivity, sustainability, crop quality, food safety, environmental protection, on-farm quality of life and rural economic development. Under exploited fruits have tremendous potential under arid and semi-arid ecosystem of western India.

14.7 REFERENCES

Chovatia, R.S. and Singh, S.P. (2000) Effect of time on budding and grafting sucess in *Jamun* (*Syzygium cuminii*). *I. J. Hortic.* 57 (3) : 255-58

Devi, S.P., Thangam, M., Desai, A.R. and Adsule, P.G. (2002). Studies on variability in physico-chemical characters of different *jamun* (*Syzygium cuminii* Skeels) accessions from Goa. Indian. J. *I. J. Hort.,* 59 (2): 153-156.

Singh, D. B., Saroj, P. L. and Vashishtha, B.B. (2003). Effect of time of budding in propagation of *lasoda* (*Cordia myxa*). *Prog. Hort.*, 35: 230-32.

Singh, Sanjay and Singh, A.K. (2005b). Studies on variability in *jamun (Syzygium cuminii*) in Gujarat. *Prog. Hort.*, 37(1): 144-148.

Singh, Sanjay and Singh, A.K. (2005c). Studies on variability in tamarind (*Tamarindus indica* L) from Gujarat. *Prog. Hort.* , 37: 280-284.

Singh, Sanjay and Singh, A K (2005) Studies on genetic diversity in *mahua* (*Bassia latifolia*) under semi-arid ecosystem of Gujarat. *Indian. J. Agril. Sci.,* Vol. 75 : 519-523.

Singh, Sanjay and Singh, A.K. (2006. Standarization of method and time of propagation in jamun (*Syzygium cuminii*) under semi arid environment of western India. *Indian Journal of Agricultural Sciences,* 76 :142-45

Singh, Sanjay, Singh, A K and V. V. Apparao (2006). Genetic diversity in *Chiraunji* (*Buchanania lanzan* Spreng.) under semi-arid ecosystem of Gujarat. *Indian. J. Agril. Sci.*, 76: 695-698.

Singh, Sanjay, Singh, A.K and Joshi, H. K (2008) Genetic variability for floral traits and yield attributes in tamarind (*Tamarindus indica* L.) *Indian. J. Hort.,* Vol. 65 (3) :228-31.

Singh, Sanjay, Singh, A.K. and Bagle, B. G. (2007). Propagating jamun successfully.*Indian Horticulture*, Nov-Dec, pp-31-33.

Chapter – 15

Exploitation of Genetic Resources of Vegetable Crops for Precision Farming

K. P. Singh

15.1 INTRODUCTION

Vegetables are rich sources of many essential micronutrients, including vitamin C, vitamin K, folate, thiamine, carotenes, minerals, and dietary fibre. In addition, vegetables are rich in health related phytochemicals, such as antioxidants. Such phytochemicals play an important role in reducing the risk of many chronic diseases. Therefore, it is essential to produce more vegetables to feed our ever growing population. In this context, use of genetically improved, disease resistant and high yielding varieties in precision farming may play a significant role in increasing vegetable productivity and production.

15.2 EXPLOITATION OF GENETIC RESOURCES

The genetic resources of a crop may be defined as *the sum total of hereditary material*, i.e., all the alleles of various genes present in a crop species and its wild relatives. Genetic resources are broadly grouped into two major types (i) cultivated germplasm, and (ii) wild germplasm. Alternatively, they may be termed as (i) indigenous (from the country in question), or (ii) exotic (from another country) based on their place of origin. Genetic resources consist the following five types of materials: (1) land races, (2) obsolete varieties, (3) modern varieties or varieties in cultivation, (4) breeding lines, and (5) wild forms and wild relatives.

15.2.1 Land races

These are primitive varieties, which had evolved without systematic plant breeding. They exhibit valuable genetic variability and, are adapted to the local soil and climatic conditions.

15.2.2 Obsolete varieties

These varieties were developed by systematic breeding efforts, and were once commercially cultivated but are no more grown. They, however, do have some desirable features which can be utilized for future breeding programmes.

15.2.3 Cultivated varieties

The cultivated varieties are mostly being used in breeding programmes. They form a major part of a working collection. They are good sources of genes for yield, quality, biotic and abiotic stresses, etc.

15.2.4 Breeding lines

These are lines/populations evolved in different breeding programmes. They also contain valuable gene combinations. It includes, nearly homozygous lines, mutant lines, lines derived from biotechnology programme and, now, transgenic lines.

15.2.5 Wilds species/forms and wild relatives

Mostly cultivated crop species are directly derived from wild species. The *wild relatives* include all other species, which are related to the crop species descending during their evolution. Wild relatives are generally difficult to hybridize with their cultivated counterparts. Both, wild forms and wild relatives are sources of valuable genes for insect and disease resistance, tolerance to abiotic stresses like drought, cold, salinity etc., and even for quality traits and yield.

Wild species and putative ancestral forms of vegetable crops contain valuable genes that are of immense use in vegetable breeding programme using conventional methods or modern tools of biotechnology. These genetic resources can be utilized for the development of new cultivars, strains and hybrids and also in restructuring the existing ones that lack one or the other attribute. In the beginning of civilization, the first step in development of cultivated plant was their domestication, which is a process of bringing wild species under human management. Knowingly or unknowingly, the man started selection according to thier needs. Under domestication, crop species have changed considerably as compared to that of wild species. The domestication of wild species is still an ongoing process and is likely to continue till life exists on the earth. This is because the human needs are likely to change with time. Consequently, the wild species of little importance today may assume great significance tomorrow. With the development in science new breeding methods were invented to increase productivity and production to feed fast developing world.

15.3 CENTRES OF DIVERSITY

The region possessing maximum diversity at global levels is tropical America, tropical Asia and the Mediterranean. In the tropical Asian region, both, India and China hold the maximum diversity. Since long, Indian subcontinent is well known as an important centre of origin and diversity of a large number of vegetable crops. For the underutilized, neglected and less known domesticated/cultivated vegetables, this diversity is mainly confined to seven regions, namely the Chinese-Japanese, Indo-Chinese-Indonesian, Hindustani, Mediterranean, European-Siberian, African, and South American regions where approximately 495 species of the 540 species occur (Arora, 1985)

India is 'Primary centre' for crops like egg plant, smooth gourd, ridge gourd, cucumber etc. and secondary centre for cowpea, okra, chillies, pumpkin and several cole crops (Table1)

Table 1: Centre of diversity of major vegetable crops

Gene Centre	Primary Centre	Secondary Centre
Chinese-Japanese	Egg plant, wax gourd, chinese cabbage, kangkong, welsh onion	Watermelon, amaranth
Indo-Chinese	Wax gourd, sponge gourd, ridge gourd, bitter gourd, sword bean, winged bean, taro, chayote, cucumber, bottle gourd, yam	Chinese cabbage, bottle gourd, cucumber, yam, bean, amaranth, yard long bean, Kangkong
Hindustani centre	Eggplant, wax gourd, cucumber, ridge gourd, bitter gourd, sponge gourd, hyacinth bean, drumstick, okra	Watermelon, muskmelon, roselle, bottle gourd, amaranth
Central asia	Onion, garlic, carrot, spinach	Egg plant, watermelon, muskmelon, cauliflower
Near East	Onion, garlic, leek, beet	Okra
Mediterranean	White cabbage, cauliflower, water melon, broccoli, radish	Sweet pepper, garlic, okra
African	Egg plant, water melon, melon, bottle gourd, cowpea, okra, roselle, locust bean	Onion, shallot, lima bean, white cabbage amaranth
European-Siberian	Lettuce	Onion, white cabbage, common bean, cauliflower, spinach, carrot
Central America and Mexican region	Tomato, hot pepper, pumpkin, squash, yam bean, sweet potato, common bean	–
South American region	Tomato, hot pepper, cassava, pumpkin, lima bean, chayote, amaranth and sweet potato	Common bean

Table 2 : Distribution of major vegetable crops variability to different agro-ecological regions of India.

S. No.	Agro ecological regions	Geographical ranges	Variability in vegetable crops
1	Humid Western Himalayan region	J&K, H.P. and parts of U.P.	Cucurbits, radish, carrot, turnip, peas, cowpea, chillies, brinjal, okra, spinach, fenugreek, amaranth, *Solanum khasianum, S. hirsutum, Sechium edule, Basella rubra*
2	Humid Bengal/Assam basin	W.B. and Assam	Cucurbits, radish, cowpea, chillies, brinjal, okra, spinach, beet, *Abelmoschus manihot.*, amaranth, *Solanum indicum, S. khasianum, S. surattense, Cucumis sativus* var. *vikkimensis, Edgeria dargelingensis, Melothria assamica, Momordica cochinchinensis, Sechium edule, Tuladiantha coordifolia, Basella rubra*
3	Humid Eastern Himalayan region and Bay Lands	Arunchal Pradesh, Nagaland, Manipur, Mizoram, Tripura, Meghalaya, Andaman & Nicobar islands	Cucurbits, radish, cowpea, pea, chillies, brinjal, okra, spinach amaranth, *Abelmoschus* var. *pungens, Solanum khasianum, S. torvum, S. sisymbrifolium, S. ferox, S. verbasifolium, Cucumis hystrix, Luffa echinata, Sechium edule*
4	Sub-humid Sutlej Ganga Alluvial plains	Punjab, U.P. and Bihar	Cucurbits, radish, peas, brinjal, okra, spinach beet, fenugreek, onion, garlic, *Abelmoschus tetraphyllus* var. *tetraphyllus, A. tuberculatus, Solanum indicum, S. khasianum, S. torvum, S. surattense, S. hispidum, Cucumis hardwickii,*
5	Humid Eastern and South Eastern uplands	East M.P., Orissa and A.P.	Cucurbits, radish, carrot, cowpea, chillies, brinjal, okra, spinach, amaranth, garlic, *Abelmoschus manihot, Solanum surattense, S.torvum*
6	Arid Western plains	Haryana, Rajasthan and Gujrat	Cucurbits, cauliflower, radish, carrot, peas, cowpea, chillies, brinjal, okra, spinach beet, fenugreek, onion, garlic, amaranth, *Abelmoschus tuberculatus, A. ficuleneus, A. tetraphyllus, Solanum torvum, S. nigrum, Citrullus colocynthes*
7	Semi Arid Lava Plateau and Central Highlands	Maharasthra and Western M.P.	Cucurbits, cauliflower, radish, carrot, cowpea, chillies, brinjal, okra, spinach, fenugreek, amaranth, onion, *Solanum surattense, S. torvum, S. nigrum, S. khasianum, Cucumis setosus, Luffa acutangula* var. *acutangula*
8	Humid to Semi Arid Western Ghats and Karnataka Plateau	Karnataka, Tamilnadu, Kerala and Lakshad weep islands	Cucurbits, chillies, brinjal, okra, *Abelmoschus crinitus, Abelmoschus angulosus, A. ficulneus, A. moschatus, A.. tetraphyllus, Solanum trilobatum, S. indicum, S. incanum, Solanum pubescens, S. surattense, S. torvum, Luffa acutangula var. acutangula, Melothria angulata, Basella rubra*

15.4 GENETIC RESOURCES OF VEGETABLE CROPS IN THE INDIAN GENE CENTRE

The distribution pattern of the wild species in different botanical/phyto geographical regions and the areas of their concentration where rich diversity of wild species still continues to perpetuate, are of special significance for undertaking programmes on collection as well as for *in-situ* conservation of biodiversity. The important families possessing wild genetic diversity are Brassicaceae (*Brassica*), Malvaceae (*Abelmoschus*), Leguminoceae (*Canavalia, Lablab, Trigonella, Vigna*), Cucurbitaceae (*Citrullus, Coccinia, Luffa, Momordica, Neoluffa, Trichosanthes*), Solanaceae (*Solanum*), Amaranthaceae (A.), Dioscoreaceae (*Dioscorea*), Amaryllidaceae (*Allium*), Araceae (*Alocasia, Amorphophallus, Colocasia*). The bulb of greater galangal (*Alpinia galangal* family Zingiberaceae) can also be eaten raw. In the Garwhal areas, *Cornus capitata* and *C. ornus controversa* (family Cornaceae) are eaten raw. In Rajasthan, the whole plant of *Gisekia pharnaceoid*e (family Mulluginaceae) is widely consumed during food shortage but in the South and West (Deccan region) the leaves are used as greens, as are the leaves of *Glinus trianthemoides*.

15.4.1 Genetic resources of solanaceous vegetables

Tomato, brinjal and chillies are most important solanaceous vegetables, widely grown all over the country throughout the year for their edible fruits.

15.4.1.1 Tomato (solanum lycoprsicum)

Tomato is a warm season crop, sensitive to frost and typically cultivated in sub-tropical and mild cold climate. It is a native of South America and Mexico, however, domestication took place in Mexico (Baswell, 1949) and now tomato is grown all over the world. The greatest variability occurs in Mexico, Central America and Coastal Peru. The wild species of *Solanum (Lycopersicon)* have shown to be a valuable source of resistance to various diseases and pest, which is vital requirement in any breeding programme. *S. pimpinellifolium (Lycopersicon pimpinellifolium)* is resistance to *Fusarium* wilt, early blight and leaf curl disease and is a rich source of ascorbic acid; *S. cheesmanii (L. cheesmanii)* is resistance to salinity and a source of high total soluble solids (TSS); *S. chilense* (*L. chilense)* is tolerant to leaf curl disease; *S. chmielewskii (L. chmielewskii)* has high content of TSS; S. *habrochaites* (*Lycopersicon hirsutum)* is an excellent source of resistance to several pest and to leaf curl disease; *S. arcanum* (*L. peruvianum)* is resistant to root knot nematode, tobacco mosaic virus, leaf curl and has high ascorbic acid content and *L. chilense* is good source of drought resistance.

15.4.1.2 Brinjal (Solanum melongena L.)

Brinjal or egg plant being the crop of Indian Origin has developed some secondary variability in China (Vavilov, 1926). A large number of land races/ traditionally grown cultivars have been developed in different agro-ecological zones of India and these land races possess valuable genes for resistance to biotic and abiotic stresses and adaptation to various environments. Maximum diversity and distribution of *Solanum* species was noted in southern India, foothills of Himalaya and northeastern region. The widely distributed species in the region includes *S. torvum, S. indicum, S. insanum, S. surattense, S. pubescens* and *S. Khasianum.*

15.4.1.3 Chillies and bell pepper

Capsicum species (Chillies or pungent pepper and Bell pepper or *Simla mirch*; *Capsicum annuum* L.) are usually grown in warm to hot and humid climate throughout the country. The large-fruited bell pepper originated from tiny pungent and pointed fruited wild species *C. annuum.* It is considered to be the cool season crop and grown in sub-tropical and temperate regions. There are four other species (viz. *C. frutescens, C. chinense, C. baccatum and C. pubescens*), which are commonly grown in southern and Central America except *C. frutescens,* which is also grown in some parts of USA. Due to continuous selection, long history of cultivation and popularity of the crop, sufficient genetic variability has been generated and now India is treated as secondary centre of diversity for chillies. In hot chillies (*C. annuum* L.), rich variability in plant and fruit morphological traits occurs throughout the country particularly in south peninsular region, northeastern region, in foothills of Himalaya and Gangetic plains. The variability includes in plant type, fruit size/shape (long, short, pointed, smooth, wrinkled), bearing habit (fruits facing upwards, horizontal, downwards) and pungency in addition to annual and perennial types. However, in bell pepper less variability has been observed. Mostly introduced lines are grown particularly in mid hill of Himalaya and parts of hilly regions of Central India including Karnataka and Maharashtra.

15.4.2 Genetic resources of cucurbitaceous vegetable

15.4.2.1 Cucurbita species

The genus *Cucurbita* comprised of about 27 species, both wild and cultivated, mostly concentrated in the tropical and sub-tropical regions of Central and South America. In this genus, few species are of commercial importance viz. Field pumpkin (*C. moschata* Duch. Ex. Poir.), Winter squash *(C. maxima* Duch.), Summer squash or vegetable marrow (*C. pepo* L), Winter squash pumpkins (*C. mixta* Pang. syn. *C. angyrosperma*) and Malabar gourd or Fig leaf gourd *(C. ficifolia* Boucha). The vegetable Marrow (*C. pepo* L), is introduced crop which withstands in cooler climate. *C. texa* Gray grows wild in Texas (Chadha and Lal, 1993).

15.4.2.2 Cucumis species

The genus *Cucumis* comprises of about 26 species. The major crops of economic importance are cucumber (*C. sativus*), muskmelon (*C. melo*), snapmelon (*C. melo* var. *momordica*), longmelon (*C. melo* var. *utilissimus*). The Indian sub-continent is considered to be the centre of origin for *C. sativus* and centre of diversity for *C. melo* (Zeven and deWet, 1982). The wild species, *C. hardwickii* is found growing in natural habitats in the foothills of Himalayas. The free hybridization with cultivated *sativus*, with no reduction of fertility in F_2 generation suggested that *C. hardwickii* is a likely progenator of cultivated cucumber. The cucumber, *C. sativus* var. *sativus* has 3-5 lobed leaves, ovary usually 3 placentiferous with fruits oblong, obscurely trigonus or cylindric. However, *C. sativus* var. *sikkimensis* has 7-9 lobed leaves, ovary often 5 placentiferous, fruits ovoid-oblong, adapted in temperate and humid climate. Muskmelon (*C. melo*) based on the distribution of diversity can be grouped into following six sub-sets (Munger and Robinson, 1991)

15.4.2.2.1 *Cucumis melo* var. *agrestis (Kachri)*

It is Wild type with slender vines and small inedible fruits, probably synonym of *Cucumis melo var. callosus* (*C. callosus.*) and *C. melo var. trigonus* (*C. trigonus*).

15.4.2.2.2 *Cucumis melo* var. *cantaloupensis Naud. (Muskmelon or cantaloupe)*

Medium sized fruits with netted, warty or scaly surface, flesh usually orange but sometimes green, aromatic or musky flavour. Fruits dehiscent at maturity, usually andromonoecious.

15.4.2.2.3 *Cucumis melo* var. *flexuosus Naud. (Cucumis melo var. utilissumum)*

Snake melon, snake cucumber, Tar-kakari, fruits long and slender consumed at immature stage, monoecious.

15.4.2.2.4 *Cucumis melo* var. *momordica*

Snapmelon or phut is a monoecious crop grown in India and other Asian countries. It has white to pale orange, less sweet pulp. The smooth surface of the fruit starts cracking at the time of maturity.

15.4.2.2.5 *Cucumis melo* var. *conomon (sweet or pickling melon)*

It is generally andromonoecious in nature and bears small fruits with green skin, white flesh.

15.4.2.2.6 *Cucumis melo* var. *inodorous Naud. (Winter melon)*

The fruit is smooth or wrinkled surface with white or green flesh, and lacking musky odor. It is also andromonoecious in nature and usually requires more time for maturity.

Out of the six melons *C. melo* var. *agretis*, *C. melo* var. *cantaloupensis* Naud. (musk melon), *C. melo* var. *utilissumus*, *C. melo* var. *momordica*, and *C. melo* var. *flexuosus* Naud are available in Indian sub-continent.

15.4.2.3 Luffa species

The Indian gene centre has rich diversity in genetic resources of *Luffa* species. The genus comprised of 9 species in the world, of which 7 species (*L. acutangula, L.*

cylindric, L. echinata, L. graveolens, L. hermaphrodita, L. tuberosa and *L. umbellata*) is native to India. There is ambiguity with regards to *L. tuberosa* and *L. umbellate* because they are considered synonym to species of *Momordica* and *Cucurbita* respectively (Chadha and Lal, 1993). The sponge gourd (*L. cylindrica*) and ridge gourd (*L. acutangula*) has rich diversity in vine and fruit morphological characteristics throughout India particularly in northeastern region including Sikkim, West Bengal, Western, Central and Southern India. *L. hermaphrodita* considered to be originated from *L. graveolens*, is another potential species distributed in parts of north-central India. *L. acutangula* var. *amara* grows in peninsular India, is a wild relative of (*L. cylindrica*) and *L. echinata* in natural habitats is in Western Himalayas, central India and Gangetic plains.

15.4.2.4 Lagenaria species

Under genus *Lagenaria*, 6 species are reported, of which *L. abyssinica, L. siceraria*, and *L. leucantha* are common. Among them *L. siceraria* is generally cultivated in all tropical parts of the world, especially in India and few African countries. Remaining species are wild, perennial and diocious in nature. Two wild species,. *L. abyssinica*, and *L. bravifolia* are perennial in nature. There are suggestions that *Lagenaria* occurs in wild form in South America and in India (De Candolle 1982).

15.4.2.5 Citrullus species

The genus, *Citrullus* has two species of economic importance *C. lanatus* (Thunb.) (Watermelon) and *C. vulgaris* Schrad var.*fistulosus*) (round gourd). Based on the evidence (DeCandolle 1982) and linguistic data (Filov and Vilenskaya, 1972), *C. lanatus* is considered to be originated in Africa and reached India in pre-historic times. Cultivation of large watermelon *Citrullus lanatus* var. *citrodes*, is comparatively recent and Soviet varieties grown today still have shape of their African ancestors (*C. lanatus* var. *caffer*). Shimotsuma (1963) reported that *C. vulgaris, C. colocynthis, C. ecirrhosus* and *C. nandinianus* are related and cross compatible with each other. It is considered that *C. colocynthoides* is a probable ancestor of watermelon.

15.4.2.6 Trichosanthes species

Trichosanthes has 22 species and is reported to be of Indian origin/ Indo-Malayan region. Among these, *T. anguina L.* (snake gourd), *T. dioica Roxb.* (pointed gourd) are cultivated throughout the country. The major zone for distribution of diversity for *T. dioica* is north-central and northeastern India including West Bengal whereas for *T. anguina* is distributed throughout the country. However, the rich diversity in *T. anguina* has been observed in north-eastern states, West Bengal, Malabar Coast and in Eastern ghats in low and mid hills.

15.4.2.7 Momordica species

The genus *Momordica* has 60 species of which 7 species have been reported to occur in India. The cultivated species is *M. charantia(bitter gourd)* is grown all over the country in tropical and sub-tropical climats. *M. dioica* (Kartoli) grows all over West Bengal, Assam, parts of Bihar and adjoining area; *M. cochinchinensis* (kakrol or sweet gourd) is most popular particularly in Tripura, Assam and West Bengal. Occurence of other species like *M. cymbalaria* is confined mainly in Western Ghats, Maharashtra, southwards and eastern peninsular tract, *M. denudate* in peninsular tract, *M. macrophylla* in northeastern region and *M. subangulata* in North-eastern hills, Eastern Ghats and Deccan penninsula.

15.4.2.8 Benincasa species

Benincasa hispida Cogn. known as ash gourd, white gourd or wax gourd, is reported to be a native of Java and Japan. It was domesticated in India during pre-historic time. It is widely grown all over the country in tropical and sub-tropical climates and possesses variability in fruit morphological characteristics and quality.

15.4.2.9 Coccinia species

Coccinia has about 35 species distributed in tropical Africa and Asia. Out of these, only one species i.e. *C. grandis* is under cultivation. The related species are *C. histella and C. sessilifolia,* which are widely cultivated in several countries like Africa, Central America, China, Malaysia, Australia and other tropical Asian countries (Pier, 2001).

15.4.2.10 Sechium species

Genus *Sechium* is said to be originated in mountainous region of America and Mexico. It includes *Microsechium compositum, Microsechium gintonii, S. compositum, S. edule, S. hintoni* and *S. jamaicense.* Principally it was confined to Tropical and sub-tropical climates and particuarly in mid hill conditions. Maximum variability occurs in Sikkim. In Meghalaya and Mizoram it is grown on commercial scale and is popular in Darjeeling Hills, also.

15.4.3 Leafy vegetables

The Indian sub-continent is considered to be the place of origin for a number of leafy vegetables like *Beta vulgaris* var *bengalensis*, *Basella rubra*, *Cichorium endivia*, *Enhyndra fluctuant*, *Lactuca indica*, etc. Most of the leafy vegetables possess good amount of diversity, which are acclimatized under extreme variation in altitude with long history of its cultivation. The wild relatives of different leafy vegetables possess valuable gene particularly for resistance to biotic and abiotic stress conditions.

15.4.3.1 Beta species

Beta species includes *Beta adanensis, B. altissima, B. atriplicifolia, B. bengalensis, B. bourgaei, B. brasiliensis, B. campanulata, B. chilensis, B. cicla, B. corolliflora, B. lomatogona, B. macrocarpa, B. macrorhiza, B. maritime, B. maritima subsp. danica, B. maritima* var. *atriplicifolia, B. maritima* var. *erecta, B. maritima* var. *prostrata, B. nana, B. orientalis, B. palonga, B. patellaris, B. patula, B. perennis, B. procumbens, B. trigyna,*

B. trojana, B. vulgaris, B. vulgaris cv. conditiva, *B. vulgaris* cv. saccharifera, *B. vulgaris* f. *rhodopleura, B. hybrida, B. vulgaris* subsp. *adanensis, B. vulgaris* subsp. *cicla, B. vulgaris* subsp. *flavescens, B. vulgaris* subsp. *lomatogonoides, B. vulgaris* var. *glabra, B. vulgaris* var. *cicla, B. vulgaris* var. *crassa, B. vulgaris* var. *atriplicifolia, B. vulgaris* subsp. *macrocarpa, B. vulgaris* var. *altissima, B. vulgaris* subsp. *vulgaris, B. vulgaris* subsp. *provulgaris, B. vulgaris* subsp. *patula, B. vulgaris* subsp. *maritime, B. vulgaris* subsp. *orientalis, B. vulgaris* var. *erecta, B. vulgaris* var. *flavescens, B. vulgaris* var. *foliosa, B. vulgaris* var. *grisea, B. vulgaris* var. *macrocarpa, B. vulgaris* var. *maritime, B. vulgaris* var. *orientalis, B. vulgaris* var. *perennis B. vulgaris* var. *pilosa, B. vulgaris* var. *prostrata, B. webbiana, B. x intermedia and B. vulgaris* var. *trojana*. The spinach beet also known as Indian spinach beet (*B. vulgaris* var. *bengalensis* Hort.) is a close relative of beet root. It is grown in the plains and hills of India from Punjab to northeastern region including West Bengal, Orissa, Madhya Pradesh., Rajasthan, Gujarat and Maharashtra.

15.4.3.2 Amaranthus species

The origin of various species of cultivated amaranth is complicated because the wild ancestors are pan tropical cosmopolitan weeds (Mohideen and Irulappan, 1993). *Amaranthus spinosus, A. hybridus* and *A. dubius* are tropical types. *Amaranthus retroflexus, A. viridis, A. lividus* and *A. graecizans* are more temperate hot season weeds. In all probably, it might have originated in India and have spread to neighboring countries by traders, buddhist monks and others. In India, a number of domesticated forms are available all over the country especially in northeastern states, West Bengal, Orissa, Tamil Nadu, Andhra Pradesh, Karnataka and Kerala. Distribution pattern of *A. blitum* (weed throughout India), *A. caudatus* (wild throughout India), *A. gangeticus* (weed warmer parts of India, variants as *A. mangostanus*), *A. paniculatus* (warmer parts of India), and *A. spinosus* (weed throughout India) are reported. Wide ranges of variability in plant morphological characteristics are available in different parts of the country. In Orissa and northeastern states a special type is grown where stem is consumed after full growth and different genotypes are grown in different parts of the year.

15.4.3.3 Spinacia species

Genus *Spinacia* belong to the family chenopodiaceae and said to originate in southwest Asia. There is well description about four types of plants with reference to sex expression, which are: (i) extreme males, (ii) vegetative males, (iii) monoecious, and (iv) females. The monoecious plants may predominantly staminate, pistillate or pure pistillate early but with some staminate flower late or almost equally staminate or pistillate throughout the season. The greater variability can be spotted in the plains and in foothills. Variability occurs in plant morphological characteristics, seed type and foliage yield attributing component.

15.4.3.4 Chenopodium species

Genus *Chenopodium* belong to the family Chenopodiaceae. *Chenopodiaceae album* as found as weed in plains and *C. murale* in hills upto 3000m. In both the species maximum variability can be observed in upper Gangetic plains, extending to northern hills; elsewhere only sporadic distribution.

15.4.3.4 Basella species

Climbing spinach, Malabar spinach or *Poi* is grown all over the country particularly in the plains and mid-hills during hot humid season The genus *Basella* includes *B. alba, B. cordifolia, B. rubra, B. tuberosa* and *B. vesicaria*. Rich variability is found in plant morphological characteristics including stem fleshiness, leaf shape and size in South India, northeastern region including Assam, West Bengal and Orissa

15.4.3.5 Trigonella species

The genus *Trigonella* is having 22 species, which are having different chromosomes numbers. The 15 species having 2n=16 chromosomes includes *T. anguina, T. arabica, T. bolansae, T. calliceras, T. coerulea, T. corniculata, T. cretica, T. foenum-graecum, T. gladiota, T. glomerate, T. gordej, T. hamosa, T. lypokyi, T. monspeliach, T. radiata:* three species have 2n=18 (*T. stellata, T. striata, T. ornithopodioides*): two species have 2n=44, (*T. geminiflora, T. grandiflora*) and another species have 2n= 44 (*T. polycerata*). Two species of *Trigonella* are of economic importance. These are *T. foenum-graecum* L., the common *methi* and the other is *T. corniculata Kasuri or Champa methi*. It is found growing wild in North Western parts of India (Bailey, 1950)

15.4.3.6 Portulaca species

Parslane (*Portulaca oleracea* L.) is said to be indigenous to Himalaya. It occurs in several forms during hot weather, in the cultivated and wild state. The cultivated forms have broad leaves whereas in wild; both broad and small leaves types are seen growing. It is tiny creeper with small, succulent fleshy leaves and tender stems, which is cooked like spinach.

15.4.3.7 Brassica species

Chinese cabbage (*Brassica pekinensis* and *Brassica chinensis*) is indigenous to China and eastern Asia where it has been in cultivation since 5th century. There are two types of Chinese cabbage-one that heading type (*Brassica pekinensis*) and other open leaf type (*Brassica chinensis)*. The former forms an erect, moderately compact, usually cylindrical heads, while the later develops clusters of succulent leaves. Rich diversity in plant type occurs in foot hills of Himalaya and northeastern states including Sikkim.

15.4.4 Allium species

The Alliums includes onion, garlic and their relatives like leeks, shallots and chives that are considered prized vegetables due to food, medicine or religious purposes. The genus *Allium* is represented by about 750 species widely distributed over the warm -

temperate and temperate zones of the northern hemisphere. Regions of high species diversity occur in Turkey, Iran, North Iraq, Afghanistan, Soviet, middle Asia and West Pakistan and a second less pronounced centre of species diversity occurs in Western North America. About 40 species of *Allium* are reported in India of which seven are cultivated, the main being onion and garlic. Much diversity has built up in India in these crops as they have been cultivated from ancient times and selections have been made in different agro-ecological zones. Both these crops are known only in cultivation and while no known wild ancester is known for *A. cepa, A. longicuspis,* endemic to central Asia, has been proposed as a progenitor for *A. sativum*. Within *A. cepa* two horticultural varieties are recognized: the dry bulb onion and the multiplier onion (shallots). In *A. sativum* despite it being propagated exclusively by the division of ground cloves or aerial bulbils, a wide range of variability is available in several traits. The leeks (*A. ampeloprasum*) succeed fairly well in India thriving at elevations above 300m and often remain fit for use until the end of rainy season. The chives (*A. shoennoprasum*) are distributed in the northern Himalayas from Kashmir to Kumaon. Cultivated and wild types interbreed and there is plenty of variation in cultivated forms. Attention is being given on these and other wild species, as they possess many disease resistance genes potentially useful in improvement programme. *A. ampeloprasum* for white rot resistance, leek yellow stripe virus, leek rust; *A. roylei* for downey mildew; *A. fistulosum* for pink rot, smut and onion fly, and *A. tuberosum* for *Fusarium oxysporum* and leek moth are particularly noteworthy.

15.4.5 Raphanus species

Genus *Raphanus* includes *R. caudatus, R. landra, R. lyratus, R. pterocarpus, R. raphanistrum, R. raphanistrum* subsp. *landra, R. sativus, R. sativus* var. *caudatus, R. sativus* var. *longipinnatus, R. sativus* var. *niger, R. sativus* var. *oleiferus, R. sativus* var. *oleiformis, R. sativus* var. *sativus, R. violaceus* and *Brassicoraphanus* species.

15.4.6 Daucus species

Genus includes several species like *D. aureus, D. bicolor, D. bocconei, D. brachiatus, D. broteri, D. capillifolius, D. carota,, D. carota* cv. *atrorubens, D. carota* subsp. *Carota, D. carota* subsp. *commutatus, D. carota* subsp. *drepanensis, D. carota* subsp. *gadecaei, D. carota* subsp. *gummifer, D. carota* subsp. *hispanicus, D. carota* subsp. *hispidus, D. carota* subsp. *maritimus, D. carota* subsp. *maximus, D. carota* var. *atrorubens, D. carota* var. *boissieri, D. carota* var. *commutatus, D. carota* var. *sativus, D. crinitus, D. durieua, D. gingidium,* etc. In these genus two types i.e., Asiatic and European types are more popular.

15.4.7 Abelmoschus species

The centre of diversity of *Abelmoschus* includes West Africa (Benin, Togo, and Guinea), India, and southeast Asian countries, i.e., Burma, Indochina, Indonesia, and Thailand Van Borssum Waalkes, 1966; Siemonsma, 1982a, b, Hamon and Hamon, Bisht *et al.*, 1995). *A. esculentus* is the commercially cultivated species of okra. However, in West Africa *A. caillei* is also cultivated for its fresh fruit. The geographical distribution of *A. caillei* is from northern Guinean extending to southeast Cameroon. *A. esculentus* is cultivated in most humid parts of Guinea, mostly in an intercropped system with *A. caillei*. The genus *Abelmoschus* comprises nine species (Anon, 1991) and amomg them *A. moschatus* and *A. manihot* are semi wild, show a greater diversity than the cultivated forms (Hamon *et al.*, 1991) *A. moschatus* is morphologically and genetically the most different from other species (Hamon and Yapo, 1986) with probably the widest geographical distribution, which extends from the South Pacific Islands through Indochina and India and over to the central and western parts of Africa. The cultivated form (*A. moschatus* var. *moschatus*), better known as "*musk mallow*," "*Jew's mallow*," or "*ambrette*" in Africa, Asia and America has fragrant seeds used for making perfume. The subspecies *biakensis* is endemic to Papua New Guinea (Charrier, 1984). The cultivation of *A. manihot* is mainly in south-east Asia and is known under the vernacular name "*qibika*", which is also found in India, northern Australia and less frequently in the American continent and tropical West Africa. Its cultivation is exclusively for leaf consumption. The species *A. tetraphyllus* (Anon, 1991) *A. ficulneus*, *A. crinitus*, *A. angulosus* (Van Borssum Waalkes, 1966), and *A. tuberculatus* (Pal *et al.*, 1952) are truly wild. *A. ficalneus* is found from the northern parts of Australia to south-east Asia and from India to most parts of Africa. *A. tetraphyllus, A. crinitus, A. angulosus* and *A. tuberculatus* are exclusively of Asian origin. *A. tetraphyllus* var. *tetraphyllus* and var. *pungens* are endemic to south-east Asia with extension of the former to Papua New Guinea and New Ireland. *A. crinitus* is found at low altitudes in India and south-east Asia and *A. angulosus* at high altitudes in India, Sri Lanka, Indochina, and Indonesia. *A. tuberculatus* is endemic to the northern and western parts of India (Charrier, 1984; Velayudhan and Upadhyay, 1994).

Based on different floras, floristic accounts and other published evidence expressed main areas of concentration of wild species are as follows:

- Semi-arid tracts of north and northwestern India are for *A. ficulneus* and *A. tuberculatus.*
- *Tarai* range and the foothills of the Himalayas are for *A. crinitus, A. manihot* and *A. tetraphyllus or A. tetraphyllus* var. *pungens.*
- Western and Eastern Ghats and also peninsular tracts are for *A. manihot* (including *tetraphyllus* type), *A. angulosus* mainly confined to hilly tracts up to 2000 m in south India with only meager distribution elsewhere.
- Northeastern regions are for *A. crinitis* and *A. manihot*, mostly var. *pungens.*

In some of the wild taxa, intra-specific variation exists and has been taxonomically identified as in the *A. manihot* var. *tetraphyllus* and *pungens*, in *A. angulosus* var. *purpureus* and *grandiflorus*, in *A. moschatus* var. *biakensis* and *rugosus*, and in *A. tuberculatus* var. *deltoidefolius*.

Table 3 : Promising Germplasm of Vegetable Crops registered at NBPGR.

Botanical name	Crop	National identity no.	Institute	Unique Features
Abelmoschus esculentus	Okra	IC526802 (INGR 5026)	IIVR, Varanasi	New plant type (Dwarf, bushy with short internodal length)
		IC565533 (INGR 9125)	IIVR, Varanasi	Thin (0.9 cm) and long fruited (26 cm)
Canavalia gladiate	Sword been	IC427811 (INGR4056)	CIAH, Bikaner	Drought tolerant, alternative source of vegetable for arid zone.
Capsicum annum	Chilli	IC 395318 (INGR 3077)	IIHR, Bangalore	CMS
		IC 296662 (INGR 4052)	IIHR, Bangalore	CGMS line with good general combining ability (GCA)
		IC 296664 (INGR 4053)	IIHR, Bangalore	CGMS line with good GCA in different genetic background.
		IC 296666 (INGR 4054)	IIHR, Bangalore	Fertility restorer line of MS1 and MS2
		IC 296667 (INGR 4055)	IIHR, Bangalore	Fertility restorer line.
		IC 399066 (INGR 3092)	DRL, DRDO, Tezpur	High pungency
C. annum var. grossum	Chilli/ sweet pepper	IC 296760 (INGR)	SKUAST, Srinagar	Early, high yielding (207.95 q/ha), export quality, superior shelf life with high nutritive value.
C. annum	Chilli	IC 526794 (INGR 5024)	IIHR, Bangalore	Fertility restorer
		IC565015 (INGR8095)	NBPGR, RS Hyderabad	Resistant to thrips and mites
		IC505489 (INGR8097)	NBPGR, RS Hyderabad	Resistant to thrips and powdery mildew
		IC569194 (INGR9029)	Navaneeta Evergreens, Anandapuram Mandal, Tarulawada, Visakhapatnam District Andhra Pradesh	Erect cluster bearing de-stalking nature of fruits, low pungency and bright colour.
C. frutescens		IC553284 (INGR7039)	IIVR, Varanasi	Resistant to leaf curl virus disease
Cucumis callosus	Kachri	IC299696 (INGR98014)	CIAH Bikaner	High yielding, drought hardy with large fruits, suited for italicize
C. melo	Muskmelon	IC557426 (INGR8043)	PAU, Ludhiana	Fruit wall round, light brown
		IC557706 (INGR8043)	IARI, New Delhi	Source of resistance to Cucumber Green Mottle Mosaic Virus (CGMMV) and Downy mildew
C. sativus	Cucumber	IC296699 (INGR98017)	CIAH Bikaner	High yielding and long fruit
		IC296700 (INGR98018)	CIAH Bikaner	High yielding, small fruit, drought hardy and high temperature insensitive

Unique traits of Germplasm of Vegetable Crops registered at NBPGR.

Crop	Line	No.	Unique traits
Pointed gourd	IIVR PG-105	INGR-03035	Seedless fruits, obligate parthenocarpic, long duration fruiting (IIVR, Varanasi)
Bitter gourd	GY-63	INGR-03037	Gynoecious line with high yielding ability and attractive fruits (IIVR, Varanasi)
Watermelon	RW-187-2	INGR-01037	High yielding ability with yellow coloured flesh (ARS, Durgapura, Jaipur)
	RW-177-2	INGR-01038	Leaf mutant with simple unlobed leaves (ARS, Durgapura, Jaipur)
Bottle gourd	Androman-6	INGR-99009	Andromonoecious sex form of bottle gourd (NDUA&T, Faizabad)
	PBOG-54	INGR-99022	Segmented leaf genotype
	Cut Leaf		GB Pant Univ. of Agril. & Tech., Pantnagar
Potato	MP/99-322	INGR 04109	High starch and dry matter, low amylose and resistant to late blight
Chillies	MS1 and B line	INGR-04052	CGMS line with good GCA (IIHR, Bangalore)
	MS3 and B line	INGR-04053	CGMS line with good GCA (IIHR, Bangalore)
	IHR14 (PMR-14)	INGR-04054	Fertlity restorer line (IIHR, Bangalore)
	IHR-11-2 (IHR11#2)	INGR-04055	Fertlity restorer line (IIHR, Bangalore)
	MS-2A & 2B	INGR-03077	CMS line
	Capsicum Selection-2 (Nishat-1)	INGR-00001	High yielding with early maturity, export quality, superior shelf life and high nutritive value
	DRLT-1107	INGR-03092	High pungency
Brinjal	Hisar Jamuni (H-9)	INGR-99037	Intermediate, semi-spreading, medium-tall, thornless, better ratoon for spring crop, fruit colour is retained for a long time, less seed content, serves purpose of long and round types.
	IIHR-3 (96-2-1)	INGR-03074	Resistance to bacterial wilt
Cucumber	AHC-2	INGR-98017	High yielding and long fruited types
	AHC-13	INGR-98018	High yielding, unique, small fruited types, drought hardy, bear fruits even under high temperature in arid region with limited irrigation
Kachari	Kachari AHK-119	INGR-98013	High yielding and drought hardy
Fenugreek (Methi)	Hissar Methi-346	INGR-01012	Downy mildew resistant genotype with quick germination, faster initial growth, long pod, bold seed with green tan seed coat color and light green narrow leaves
	Hissar Methi-350	INGR-01012	Powdery mildew resistant and Downy mildew tolerant genotype. Plant is tall erect; multi

contd.

			branched with whitish green stems and dark green narrow leaves.
Peas	PMR-27	INGR-03078	Tendril type with resistant to powdery mildew
	DPP-62	INGR-03079	Yellow wrinkled seeds with resistance to powdery mildew
Round melon	HT-10	INGR-99038	Intermediate semi-spreading vine type round melon, tolerant to downy mildew and root rot wilt complex.
Snapmelon	AHS-10	INGR-98015	High yielding and drought hardy
	AHS-82	INGR-98016	High yielding and drought hardy
Tomato	Hisar Lalit	INGR-03036	Resistant to root knot nematode
	IIHR-2195	INGR-03076	Resistant to tomato leaf curl virus and bacterial wilt
Soybean	P-1366	INGR-01035	Vegetable type

With the advent of the regime of ownership and IPR in relation to PGR, it has become imperative to develop mechanisms to ensure that developers of plant get due credit. Similarly, there is no mechanism to recognize scientists associated with the development of improved germplasm/genetic stocks (new sources of resistance, male sterility, varied types of mutants, cytogenetic stocks. etc.). Lack of formal recognition of such useful materials and the role of scientist in development of these materials, discourages them from sharing the valuable materials with other workers. Consequently, most of such valuable material remains underutilized or is lost.

15.5 UTILIZATION OF GENETIC RESOURCES

The modern plant breeding methods have their bases in the genetic and cytogenetic principles. The science of genetics began with the rediscovery of Gregor John Mendel's paper in 1900 which was originally published in1866. Mendel's laws of inheritance provided the foundation of crop improvement. Realization that the chromosomes are the carriers of genes has led to the development of specialized plant breeding methods of chromosome engineering. A noteworthy development resulted from the studies of G.H. Shull on maize, who found that inbreeding produced a considerable loss of vigour, and put forth the idea of hybridization and coined the term heterosis. Selection of desirable plants, especially during F2 and subsequent generation is largely dependent on the skill and judgment of the breeder. However, the goal of all plant breeding activities has been always to develop an improved crop variety which is utilized by farmers for commercial cultivation. The desired changes in genotypes of crop species and consequent benefits to farmers are brought about by a series of interrelated and largely interdependent activities. These activities include viz., creation of variation, selection, evaluation multiplication and distribution. However, selection is effective in altering a species only when genetic variability exists in the populations of the species.

Mendelian variation (generated mainly by gene mutation), site-specific hybridization, and polyploidy are the three major ways to bring genetic variability in various crop species. The other developments in the science of crop breeding are totipotency and recombinant DNA technology which have enabled the transfer of desirable genes from any source into plant system.

Diseases, insect pests, and high temperature are threats to vegetable growers. This, in turn, reduces the availability of vegetables for consumers. Natural genetic resistance is the most effective and environment friendly solution. Genes from wild plants are introduced into crops through crossing techniques. Sometimes, these hybrids are multiplied via tissue culture. Plants with resistance genes are identified through molecular markers or other disease screening tests. Resistance is combined with other desirable horticultural traits such as high yield. The whole process takes about 5-10 years depending upon the traits to be incorporated. The prized genetic materials along with their wild and close relatives are used by plant breeders for development of resistant varieties/ hybrids. They also constitute a priceless reservoir that contains genes conferring better adaptation to stress environments and resistance to diseases and pests.

Table 4 : List and description of selected genotypes utilized in crop improvement programme of *Capsicum* spp. at IIVR Varanasi.

S.N.	Genotype (sp.)*	Remarks, especially on origin/pedigree, degree of pungency, reaction to leaf curl
1.	Bhut Jolokia (*f* x *c*)	Interspecific derivative, landrace of north east India, world's hottest known so far , highly resistant to leaf curl (HR), partial restorer
2.	BS-35 (*f* x *c*)	Interspecific derivative, landrace of north east India, very hot leaf curl symptomless, partial restorer
3.	COO-309 (*f*)	Introduced from AVRDC, originally collected from Taiwan, highly pungent, HR, partial restorer, resistant to tomato etch virus
4.	Lankamura Tripura Collection (*f* x *c*)	Interspecific derivative, landrace of north east India, highly pungent, leaf curl resistant (R) partial restorer
5.	CO-5635 (*b*)	Introduced from AVRDC, originally collected from Brazil, non-pungent, susceptible to leaf curl (S), non-crossable
6.	COO-304 (*c*)	Introduced from AVRDC, originally collected from the United States, mild pungent, highly resistant to leaf curl (HR), resistant to tomato etch virus, non-crossable
7.	IC-383072 (*f*)	Collected by NBPGR, India, highly pungent, HR, partial restorer
8.	GKC-29 (*a* x *c*)	Interspecific derivative, landrace of north east India, highly pungent, SL, non-crossable
9.	Punjab Lal (*a*)	Improved variety developed by PAU, India, pungent, R, restorer
10.	NMCA-40008 (f)	Introduced from NMSU, USA, highly pungent, Tabasco, R, partial restorer
11.	NMCA-50003 (*a*)	Introduced from NMSU, USA, non-pungent, Yellow Wonder, moderately susceptible to leaf curl (MS), maintainer
12.	AMK-11 (*a*)	Collection from Dharwad, India, pungent, S, fruit Mulakka type, restorer
13.	ISC-9 (*a*)	Introduction from Israel, pungent, MS, partial restorer
14.	EC-519636 (*a*)	Introduced from AVRDC, mild pungent, S, restorer
15.	Sel-4 (*a*)	Developed at Solan, India, improved sweet pepper line (non-pungent), S, maintainer
16.	Solan Local (*a*)	Developed at Solan, India, improved sweet pepper line (non-pungent), S, maintainer
17.	AKC-89/38 (*a*)	Developed at PDKV, Akola, India, round fruited, upright bearing, pungent, MS, restorer

contd.

18.	DSL-2 (*a*)	Collected from north eastern region, pungent, MS, restorer
19.	VR-339 (*a*)	Developed by IIVR, Varanasi, India, pungent and MS, restorer
20.	0337-7545 (*a*)	AVRDC, pungent, MS, restorer
21.	IVPBC-535 (*a*)	Introduced from AVRDC, purified and identified for oleoresin extraction at IIVR, non-pungent, S, resistant to PVY, BW, partial restorer
22.	VR-338 (*a*)	Developed by IIVR, mild pungent, S, restorer
23.	Japani Longi (*a*)	Improved population from IIVR, pungent, moderately susceptible to leaf curl
24.	Taiwan-2 (*a*)	Introduced from AVRDC, highly pungent, MR, restorer
25.	KTPL-19 (*a*)	Introduced from Spain at IARI, non-pungent, suitable for oleoresin extraction, MR, partial restorer
26.	NuMex Twilight (*a*)	Introduced from NMSU, USA, ornamental type, mild pungent, S, restorer
27.	NuMex Senteniales (*a*)	Introduced from NMSU, USA, ornamental type, mild pungent, S, restorer
28.	DC-16 (*a*)	Collected from north east India, purple fruit, pungent, MS, restorer
29.	CM-334 (*a*)	Mexican landrace, pungent, resistant to PVY, PeMV, *Phytophthora* and nematodes, MS, restorer
30.	Pant C-1 (*a*)	Developed at Pantnagar, India, pungent, MS, strong fertility restorer
31.	Pusa Jwala (*a*)	Improved variety developed at IARI, India, very common in northern India, pungent, S, restorer
32.	PKM-51 (*a*)	Improved variety developed at Periyakullam, India, pungent, moderately resistant to leaf curl (MR), restorer
33.	PDC-50 (*a*)	Local collection, pungent, MS, restorer
34.	LCA-235 (*a*)	Variety developed at Lam, India, pungent, MS, restorer
35.	California Wonder (*a*)	Originally collected from IARI, popular sweet pepper variety, HS, fertility maintainer
36.	Local Tripura (*a*)	Landrace collected from north east India, pungent, MS, restorer
37.	TC-6903 (*a*)	Introduced from AVRDC, originated in Mexico, non-pungent, S, maintainer
38.	EC-119457 (*a*)	Introduced from Taiwan, non-pungent, MR, restorer
39.	MS-12 (*a*)	Nuclear genic male sterile line in the background of Punjab Lal, developed by PAU, India, pungent, MR
40.	KDCS-810 (*a*)	Improved variety developed in Kalyanpur, India, pungent, MR, restorer
41.	EC-458206 (*a*)	Introduced from AVRDC, sweet pepper, HS, maintainer
42.	NCCH-143 (*a*)	Developed at Nirmal Seed Company, India, Sweet pepper hybrid, MS
43.	No. 8 (*a*)	Developed at Solan, India, sweet pepper, S, maintainer
44.	CCA-4261 (*a*)	Introduced from AVRDC, pungent, S, cytoplasmic male sterile (CMS) in Tumpang (Korean) back ground
45.	PBC-473 (*a*)	Introduced from AVRDC, pungent, S, Restorer of CMS
46.	AC Assam (*a*)	Collection from north east India, pungent, MS, restorer
47.	Kashi Anmol (*a*)	Improved variety developed by IIVR, India, pedigree: Pant C1 x MI-2, pungent, MS, restorer
48.	9915-15 (*a*)	Introduced from AVRDC, non-pungent, MR, restorer

- a = annuum, b = baccatum, c = chinense, f = frutescens;
- SL: symptomless
- MS:moderately susceptible
- MR: moderately resistant
- S: susceptible
- PeMV: pepper mottled virus
- NMSU: New Mexico State University

15.5.1 Cucurbits

Table 5 : Some promising recent introductions in cucurbits.

Crop / Vegetable	Accession number	Salient features	Country
High yielding and desirable fruit type			
Cucumber	EC 398030	Early and determinate type	China
Watermelon	EC 380989 to 92	Yellow flesh /early maturity	USA
	EC 393240 to 43	Red flash/round and small	USA
Summer squash	EC-380995	Unique, small, oblong early maturing, the flesh comes out in strings, can be baked or used in breads	USA
	EC 380996	Hull-less seeded pumpkin; seeds can be popped and eaten like snack, high in nutrition and good source of Zinc.	USA

Table 6 : Male sterile/ restorer lines and hybrids

Cucumber	EC 382737 to 39	Gynoecious lines	USA
	EC 329300	Gynoecious (G4)	USA
Muskmelon	EC 348140 to 43	F1 Hybrid (Early and Sweet)	USA
	EC 382726 to 36	Male sterile gene (Ms2 to Ms5)	France
Germplasm for abiotic resistance			
Muskmelon	EC350603	Sulphur	USA
	EC 342751-54	Heat	Japan
Germplasm for biotic resistance			
Watermelon	EC 383809	Fusarium wilt	USA
	EC 378523-24	Fusarium wilt	USA
	EC 333659-69	Wilt	USA
	EC 217073-74	Fusarium & Gummy stem blight	USA
	EC-382753	Breeding line SSDL resistant to Fusarium wilt and anthracnose	USA
	EC 402549	Multiple disease resistant to race-2, anthracnose, Fusarium wilt and gummy stem blight with early maturity	USA
Muskmelon	EC 350602	Aphid	USA
	EC 350603	WMB	USA
	EC 330956	Multiple disease	USA
	EC 399866 and EC 399212	Resistant to downy mildew, Powdery mildew and CMV	UK

contd.

Cucumber	EC 320526	Multiple diseases	USA
	EC-399914 to 37	Resistant to CMV, downy mildew	UK
	EC-398966 and 67	Resistant to angular leaf spot	USA
	EC- 398968 to 70	Resistant to anthracnose	USA
	EC- 398971 to 73	Resistant to fruit rot	USA
	EC- 398974 to 90	Resistant to downy mildew	USA
	EC- 398991 to 9007	Resistant to leaf spot	USA
Other Cucumis spp.	EC-382500 to 69	Nematode resistance	USA

Table 7 : Direct introductions of cucurbits commercialized in India.

Crop	Introduced variety	Origin
Watermelon	Asahi yamato	Japan
	Sugar Baby	USA
	New Hampshire Midget	USA
	Improved Shipper	USA
	Dexielce	USA
Cucumber	Japanese Long Green	Japan
	Straight eight	USA
	Poinsette	USA
Summer squash	Australian Green	Australia
	Patty pan	USA

Table 8 : Major biotic stresses and their sources of resistance.

Crop/ biotic stress	Resistance sources
Muskmelon	
Powdery mildew	PMR-45, PMR-450, PMR-5, PMR-6, PI-124111
Downey mildew	MR-1, PI 414723, DMDR-1, DMDR-2
CGMMV	DVRM-1, 2, *Cucumis africanus, Cucumis ficifolium, Cucumis anguria*
CMV	PI 161375
WMV	PI 414723
Fruit fly	*Cucumis callosus*
IPM, DM, MMV	MR-12
Nematode	*Cucumis metuliferus*
White fly	*Cucumisasper, Cucumis denteri, cucumis dipsaceus, Cucumis sagittatus*

contd.

Watermelon	
Fusarium wilt	Summitte, Conqueror, Charlistm gray, Dixilee, Crimson sweet
Anthracnose	Fair, Charleston gray, congo, PI 189225
Bottle gourd	
CMV, SMV, WMV	PI 271353
Fusarium wilt	Taiwan variety Renshi
Cucumber	
Anthracnose	PI 175111, PI 175120, PI 179676, PI 182445, wise 2757 (USA)
Downey mildew	B-184, B159, wise 2757 (USA)
Powdery mildew	PI 200815, PI 200818, *Cucumis hardwikkii*, wise 2757 (USA)
CMV	Wisc SMR-12, SMR-15, SMR-18, wise 2757 (USA)
Pumpkin	
PM and Viruses	*Cucurbita lundelliana, Cucrbita martenezii*
ZYMV, WMV	*C. ecuadorensis, C. faetidistima, Cucrbita martenezii*

DM-downy mildew; MM- muskmelon mosaic virus; CGMMV- cucurbits green motel mosaic virus; CMV-cucumber mosaic virus; WMV- watermelon mosaic virus; SMV-squash mosaic virus

15.6 SUCCESS STORIES

Tomato : Tomato has been exploited as a classical experimental material for research. Development of resistant lines using resistant sources from different species of tomato could be an ecofriendly and cost-effective approach to minimize losses caused by various harmful organisms. Resistance genes from different tomato progenitors, *S. chilense, S. peruvianum,S. pimpinellifolium* and *S. herbrochaites* have been extensively used for the development of resistant lines against TYLCV. *Solanum pimpinellifolium* is being used to develop resistant varieties against bacterial wilt. The genotype, Hisar Anmol (H-24; moderately resistant to ToLCV) has been developed by inter-specific crosses *L. hirsutum f. glabratum* line B6013 and *L. esculentum*. A number of varieties resistant to wilt viz., Utkal Pallavi, Utkal Kumari, Utkal Urbashi, Utkal Pragyan Mukthi, Swarna Baibhaw, Swarna Naween, Swarna Lalima, Swarna Sampada has been developed for commercial cultivation. The varieties Arka Vardhan and Hisar Lalit resistant to nematode have been developed. Besides, multiple disease resistant varieties Arka Ananya (ToLCV and bacterial wilt), Arka Samrat (ToLCV, BW and early blight) and Arka Rashak F1 hybrid (ToLCV, BW and early blight) have been developed. Other advances in tomato crop improvement have been the synthesis of various pre-breds superior tomato cultivars in which substituted whole chromosome or substituted chromosome segments have been added from wild species. Breeders can select pre-breds with desired traits for attaining their goal.

Since 1940, tomato breeders have relied heavily on wild species as important plant genetic resources for economically important traits. At present resistance introgressions for more than 40 diseases have been achieved using wild tomato resources. In addition, tolerances to environmental stresses as heat, cold, drought, submergence and salinity have been ascertained. Transfers of such introgressions from wild species would be critical in the present climate change scenarios.

Brinjal: The bacterial wilt resistance is governed by a single dominant gene R ps. Wild species viz., *Solanum melongena* var. *incanum, S. xanthocarpum, S. toxicarium, S. nigrum* and *S. sisymbriifolium* are reported donors for bacterial wilt. Breeding for resistance to bacterial wilt in brinjal resulted in the selection of Swarna Pratibha, Swarna Shymal, Swarna Ajay, Swarna Abhilamb, Utkal Jyoti, Arka Nidhi, Arka Anand. BB-44 variety has been bred for resistance to Fusarium wilt and drought.

Phomopsis fruit rot is recessive and polygenic. The resistant sources are wild species like *S. xanthocarpum*, *S. indicum*, *S. gilo*, *S. khasianum, S. torvum, S. sisymbriifolium* and *S. testiculatum*. Among cultivated varieties Florida market, Florida Beauty, Muktakesi, Bargan, White Gih and Pusa Bhairav are tolerant and can be used in crop improvement of phomopsis blight resistance.

***Capsicum* spp.:** The varieties Punjab Lal (resistant to vCMV and leaf curl), Punjab Surkh (resistant to viral diseases), Utkal Ava (resistant to bacterial wilt), Pant C1 (tolerant to leaf curl and mosaic) have been bred for commercial cultivation.

Okra: Two major viral diseases, YVMV, and OLCV infest the okra crop regularly and cause severe yield loss. Other diseases such as *Cercospora* leaf spot and powdery mildew appear sporadically, but have not attracted the attention of breeders for breeding. The transfer of resistance from wild relatives has been hampered by sterility problems. A. callei and A. tetraphyllus have been utilized in the improvement of okra. Through inter-specific hybridization, it was possible to transfer symptomless carrier type of tolerance to YVMV from *A. callei* (Parbhani Kranti, Punjab-7 and Punjab Padmini) and *A. tetraphyllus* (Arka Anamica and Arka Abhay) to the conventional okra varieties.

Vegetable Cowpea: Cowpea golden mosaic virus (CGMV) is a major disease that infests vegetable cowpea. Genotypes, CNC-043, TVu-612, CE-315, BR1-Poty, Arka Samrudhi and Arka Garima (IIHR), Cowpea-263 (PAU), BC-244002 (BCKV, Kalyani), NDCP-3 (NDAUT) are resistant to CGMV. Resistant varieties have been developed at IIVR Varanasi (Kashi Gauri, Kashi Unnati, Kashi Kanchan and Kashi Sudha) by utilizing these varieties (Ferry and Dukes, 1995).

French Bean: Arka Anoop is resistant to both rust and bacterial blight. Lines, SP6c, SP17B, SP61, SP180D, SP377C have been developed, which are resistant to BCM, BYM, CYV, and BBW. Scully *et.al,*. (1995) developed Bellersnap and Tender Cream varieties of cowpea, which are resistant to root knot, and *Cercospora* leaf spot.

Vegetable Pea: In vegetable pea, powdery mildew is a major disease causing heavy yield loss. Early varieties grown for table purpose escape disease incidence. In addition, rust also infests pea crop and causes significant yield loss. Varieties, PRS-4, JP-4, JP-83, NDVP-4, and FC-1 are resistant sources for powdery mildew. The varieties, Kashi Mukti, Kashi Shakti, Arka Ajit, DPP9411, are resistant to powdery mildew. The varieties, Arka Karthik and Arka Sampoorna are resistant to both rust and PM.

If one analyzes the vegetable crop improvement since beginning through various means, it appears that since beginning from domestication, selection hybridization, heterosis, polyploidy, back cross breeding and DNA recombinant technology are the various ways involved in vegetable crop improvement. To gain the benefit of genetic modification to resource poor families, it is important that pre-breeding centres are established. The novel genetic combinations produced as such Pre-breeding centres should be used for developing location specific varieties through participatory breeding with farming families. This will help to combine genetic efficiency with genetic diversity at field levels. Also it brings wisdom of past and innovations of the present into synergetic blend.

15.7 REFERENCES

Anonymous (1991) Report on international workshop in okra genetic resources, October, 1990, IBPGR, Rome, pp. 2-3 and 8-12.

Arora, R.K. (1985) Genetic resources of less known cultivated food plants. *NBPGR Sci. Monogr.* No. 9:125.

Bailey, L.H. (1950) The Standard Encyclopedia of Horticulture. The MacMillan Co., New York.

Bisht, I.S., Mahajan, R.K. and Rana, R.S., (1995) Genetic diversity in South Asian okra (*Abelmoschus esculentus*) germplasm collection. *Ann. Appl. Biol.*, 126: 539-550.

Chadha, M.L. and Tarsem, Lal, (1993) Improvement of cucurbits. In *Advances in Horticulture*, Vol. 5 Vegetable Crops. (Eds.) K.L. Chadha and G. Kalloo, *Malhotra Publ. House, New Delhi*

Charrier, A., (1984). Genetic Resources of the Genus *Abelmoschus* Med. (Okra, English translation). IBPGR, Rome

De Candolle, A. (1982) Origine des plantes cultivers. Paris (English Translation, 1886, Kogan Paul).

Filov, A.I. and Vilenskaya, G.M. (1972) Cultivated cucurbits in different languages of the world. *Trudypo Prikladnoi Botanike Genetike L Selektsii*, 47: 138-150.

Hamon, S. and Hamon, P., (1991) Future prospects of the genetic integrity of two species of okra (*Abelmoschus esculentus and Abelmoschus caillei*) cultivated in West Africa. *Euphytica*, 58: 101-111.

Mohideen, M. K. and Irulappan, I. (1993) Improvement of Amaranthus. In *Advances in Horticulture.* Vol.5, Vegetable crops. 305-323. (eds) K.L. Chadha and G. Kalloo, Malhotra Publ. House, New Delhi.

Munger, H.M. and Robinson, R.W. (1991) Nomenclature of Cucumis melo L. Cucurbit Genet. Coop. Rep. 14:33

Pal, B.P., Singh, H.B. and Swarup, V., (1952) Taxonomic relationship and breeding possibilities of species of Abelmoschus related to okra (*Abelmoschus esculentus* L.). *Bot. Gaz.*, 113: 455-464

Pier, (2001) " Invasive plant species: *Coccinia grandis*" Pacific Island Ecosystems at risk. United States Department of Agriculture, Natural Resources Conservation Services. The Maui Weekly Wrap. *Maui News*, p 10-18.

Siemonsma, J.S., (1982a) La culture du gombo (Abelmoschus spp.) legume fruit tropical avec reference speciale a la Cote d'Ivoire. Thesis Univ. Wageningen, the Netherlands.

Shimotsuma. (1963) Cytogenetics and evolutionary studies in genus Citrullus, Seiken Zin Rept. Kihara Inst. *Biol. Res.* 15: 24-34.

Van Borssum Waalkes, J., (1966) Malesian Malvaceae revised. Blumea, 1: 1-251.

Vavilov, N.I.(1926) Studies on the origin of cultivated Plants. *Bull. Appl. Bot.* 26 (2):1-248.

Velayudhan, K.C., and Upadhyay, M.P., (1994) Collecting okra, eggplant and their wild relatives in Nepal. *Pl. Genet. Resour. Newslett.* 97: 55-57

Zeven, A.C. and J.M.J. de Wet, (1982) Dictionary of cultivated plants and regions of diversity. Wageningen p. 259.

Chapter – 16

LED Lighting in Horticulture

Mam C. Singh

16.1 INTRODUCTION

The world floriculture business has reached to almost over a $ 40 billion mark. It is continuously going through the rapid transitional changes to have more unique and competitive products of wide choice to the consumers. Netherland is ahead in floriculture trade. Throughout the year, a large number of ornamental commodities are imported in Holland and exported back to the countries for its retail and consumption mostly to the countries in the Gulf, Middle-East Asia and America. According to the statistical estimates worldwide, a large quantity is consumed in US and the other countries. For an example, the US alone consumes 100 lacs of cut flowers a day and approximately 4000 lacs every year, though it ranks only 17th in per capita cut-flower consumption. The average Swiss spends Rs 4500 a year on flowers, nearly four times than that spent by the average American besides the fact that the consumption of fresh cut flower and foliage plant is on manifold increase every year resulting in the competition worldwide for growing of exportable eco-friendly products. At the same time there is great environmental consciousness towards transforming floriculture with "natural and unspoiled" system of having the exportable produce into a mass and its ease to transport to any distant market. The system relies greatly on the industry with extensive network of breeders, greenhouse managers, field workers, auction houses, sales representatives, shippers, and florists to offer consumers a variety of choices matching their taste. This is the fact that in the recent times, the commercial growers don't rely on the nature to introduce the genetic mutation that will make them millionaires but depends on the art and scientific knowhow to create a novelty in its own. This has resulted in the seriousness and painstaking effort of the growers to produce the plants in new forms and shape to put forward a new plant with extra-ordinary marketable characteristics every year, e.g. compact chrysanthemum, *kalanchoe*, bromeliads, orchids, etc (Plate1) are some of the common products with an unusual and eye-catching hue in the European markets.

Similarly, the consumers can purchase gerberas with blooms as wide as salad plates and roses with yard-long stems perfectly suited for valentine-day gifts which mainly they get produced under the specific and artificial environment created for a specific growth stages of the plant.

16.2 THE PRINCIPLE OF LEDS USEFULNESS IN PLANT'S GROWTH

16.2.1 Plant growth is spectrum dependent

Since plants only use a tiny percentage of the full light spectrum and chlorophyll being green acts as a "Sun-block" to remove 95 per cent of the unwanted and damaging light wavelengths due to higher irradiance at high temperature (Hughes and Cockshull, 1972) resulting in poor growth. On the other hand, red and blue (LED) lighting have such narrow band widths the plants without green chlorophyll can still survive and will be safe against the harmful light wavelengths. It is further to note that nearly 100 per cent of the light an LED grow light emits is completely absorbed by the plants.

Plate 1. Compact ornamental plants grown with LEDs grow lights

LED's can also be used for 24 hours a day without stressing plants since the Chlorophyll does not have to do battle with unwanted light-waves as they produce a very small amount of hurt over the plants canopy (fig.1). If we measure the quantum of LED grow lights in Einstein's it means that how many photons of light strike an area per second which sometimes is just vague in measuring the quantum absorbed in the plants as tiny fraction. However, the irradiance power factor or amount of light that actually gets used in plant growth or chlorophyll production as phytochrome response (Cathey, 1974) can be measured as µmol m-1s-1 in the conditions when the lights are hardly 95 per cent growth efficient. Also in that case the 95 per cent light is wasted since plants only use very narrow bandwidths of the total light energy.

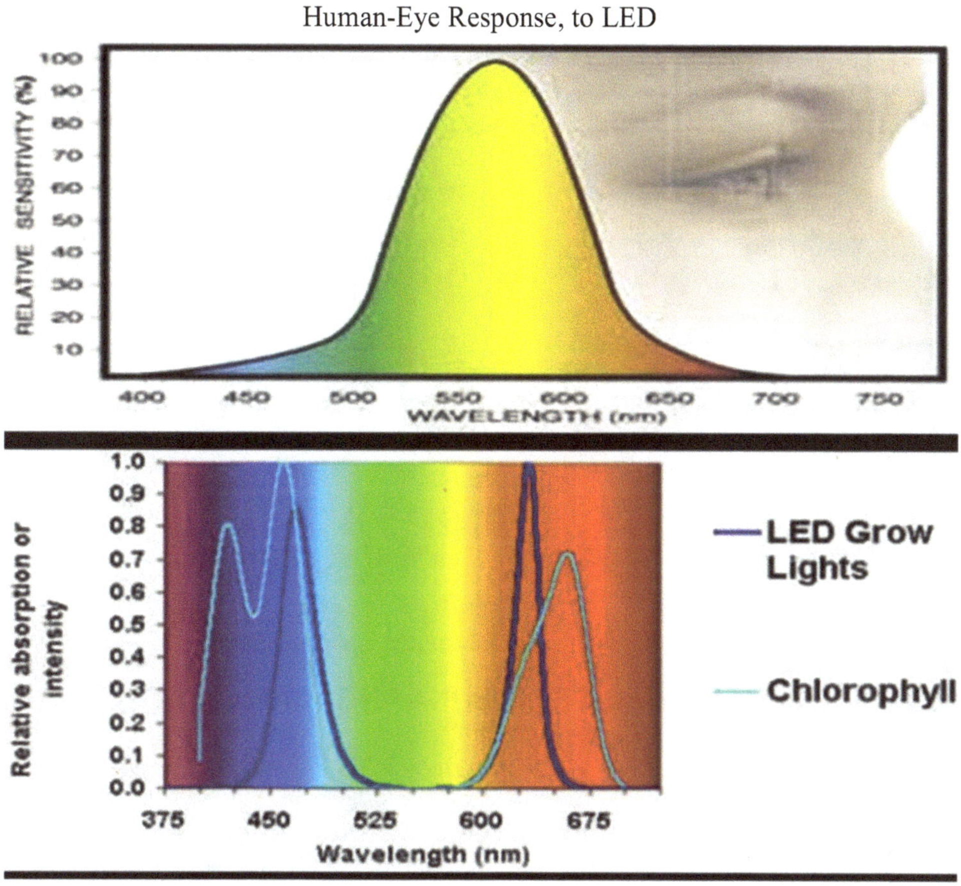

Fig 1. The usage spectrum by plants for photosynthesis

16.3 LEDS AND ITS UTILITY

A light emitting diode (LED) is a semiconductor diode that emits incoherent narrow-spectrum of light when electrically biased in the forward direction in the form of an electroluminescence. This light source is usually a small area (plate 2), often with extra optics added to the chip that shapes its radiation pattern and flow over the plant canopy in different colours. The colour of the emitted light depends on the composition and condition of the semiconducting material used which can be infrared, visible or near-ultraviolet.Infect, the light emitting diodes are rectifying semiconductor device which converts electrical energy into electromagnetic radiation. The wavelength of the emitted radiation ranges from the near-ultraviolet to the near-infrared, that is, from about 400 to over 1500 nanometers. This light can be used as a regular household light source and most commercial light emitting diodes, both visible and infrared, are fabricated from type III-V semiconductors. Wherein, these compounds contain elements such as gallium, indium, and aluminum from column-III (or group 13) of the periodic table, as well as arsenic, phosphorus and nitrogen from column-V (or group 15) of the periodic table.

There are also LED products made of II-VI (or group 12-16) semiconductors, for example Zn, Se and related compounds. If we combine them together, these semiconductors possess the proper band-gap energies to produce radiation at all required wavelengths. Most of these compounds have direct band gaps and therefore, these are efficient in the conversion of electrical energy into radiation.

What is a Blue LED ?

L.E.D. = Light Emitting Diode

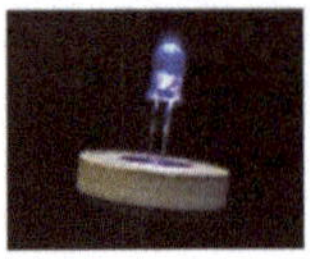

Blue LED

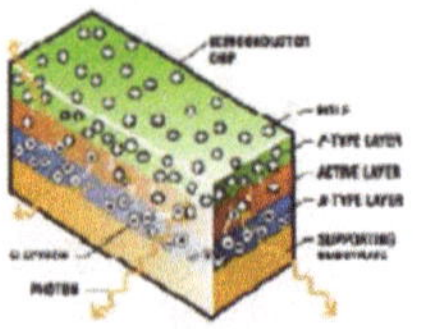

An LED produces light by combining positive and negative charges inside a Gallium nitride (GaN) crystal.

"The Promise" Energy Usage Comparison

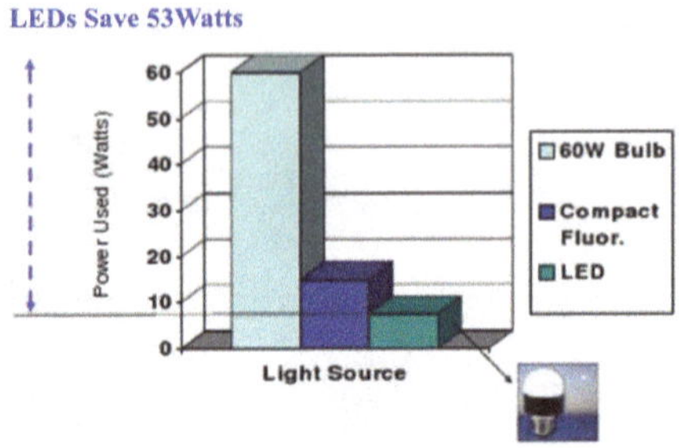

"Best" White LED and Compact Fluorescent vs. 60Watt Light Bulb Comparison

Plate 2. Useful LEDs source and energy usuage

16.4 COMMERCIAL USE OF LEDs IN LANDSCAPE HORTICULTURE

LED lights can be successfully incorporated into a landscape in a variety of ways. Specifically, LED strips are effective for lighting long paths, stairs or pools, thanks to their lengthy design. (Annon., 2010). Throw down a long strip, and a large amount of real estate can be covered and properly lit. Mounted LED fixtures that are strategically placed can illuminate a specific spot on a lawn. LEDs can also be installed directly into other devices, such as lanterns, for enhanced lighting, Also, thanks to their immense flexibility; LEDs can be brighter or softer depending on your security needs or the mood that needs to be set. Voltages and lamps for the LED are highly configurable. In terms of a landscape job, this means that non-intrusive, low-light bulbs can be used for illuminating footpaths and sidewalks, while brighter lights can be housed just outside the home, or on a stoop. Additionally, lights of wildly different colors can be used to spice up a landscape and further enhance a desired mood. Above all, if we see the aesthetics in landscape lighting even a nighttime lighting is becoming common in and outdoor, almost at no cost is elegance beyond imagination (Plate 3.1).The following are kind of styles recently being used in gardening.

Plate 3.1 Garden Solar LEDs lights

Plate 3.2 Solar powered LED garden stick lights

LED garden stick lights are being used mainly for energy saving, long working life and easy "stick and play" installation along with weather-proof and super bright LED light properties. These garden lights look like star at night or fireflies shuttling in your garden. These lights serve as great supplement for garden or home decoration These LED garden stick lights (Plate 3.2). can be solar powered as super bright LED color changing light having almost no electric cord wiring, long working hours in an aluminum stick body and durable weather-proof super bright LED light bulbs. These lights normally are built in two solar panels and with one AAA Ni-Nh rechargeable battery and can light up 10 hours after each full recharge. These are best fit for yard decoration and perfect gifts for Christmas and holiday seasons. The most valuable advantage of this kid of lights in the garden is that they can be put simply in water as plug and play, easy to relocate and need no hardware installation. These make a great ease for display in pond, party event or it can be sit on top of the regular fountain to make a beautiful lighted water spray.

Plate 3.3 Garden Solar powered LEDs grow lights

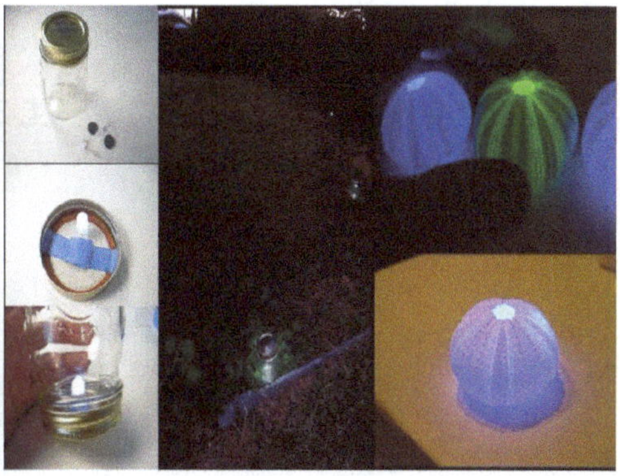

Plate 3.4 Temporary and inexpensive garden LEDs lanteerns

The solar power LED lights are large variants of solar garden lights e.g., solar fairy lights with changing colours and having almost 100 LED bulbs. These bulbs are also fixed as solar powered illuminated rock lights to provide an ambient lighting for the garden or patio. These garden lights save energy while the lighting unit illuminates automatically (Plate 3.3.) In general, solar garden lights, as with many things, new solar LED technology allows light discharge of upto 12 hours per night. The individual advantage is that these solar energy LED garden lights require no wiring as they charge through the daytime from the sun's rays (they require a sunny position, although will gain some charge on an overcast day). The solar based LED lights come on automatically at dusk, although have a manual over-ride switch. The Solar LEDs panels are also highly efficient, mono-crystal or multi-crystal silicon solar based panels and have rechargeable battery (Ni-Cd/Ni-MH), preferable and adjustable LED color e.g., super bright white, yellow, red, green, blue etc (Keeler and Nicole, 2010) and the all lights in panel are in automatically turned on mode.

The new fancy LED lights have been developed in a matching shape of sea objects to give an impression of sea-life by illumination objects in a open garden corners. These lights look like tiny sea urchin shells and often are available at beach shops in US and Europe. These water garden landscape lights are thin and lighten as eggshells. However the sea urchin lamps were available before, but they've always been made with large (i.e., sturdy) and colorful ones (Plate 3.4). In contrast, these tiny ones might be better to hang around Christmas lights like little paper lanterns. These when lightens look like living urchins and one will always be surprised by how elegant and mobile they are. These lights very well fit in the 10 mm LEDs holders and require a slightly smaller battery than lamps (CR2032) to fit in here, may be a low input bulb (CR1632) can be suitable one for tiny illuminated sites.While using solar power based LED lights in the garden ultra-low cost temporary garden lights using LEDs, lithium coin cells, and mason jars can be made to start with for which we need LEDs and CR2032 lithium coin cells one each per jar. The best kind of LED for this design is an ultra bright LED with a diffused lens available in commercial stores and they are effective in casting the light by LED chip every direction. One may also opt for an easy access, with 10mm diffused white LEDs and there are still similar LEDs available elsewhere as well.

16.5 COMMERCIAL USE OF LEDS IN PLANT PROPAGATION

Among all the light intensities intercepted by the plants red and blue light is essential for plant growth as it is a fact that the leaves perform the process of photosynthesis and red and blue spectrum or wavelengths are most essential for photosynthesis. The blue light is absorbed to its maximum in the spectrum region of 430-450nm while the red in the spectrum with 650-670nm. It has been estimated that the ideal balance for photosynthesis is 92 per cent red LEDs and 8 per cent blue LEDs if it is practised artificially. In this case the blue lights have a smaller influence than those of the red lights and it makes easy to select anything between 1 per cent and 20 per cent of blue light, depending on requirements of the plants for their growth and developments. The commercial LED lights are available as per colour and the frequency of light used by plants for efficient and healthy growth so as to provide an efficient photosynthetic energy than the use of conventional grow lights (Plate 4.1). In general, red light being in the vicinity of a plant in the first peak of light absorption spectrum i.e. 660 nm it contributes maximum to the plant photosynthesis. However, the red light, when combined with blue light, encourages flowering and most often used in higher doses to stimulate flowering, seeding and fruiting in certain plants of economic importance (Plate 4.2). Whereas, the blue light is responsible primarily for vegetative growth (leaf) as in fect the chlorophyll has its second distinct absorption peak in the vicinity of blue light wavelengths (450nm). The blue light is indispensable to the morphologically healthy growth of plants.

Plate 4.1 Plants exposed to LEDs grows faster

Plate 4.2 LEDs grown copact ornamental foliage plants

16.6 LEDS APPLICATION IN GREENHOUSE HORTICULTURE

In majority of the flower growing countries, conditions are perfect and the days are always 12 hours long; sunlight is intense, and nights are cool. However, the most commercial sectors are now moving to the still better climatic conditions, into the developing countries. The flowers are now the third most important product, after oil and bananas in gulf countries and despite the hard economics the flower business is thriving well and there are several flowering plants those respond to the natural short day (light) conditions. To cut short their vegetative phase into reproductive with sufficiently elongated stems and quality flower buds we may have provision of providing the plants an extra photoperiodic as well as photosynthetic light to make them accessible to enough exposure, especially in winter.

Plate 5.1 LEDs application has increased 20% stem yields

Plate 5.2 An ensured higher productivity in greenhouse cultivation

It is evident that every plant contains a photosensors that respond to the specific part of the spectrum from the ultra violet radiation (UV) to 740 nm spectrum which is known to control various aspects of plant growth and flowering. In last decade, several researchers worldwide have studied the effects of light on plants to explore that which part of the daylight provides most suitable spectrum for photosynthesis. Its further more critical that how low intensities of LED lights (on the order of 1 W/m2 or 100μmoles m-2sec-1in total) can be used to control the desired characteristics in the plants for final consumer as a finished product whether be it size, shape and colour of the foliage or uniform compactness and architecture of the plant as whole. It has been already standardized that the plants use light from 400 to 700nm for photosynthesis, with some wavelengths having the greatest effect at certain times of the day and theirs at various stages of a plant's growth. The spectrum can affect properties such as compactness of shape and time to flowering and final yields (Plate 5.1). Not only that in such plants grown for food or medicinal use, it can change the plant's chemical content, which can affect flavour, colour, and nutritional and/or pharmaceutical value.

It is very important to optimize the lighting requirement for a given effect on the plants. For example, the various scientists in Europe and US have revealed that a particular red wavelength causes a chrysanthemum plant to flower four weeks earlier (Ieperen *et. al.*, 2009) than it does under the same conditions without the additional LED lighting using daylight. It has also been reported that the system could reduce the cost to an economically viable level and is affordable with present-day LEDs. The application of LEDs is commercially highly efficient with available narrow emission spectra across the relevant wavelength range over a plant canopy (Plate 5.2) resulting in an flower bud induction without disturbing the dark period in photo-sensitive plants while using smart LEDs in a mixture of 80 per cent red and 20 per cent blue (Singh *et al.*, 2009). Recently introduced method for commercial greenhouse is use of plastic sheeting with special reflectance spectra and placed as reflective mulch on the floor or on benches where pot plants are grown (Plate 5.3) as its obvious that a white plastic mulch can increase light intensity at the plant while absorbing UV. On the other hand, the red plastic mulch is often used to improve the flavour in soft fruit. Whereas the silver plastic mulch have been found to reflect throughout the near-UV, visible and near-IR lights, are commonly being used on large scale in the greenhouses to improve grape quality and increased productivity.

16.7 ADVANTAGES AND DIS-ADVANTAGES OF LEDs

16.7.1 Advantages

1. The LEDs are more efficient to produce light per watt than incandescent bulbs.
2. These can emit light of an intended colour without the use of colour filters that is required in traditional lighting methods. Thus, the LEDs are more efficient and can lower the initial costs
3. The size of LEDs can vary from very small (>2 mm2) to a large size and can be easily populated onto the circuit board

4. Frequency of their lighting on and off make them convenient to use for lighting up very quickly as a typical red indicator LED can achieve full brightness in microseconds
5. The LEDs are ideal for use in applications that are subject to frequent on-off cycling, unlike fluorescent lamps that burn out more quickly when cycled frequently, or HID lamps that require a long time before restarting
6. The LEDs can very easily be dimmed either by pulse -width modulation or lowering the forward electric current
7. In contrast to most light sources, LEDs generate light and waste heat by different mechanisms - respectively radiation and conduction and suitably designed luminaries can produce a relatively cool light stream
8. The LED lights are mostly failing by dimming over time, rather than the abrupt burn-out of incandescent bulbs
9. The LEDs can have a relatively long useful life and estimated to have up to 35,000 to 50,000 hours, though time to complete failure may be longer. If we compare with the fluorescent tubes, which are typically rated at about 10,000 to 15,000 hours, depending on the conditions of use. Whereas, the incandescent light bulbs can be rated only at 1,000-2,000 hours
10. The LEDs, being solid state components are difficult to damage with external shock, unlike fluorescent and incandescent bulbs which are fragile
11. The solid package of the LED can be designed to focus its light but the incandescent and fluorescent sources often require an external reflector to collect light and direct it in a usable manner
12. The LEDs do not contain mercury unlike the fluorescent lamps and avoid the environmental pollution

Plate 5.3 Quality foliage plants fetches very high price in market

16.7.2 Dis-advantages of using LEDs

1. LEDs are currently more expensive on an initial capital cost basis, than most conventional lighting technologies
2. The performance of LEDs are largely dependant on the ambient temperature of the operating environment as over-driving in high ambient temperatures may result in overheating of the LED package, eventually leading to device failure
3. These are voltage sensitive and must be supplied with the voltage above the threshold, a current below the rating involving the series resistors or current-regulated power supplies
4. The whitest lights (LEDs) have spectra that differ significantly from a black body radiator like the sun or an incandescent light. However, the color rendering properties of common fluorescent lamps are often inferior to that now available with white LED lights
5. LEDs do not approximate a "point source" of light and cannot be used in applications needing a spherical light field. These are not capable of providing divergence below a few degrees
6. There is increasing concern that the blue and white LED lights are now capable of exceeding safe limits but still the eye safety specifications are needed to use as the LEDs can cause more light pollution than other light sources if not shielded
7. Therefore, it is very important that the LEDs are fully shielded when used outdoors as the blue LEDs is scattered about 2.7 times more by the Earth's atmosphere

16.8 REFERENCES

Anonymous.2010.http://www.articlesnatch.com/Article/Lighting-Your-Landscape-With-Led-Lights/1063664#ixzz0ogNN6eFh

Cathey HM.(1974) Participation of phytochrome in regulating internode elongation of chrysanthemum moriflolium (Ramat) Hemsl. *Journal of the American Society of Horticultural Science* 99 : 17-23.

Hughes AP and KE Cockshull. (1972) Further effects of light intensity, carbon dioxide concentration, and day temperature on growth of *chrysanthemum morifolium* cv. Bright Golden Anne in controlled environments. *Annals of Botany,* 36 : 333-350.

Ieperen, Wim van; Singh, M.C. and Ep Heuvelink.(2009) Flowering Preserved in Short-day Chrysanthemum plants grown under Long-Day conditions using PAR with a variable spectrum over the day. In: *Proceedings of 6th International Symposium on Light in Horticulture.* November 15-19, 2009, Tsukuba, Japan. p. 92.

Keeler, Bill and Nicole Slein. (2010) Light is a registered trademark of Luminus Devices Schwartz Communications, Luminus Media, 781-684-0770. www.luminus.com

Singh, M.C., Wim van Ieperen and Ep Heuvelink.(2009). Flower induction in *Chrysanthemum morifolium* under long days using LEDs. (M.Sabu Award Paper).In: International Symposium on "*Angiosperm Systematics* and *Phylogeny: Retrospect's and Prospects*", November 12-14, 2009. NBRI (CSIR), Lucknow, UP.

Chapter – 17

Cultivation of Anthurium Through Precise Technology

S. L. Chawla, A. R. Gaikwad, Alka Singh and L. K. Dashora

17.1 INTRODUCTION

Flower consumption is growing at an alarming rate of 20 per cent per annum at global level and India is, today, one of the world's largest consumer base and fastest growing retail destinations. Commercial floriculture is today a potential money-spinner and an economically viable agri-business.

Cut flower trade in terms of export oriented units dealing with rose and gerbera is flourishing. Other cut flowers like gladiolus, tuberose, lilum, chrysanthemum, etc. are also being cultivated along with traditional loose flowers like jasmine, rose, crossendra, marigold on a higher scale. However, growing awareness of Indian consumers and popularity for cut flowers is creating demand for diversified flower crops. Further, trade of some flowers in the market is responsible for gluts. Thus, there is an urgent need to diversify the floriculture industry by introducing and adoption of anthurium as a cut flower with unique shape and high longevity.

Precision farming is an advanced approach for the flower growers to enhance their income. It mainly includes information and technology of farm management system to identify, analyze and manage variability within fields for optimum profitability, sustainability and protection of the land resource. Precision farming is applied for increasing efficiencies that can be realized by understanding and dealing with the natural variability found within a field.

Anthurium is taking a good place in market with high value. Farmers are undertaking anthurium cultivation and in this direction, precise farming technology will help the farmers to grow the crop with good quality and higher production per unit area.

Anthurium a herbaceous tropical plant, belongs to family Araceae and is highly praised flowering plant for its colourful long lasting unique flower and shining foliage.

It is an excellent plant for interior decoration. The flower remains in high demand especially for flower arrangement. The most suitable part of anthurium is modification of the leaf and botanically known as 'spathe'. The heart shaped spathe appears on a long flower stalk with a cute spadix. This beautiful plant blooms almost continuously under suitable conditions. It is epiphytic in nature with creeping, climbing or arborescent stems including lots of aerial roots that aid in taping water and nourishment. In India, growing of anthurium has been a hobby earlier but at present this has become an important export oriented crop. It requires a highly organic and aerated medium with good water retention capacity. However, the secret of success for commercial cultivation is to have good drainage in medium and nutrient and environment management.

The genus anthurium has around 600 species. The most popular and important species known for their beautiful inflorescence are *A. andreanum and A. scherzianum* and for their excellent foliage are *A. magnificium, A. digitatum, A. crystallinum, A. watermaliense, A. clarinervium* and *A. hookeri.* Among them, *Anthurium andreanum* is also known as Oil cloth flower, Tail flower or Painter's Palette.

17.2 TYPES OF ANTHURIUMS

1. **Standard:** These are common heart shaped spathe lobes ranging from 12x11 to 20 × 18 cm.
2. **Obake:** These have two-tone coloration i.e. bi-colour effect of green with a major spathe colour. In some varieties there is loss of green colour in summer. Size of spathe ranges from 8 × 7 cm to 28 × 23 cm.
3. **Tulip:** These are upright cupped spathes and straight and erect spadix. Size of spathe is about 10 × 6 cm. These are mostly hybrids.

Cut Flower Varieties

S.N.	Colour	Varieties
1	Red	Jewel, Hawaiian Red, Ozaki, Cancan, Tropical, Fla Red, Avo Netta, Scarlet, Mauritius Red, Tanaka, Tina Red, Cherry Red, Eureka Red, Honduras, Flame, Pumesila
2	Orange	Nitta, Sunset Orange, Sun Burst, Gino Orange, Avo Gino, Mauritius
3	White	Acropolis, Manoa Mist, Trinidad, Meringue White, Lima White, Haga White, Ivory, Cotopaxi
4	Pink	Agnihotri, Passion, Marian Seefurth, Candy Queen, Candy Stripe, Pink Hawaii, Abe Pink, Sonata, Aymara
5	Obaque	Red Dragon, Fantasia, Madonna, Lambada, Senator
6	Green	Midori, Esmeralda, Akapana (whitish-green)
7	Maroon	Ceaser
8	Light Peach	Vogue

Pot Varieities (Obake): Patino (Pink), Rosario (Peach), Pumesilo (Red), Coralis (red), Condor (orange)

17.3 ENVIRONMENTAL CONDITIONS

Anthurium grows naturally in a shady environment, well protected from extreme weather conditions. Wind can cause heavy damage to the anthurium plants and flowers.

(1) **Temperature and relative humidity:** Anthurium can grow in a temperature range of 15°C to 35°C. Temperature of 21° to 24°C is essential for initiation of flowering. High temperature along with low humidity causes leaf scorching. It is humidity loving plant and prefers 80-85 per cent relative humidity. Mandhar *et al.,* (1998) reported 25 to 30°C temperature 80 per cent relative humidity and shade upto 75 per cent ideal for anthurium.

(2) **Sunlight:** Anthurium is a semi-shade loving plant. Too much light can cause the colour of flower to fade. The leaf can also show the scorching and burning effect. It thrives well in 50 per cent shade. per cent. Nevertheless, it can tolerate 80 per cent shade also with good growth.

17.4 PROPAGATION TECHNIQUES

Anthurium is propagated mainly by division of offshoots and cuttings. Cutting using 2-3 terminal leaver was found better in terms of number of roots (Kunisaki and Sagawa, 1971). Treatment of cutting with hot water +IBA increased the number of shoots per cutting (Hata *et al.*, 1994) Micro propagation is also a good option for obtaining large number of plants in short time. Leaf segments, spadix segments, vegetative buds, stem sections etc., are mainly used in micro propagation.

17.5 GROWING STRUCTURES FOR ANTHURIUM CULTIVATION

Structures for anthurium cultivation depend upon the agro climatic conditions of area. Generally at commercial scale anthurium cultivation is done in natural ventilated polyhouses, net houses and climate controlled green houses. In moderate climate greenhouse protects plants from low temperature (Singh, 2006).

The anthurium generally grows good under shade houses. Shade net houses of 50 to 75 per cent intensity can be used to maintain intensity of sunlight and temperature for successful cultivation. Provisions can be made for installing overhead sprinkler/misters/fogging systems by running GI wires. Misters can be installed upto plant height. Such structures are more suitable in areas where the ambient relative humidity levels are already high and frequent air exchange through the porous side nets do not result in drastic reduction in the relative humidity levels.

In areas where the relative humidity levels are low and the temperature levels are high, low cost polyhouses are more suitable for commercial cultivation. By creating a protected climate by UV stabilized polythene, the relative humidity levels can be easily maintained. The hot air can be easily expelled. With top or side ventilation, inside the polyhouse, the portion can be provided with layers of two shade nets (50 and 25%). If the light intensity is very high, a thin film of lime can be sprayed on the external surface of the polyhouse to reduce the incoming radiation.

17.6 PLANTING MEDIUM

Highly organic, well aerated medium with good water retention capacity and drainage is suitable for anthurium cultivation. The medium generally used are sugar bagasse, wood shavings, leaf mould, coarse sand, small brick pieces, coir pith, compost, charcoal, coconut husk pieces etc.

Pot planting: Earthen pots of 23 to 30 cm diameter with a minimum of two holes are preferred. Young plants require repotting every year while adults ones, in every 2 or 3 years.

17.7 BED PLANTING

The most common growing system is cultivation in beds. Land having gentle slope (3.0cm for every 100m) is more suitable so that water stagnation could be avoided. Planting is done on the beds preferably in four rows. The usual width provided for bed is 1.20 m and for the paths, 0.80 m. Planting should be sparse to facilitate air circulation and spray penetration. Beds can be arranged on platform of 3-4 feet height to facilitate good aeration in heavy soil region.

17.8 IRRIGATION MANAGEMENT

The plants are to be watered at least twice a day in summer months. A thorough irrigation has to given in the morning; irrigation in the afternoon should be tied in such a way as to leave sufficient time for the water to evaporate so that the plants will not be damp during night hours. The methods of irrigation practiced in commercial farms are sprinkler irrigation or drip irrigation through which fertilizers are also often given. Water should be preferably free from dissolved salts. Water with an EC of more than 0.5m ds/cm should not be used. The optium pH range is between 5.2 and 6.2. If the pH is too low, colour stability of the spathe will be affected. Voogt (1972) reported yield reduction when EC of water was found above 0.6 mmhos/cm.

17.9 NUTRITIONAL MANAGEMENT

Nitrogen, Phosphorus, Calcium, Potassium, Magnesium and Sulphur are the important elements required in anthurium nutrition. It is better to apply the fertilizers in smaller doses and at frequent intervals than longer doses at longer intervals. Due to this the fertilizers are applied by fertigation rather than individual application, which will help to apply fertilizers to the root zone. The pH of the nutrient water should be slightly acidic. Young plants may be sprayed with a fertilizers mixture containing higher ratio of nitrogen (3:1:1) at 0.2 per cent, twice in a week. For flowering plants, it is better to reduce the quantity of nitrogen to increase the post harvest longevity of flower. Application of 1:1:2 NPK mixture 2-3g/l twice in a week is found to be beneficial.

Organic manures dissolved in water can be sprayed on the plants once a month. For this, cowdung or neem cake mixed with 10-15 times the quantity with water and kept for three or four days can be used. Under greenhouse conditions anthuriums are fretigated with solution containing nitrogen 14ppm, potassium 176ppm, calcium60 ppm, magnesium24 ppm, nitrate91 ppm, sulphate48 ppm and phosphate31 ppm.

17.10 LEAF PRUNING

An anthurium plant has to be pruned to retain adequate number of leaves, without any adverse effect on the flower production and quality. This facilitates good plant health in view of better air circulation. Pruned leaves that are healthy can also be sold as cut foliage.

17.11 HARVEST AND POST HARVEST MANAGEMENT

17.11.1 Harvesting stage

The flowers are harvested after the unfolding of the spathe is complete and the spadix shows more than half colour change. They are harvested in the morning and are cut with long stalks. After harvest, the basal portion of the stalk of the flowers is kept in water as soon as possible to prevent drying out. If the flowers are taken to distant markets, a piece of cotton soaked in water is kept at the end of the stem to prevent desiccation. About eight flowers per plant are obtained annually under good management conditions.

17.11.2 Pulsing

Pulsing is given to anthurium flowers before shipment to extend the holding period. Pulsing treatment with BA 50 ppm for 12 hrs is known to delay initiation of spathe bluing and spadix necrosis in anthurium flowers. This treatment is also helpful in retention of spathe glow.

17.11.3 Storage

The optimum storage temperature for anthurium is reported to be between 14° and 17°C. (Paull, 1987)

17.11.4 Packaging

Soft protective materials can be put in between the spathe and spadix to prevent bruising. The spathe along with the spadix is then inserted into a polythene cover of appropriate size. This will prevent bruises on the spathe due to pressing of the spadices of adjacent spikes that causes blackening effect. The open ends of the polythene cover are stapled just outside the spathe to prevent movement of the spathe inside the cover.

The flowers should be spotlessly clean and shining. They are packed in cardboard boxes of size 60 cm × 30 cm × 23 cm. A box of this size can be hold 70-100 flowers.

17.12 DISEASES AND PESTS MANAGEMENT

Anthurium plants are extremely sensitive to several insecticides and fungicides. Hence, require great care for diseases and pests management. Some diseases and pests are given below.

17.12.1 Bacterial blight

Initial symptoms of the disease are the appearance of small scattered angular water-soaked spots near the margins of leaves. The infected tissues are killed very soon and there is a bright yellow halo around the spot. The bacterium is capable of becoming systemic. The bacterium can spread through the planting materials taken from the diseased plants, irrigation water or splashing of rain and contaminated tools

Strict sanitation should be maintained and the plants should be sprayed thoroughly with streptomycin sulphate or oxytetracycline (0.5 to 1.0 g/l) on a weekly schedule for about six weeks.

17.12.2 Anthracnose (spadix rot)

The fungus primarily infects individual flowers in the spadix. Infection starts as tiny dark spots which expand to triangular or other angular shapes. This fungus can also infect leaves, producing elongated diamond shaped lesions.

The spores spread primarily by splashing of rain and irrigation water. Insects, mites and other organisms may also spread the spores. Badly affected plant parts have to be removed and destroyed. Spraying of fungicides Mancozeb (0.25%), Captan (0.25 per cent) or Carbendazim (0.1 per cent) is effective in controlling this disease.

17.13 PESTS

Nematodes, scales, mealy bugs, thrips and mites are common pests of anthurium. Nematodes are better controlled by disinfecting the culture medium. Against scale insects, which attack stem and leaves, wiping the affected parts with a cotton wool soaked in Malathion 0.2 per cent is effective. Formulations containing pyrethrum extract can control aphids and thrips. The attack of mites is more severe in summer months and can be controlled by spraying the under surface of the leaves with Kelthane (0.2%). Snails and slugs appear during rainy season and are difficult to control. Metaldehyde based baits are effective in controlling them.

17.14 CONCLUSION

With its numerous colours, showy spathe, long vase life and growing demand, anthurium is fast becoming prime cut flower of the tropics. The high humid stretch of the tropical belt and the NE states offer great scope for growing anthuriums on commercial scale, without much infrastructure facilities, but in other areas it can be grown successfully through precision farming system. Coupled with this, more attention is required to grow, superior varieties, adopt latest technology and to resort better post harvest management, in order to develop a commercially viable anthurium industry in the country on a large scale.

17.15 REFERENCES

Hata, T.Y., Hara, A.H., Nagao, M.A. and Hu, B.K.S. (1994). *Hort Technology*, 4 : 159-162

Kunisaki, J.T. and Saogawa, (1971). *Hawali Fm. Sci.* 20 : 4-5

Mandhar, S.C. Kumaran G.S. and Prakash, D. (1998) *Souvenir and abstract.* National Seminar on anthurium production, IIHR, Bangalore P.15.

Paull, R.E. (1987). *Hort. Science* 22 : 459-460.

Singh, A.K. (2006) Anthurium. In : *Flower crops cultivation and management*, NIPA, New Delhi, P. 35.

Voogt, S.J. (1979) Vakblad Voor de Biocmistery, 24 : 133

Chapter – 18

Organic Farming in India : Challenges and Opportunities

Hari Prakash and Sanjai K. Dwivedi

18.1 INTRODUCTION

Organic farming systems have attracted increasing attention over the last one decade because they are perceived to offer some solutions to the problems currently besetting the agricultural sector. In the western world, organic farming is catching fast (Malik, 2000). Indiscrimpte use of chemical fertilizers, herbicides and pesticides have resulted in environmental and health hazards along with socio-economic problems (Padmadevi, *et al.*, 2008). India is bestowed with lot of potential to produce all varieties of organic products due to its diverse agro-climatic regions. In several parts of the country, the inherited tradition of organic farming is an added advantage. Organic farming is necessary to attain the goal of sustainable development. According to the Food and Agriculture Organization (FAO), sustainable agriculture "is the successful management of resources for agriculture to satisfy changing human needs while maintaining or enhancing the quality of environment and conserving natural resources". All definitions of sustainable agriculture lay great emphasis on maintaining an agriculture growth rate, which can meet the demand for food of all living things without draining the basic resources. Organic farming is one of the several approaches found to meet the objectives of sustainable agriculture. Many techniques used in organic farming like intercropping, mulching and integration of crops and livestock are not alien to various agriculture systems including the traditional agriculture practised in countries like India. However, organic farming is based on various laws and certification programmes, which prohibit the use of almost all synthetic inputs, treated seeds and health of the soil is recognized as the central theme of the method. Adverse effects of modern agricultural practices on the farm resources and health and over all survival of living being have been well documented all over the world. Application of technology, particularly in terms of the use of chemical fertilizers and pesticides has been serious concern. Their negative effects on the environment are manifested through soil erosion, water shortages, salination, soil contamination, genetic erosion, etc. Organic farming has the potential to provide benefits in terms of environmental protection, conservation of non-renewable resources and improved food quality. It is very important to maintain long term soil biological activity, ensure effective peak management, recycle wastes to return nutrients to the land, provide attentive care for farm animals and handle the agricultural products without the use of extraneous synthetic additives or processing in accordance with largely acceptable standard and applicable standards of destination country/ countries. Organic farming has several advantages over the conventional one apart, from the protection of both the environment and human health. Improved soil fertility, better water quality, prevention of soil erosion, generation of rural employment, etc. are some of them.

The concept of quality food has undergone a drastic change over the past few decades. It does give emphasis on the characteristics of the end product, but the process and method of production and transport are now considered equally important. Not only the importers but also the domestic retailers have their own quality specifications or standards, which in many cases are tougher than those of the government regulations. Consumers have become health conscious and are willing to pay for the clean, healthy and natural food. Many developed countries have various support programmes to help organic farming with financial incentives and technical guidance.

18.2 WORLD SCENARIO

The negative effects of modern chemical based farming system were first experienced by those countries, which introduced it initially. Harada, (1975) studied on the influence of mercury and accumlation of cadmium in food leading to *minamata* and Itai diseases. Actually, naturally, organic farming was adopted on relatively large scales during early stage. There are large number of organizations promoting the organic farming movement in European countries. The International Federation of Organic Agriculture Movements (IFOAM) and Greenpeace have studied the problems of the chemical farming methods and compared the benefits of organic farming with that of modern farming. Organic farming movements have since spread to Asia and Africa too.

IFOAM was founded in France in 1972. It spearheads and coordinates organic farming efforts world over by promoting organic agriculture as an environment friendly method. It has about 600 organizational members spread over in about 120 countries including India. IFOAM undertakes a wide range of activities related to organic farming such as administrative and policy making forums in the national and international arena; updating of production, processing and trading standards; coordination of research projects and holding of conferences and seminars etc.

The Food and Agriculture Organization (FAO) of the United Nations provides support to organic farming in the member countries. It also attempts harmonization of national organic standards, which is absolutely essential to increase international trade in organic products.

The organic food market in the world has grown rapidly in the past decade. International trade in organic foods showed an annual growth rate of about 20-22 per cent during this period. The size of the organic food markets in some developed countries in 2000 was: The US food market tops with a value of $8.0 billion, followed by Germany and Japan with $2.5 billion each while France accounts for $1.25 billion. Moreover, the organic foods command higher prices which often range from 10 to 50 per cent more than the conventional ones. Major organic products traded in the international market are dried fruits and nuts, tea, coffee, fruits and vegetables, cocoa, spices, herbs, legumes, cereals, meat, dairy products, alcoholic beverages etc. Cotton, cosmetic ingredients, toiletries and pot plants are major non-food organic products in the world markets. Organic farming is growing around the world and many country like are developing their own standards and regulations. India has adopted the National Programme for Organic Production (NPOP) with national standards.

Organic certification of farms systems is the foundation on which organic agriculture rests. Its role in production and marketing of organic foods becomes a significant corner stone in the creation of a confident consumer community. The need for a trustworthy and healthy organic market is important as the consumers are becoming health conscious. The consumers should possess knowledge and confidence that the products they buy are truly organic. Such environment can be created by an organized system of inspection and certification supported by law and regulations of the state. Organic products enjoy trust based on the label or logo approved by the certification process. The aim of organic certification is, therefore, to provide an identifiable label (NPOP logo or a symbol) and to give an assurance to the consumers that the product is truly organic. Thus, a label on an organic product conveys that the manufacturer has a license for organic production, an independent agency has inspected the production/processing practices followed by the producer, and compliance of the set standards is made. Such an assurance becomes crucial in the generation of consumer confidence in the organic products, particularly as they are costlier than the conventional ones. The marketing agents of organic products also need such an assurance to maintain fair trade practices essential to the growth and sustainability of the demand for such products. The certification and labeling laws in different countries are very rigid. Imports of organic products must conform to the country standard laid down by respective country. The ability of many developing countries to comply with these requirements is limited. A grower or a dealer in need of certification has to lodge a plan to a certification agency which is accredited for that standard. There are both private and state certification programmes. The plan must contain all details in relation to the farming or handling methods and the materials those will have to be used therein. The anticipated improvements and future aims should also be a part of the plan. The farmers or dealers of wild crops should also submit a plan mentioning that the harvesting techniques to be employed will not cause any damage both to the produce and the environment. All records related to the management practices and the materials utilized in organic production should be maintained for five years. Certification will be approved only if the crops are grown on the land free of prohibited substances for three years prior to harvest. The period of conversion is generally three years between the conventional and organic farming and the crops grown during this period cannot be labeled organic or any other label indicating that they are under conversion. Municipal solid waste compost and sewage sludge composts and the use of prohibited material as composts/ ingredients are banned. There is a check list of acceptable and banned materials which includes pest control treatments and other agricultural inputs like fertilizers and seed treatments. An evaluation of all agricultural inputs has been made for their long term effect on the environment irrespective of the fact that they are synthetic or natural.

Germany is a Leading nation in the adoption of organic farming in Europe. The history and development of organic agriculture in Germany begins from the days of bio-dynamic farming initiated in 1924. Bio-dynamic theory of farming treats farm as a living being which responds to non-material influences, called dynamic forces. The proponents of bio-dynamic farming believe that these forces can be enhanced by bio-dynamic preparations.

Most of the countries lagging behind in the adoption of organic agriculture were from Asia and Africa. In terms of both in absolute and percentage area under organic farming, Tanzania, Zambia, Ghana, India and South Africa are far behind many nations in organic agriculture.

18.3 NEED OF ORGANIC FARMING

Organic farming is farming without the addition of artificial chemicals (Gehlot, 2009). The relevance and need for an eco-friendly alternative farming system arose from the chemical and pesticide residues in food chain and ill effects of the chemical farming practices adopted worldwide during the last quarter of the last century. The methods of farming evolved and adopted by our forefathers for centuries were less injurious to the environment. People began to think of various alternative farming systems based on the protection of human health and environment with a view to have less hazard on human health ultimately by various ways like clean and healthy foods, conducive ecology, low use of natural sources, etc. Many systems of farming came out of the efforts of many experts and laymen. However, organic farming is considered to be the most suitable because of its scientific approach and wider acceptance all over the world.

18.4 INPUT-OUTPUT IMBALANCE

A crop, in its growth process, incorporates a part of the soil fertility into the parts of the plant. The roots remain in the soil. The leaves and stems are fed to the cattle/ burnt as fuel/directly returned to the soil. The consumed part by cattle and human also go back to the soil. The practice of commercial farming leads to continuous depletion of soil fertility as it doesn’t favour the build up of organic matter. The soil becomes powdery and gets eroded by wind or rain. The Earth can produce only a limited amount of biomass from a given area. The maintenance of a dynamic equilibrium balances the inputs and outputs in a region. It is essential for sustainable agriculture.

18.5 LOW PRODUCTIVITY

The productivity of cereals, millets, oilseeds, pulses and plantation crops is very low in comparison with those in other countries in the world. This is in spite of our success in improving the quality of seeds and adoption of efficient technology. The impact of green revolution is showing signs of weakness and production appears to have decreased even after an increase in the inputs used .The production of food grains in the country increased very substantially during 1960 to 2009 to reach 218.1 million tonnes from 60 million tonnes. But the decade ending 1990 and 2000 did not witness such increases and the attainment of the targeted production could not be achieved. The reasons attributed to the low productivity are the drastic reduction in soil nutrients in the areas where fertilizer is used intensively in which the organic matter is not supplemented

18.6 BENEFITS OF ORGANIC FARMING

Organic agricultural practices are based on a maximum harmonious relationship with nature aiming at the non-destruction of the environment. The developed nations of the world are concerned about the spreading contamination of poisonous chemicals in food, feed, fodder and fibre. Naturally, organic farming system is looked upon as one of the means to remedy these maladies there. However, the major problem in India is the poor productivity of soils because of the low level content of the organic matter.

The efficiency of the organic inputs in the promotion of productivity depends on the organic contents of the soil. There were many resemblances of organic farming principles in the traditional agriculture of India. But the former gives a more open and verifiable scientific foundation than the latter.

18.6.1 Safe and healthy foods

On comparing the nutritional values of both organic and conventional foods, it appears that consumption of organic food is more health safe. Such produce had better taste (Naik, 2007). Apples, pears, potatoes, corn, wheat and baby foods were analyzed to find out 'injurious' elements such as aluminum, cadmium, lead and mercury and also 'benefical' elements like boron, calcium, iron, magnesium, selenium and zinc. The organic food, in general, had more than 20 per cent less of the injurious elements and about 100 per cent more of the beneficial elements.

18.6.2 Improvement in soil quality

Soil quality is the foundation on which organic farming is based. Efforts are directed to build and maintain the soil fertility through the farming practices. Multi-cropping, crop rotations, organic manures and pesticides, and minimum tillage are the methods employed for the purpose. Natural plant nutrients from green manures, farmyard manures, composts and plant residues build organic content in the soil. It is reported that soil under organic farming conditions had lower bulk density, higher water holding capacity, higher microbial biomass, carbon and nitrogen and higher soil respiration activities compared to the conventional farms (Sharma, 2003). This indicates that sufficiently higher amounts of nutrients are made available to the crops due to enhanced microbial activity under organic farming. The effect of organic cultivation on soil fertility as reported at the farm of Central Institute for Cotton Research, Nagpur.

18.6.3 Increased crop productivity and income

Field trials of organic cotton at Nagpur revealed that during the conversion period, cotton yield was low compared to the conventional (using fertilizer and pesticides) and integrated crop management (using 50 per cent each of organic and inorganic inputs). However, the yields of organic cotton started rising from third year. Cotton yields under organic, conventional and the mixed systems were 898, 623 and 710 kg/ha respectively at the end of the fourth year of the cultivation. The yield of soybean under organic farming was also the highest compared to the other two systems .The cost of cultivation of cotton was lower in the organic farming than in the modern system ,due to the non-use of fertilizers and chemical insecticides.

18.6.4 Low incidence of pests

Counts of eggs, larva and adults of American BoUworm were far lesser under organic farming than under the conventional method (Sharma, 2003).

18.6.5 Employment opportunities

According to many studies, organic farming requires more labour input than the conventional farming system. Thus, India which faces very high labour unemployment and under employment will find organic farming an attraction

18.6.6 Indirect benefits

Several indirect benefits from organic farming are accrued to both the farmers and consumers. While the consumers get healthy foods with better palatability and taste and nutritive values, the farmers are indirectly benefited from healthy soils and farm production environment. Eco-tourism is increasingly becoming popular and organic farms have turned into such favourite spots. Protection of the ecosystem, flora, fauna, and increased biodiversity and the resulting benefits to all human and living things are great advantages of organic farming which are yet to be properly accounted for.

18.7 INITIATION OF ORGANIC FARMING IN INDIA

The International Conference on "Indian Organic Products-Global Markets" at the end of 2002 was the first to be held in India. IFOAM predicts that India and China have great potential to be organic farm produce exporters in the future. An important event in the history of the modern nascent organic farming in India was the unveiling of the National Programme for Organic Production (NPOP) in May, 2000 and the subsequent Accreditation and Certification Programme in October, 2001. The logo

"India Organic" was released in July 2002 to support the NPOP. Indian organic products are mainly exported to Europe (Netherlands, United Kingdom, Germany, Belgium, Sweden, Switzerland, France, Italy, Spain, etc.), USA, Canada, Saudi Arabia, UAE, Japan, Singapore, Australia and South Africa.

18.8 REGULATIONS

The most important step towards organic farming taken by the government was to draw a regulatory framework. It is true that the initiatives by the government to introduce organic farming by laying down regulations was taken late as many countries had already done this kind of basic work decades ago. The implementation of NPOP is ensured by the formulation of the National Accreditation Policy and Programme (NAPP). The regulations make it mandatory that all organic certification bodies should be accredited by an Accreditation Agency. The international certification agencies operating in India even prior to these regulations will also have to get accreditation under the new dispensation.

The regulations lay down the institutional arrangements for implementing the national programme for organic production. The NPOP is administered, monitored and implemented for the benefit of farmers, processors, traders and consumers. It envisages a three tiered organization under the overall guidance of the Union Government with the Department of Commerce, Ministry of Commerce and Industry as the nodal agency. Policy making and declaration of the standards for organic products, recognition of organic standards of other nations, efforts to get our standards recognized by others and coordination with other arms of the government for the successful management of the organic agriculture are the major functions entrusted to the ministry.

The agencies accredited are the Agricultural and Processed Food Products Export Development Authority (APEDA), Coffee Board, Tea Board and the Spices Board. The regulations cover exports, imports and the domestic trade of the organic products. But the government regulations are applicable to only the exports. So, an organic farm product can be exported only if it is certified by a certification body accredited for the purpose. The categories of products covered under accreditation are organic crop production, organic animal production, organic processing operations, wild products etc. APEDA had drafted organic animal husbandry, aqua culture and organic textile standards and could be finalized very soon. For better trainability APEDA has implemented trace net system, it is an online system where all possible information of organic operators is available on website.

Organic production requires certification after periodic inspections in order to ensure that all prescribed practices are followed. The inspection and certification are done by the agencies accredited to the Accrediting Agencies dealing with the commodity. As on date eighteen certification bodies are accredited in India. Inspection and certification agencies can be government departments, NGOs, trade or consumer or producer organizations. Such agencies should be registered bodies, with managements in position, declaring the persons who shall be held responsible for any miscarriage of certification and having proof of adequate field staff to undertake periodic inspections. The continued accreditation of such bodies is dependent upon their record of fidelity to the principles of organic production. They are authorised to award certificates after due satisfaction that practices conformed to those enunciated by the Accrediting Agency in relation to the item concerned. The charges levied by the certifying agencies are fixed by the Accrediting Agencies.

Accrediting agency accredit certification agency for three years and conducts surveillance audits every year. If any major discrepancies or non conformities were found then accrediting agency takes action against certification agency, it may be as severe as suspending it from carrying any certification activities.

18.9 GOVERNMENTAL SUPPORT

A National Institute for Organic Farming has been established to spearhead research in organic agriculture. The government of India constituted task force had also recommended the initiation of the postgraduate level courses in organic farming.

The agriculture ministry is promoting organic farming in the country by implementing various schemes such as National Project on Organic Farming, the National Horticulture Mission, the Technology Mission for North East and the Rastriya Krishi Vikas Yojana. 'The ICAR (Indian Council of Agricultural Research) has initiated an All India Network Project on organic farming to ensure the development and promotion of scientifically proven methodologies in organic farming..

18.10 PROBLEMS AND CONSTRAINTS

The most important constraint felt in the progress of organic farming is the inability of the government policy making level to take a firm decision to promote organic agriculture. Unless such a clear and unambiguous direction is available in terms of both financial and technical supports, from the Centre to the *Panchayat* levels, mere regulation making will amount to nothing. The following are found to be the major problem areas for the growth of organic farming in the country:

18.10.1 Lack of awareness and shortage of skilled manpower

It is a fact that many farmers in the country have only vague ideas about organic farming and its advantages as against the conventional farming methods. Use of bio-fertilizers and bio pesticides requires awareness and willingness on the part of the farming community. Knowledge about the availability and usefulness of supplementary nutrients to enrich the soil is also vital to increase productivity.

Farmers lack knowledge of compost making using the modern techniques and also its application. The maximum they do is making a pit and fill it with small quantities of wastes. Often the pit is flooded with rainwater and result is the top of the compost remains under composted and the bottom becomes like a hard cake. Proper training to the farmers will be necessary to make vermicompost on the modern lines.

Attention on the application of composts/organic manure is also lacking. The organic matter is spread during the months when the right moisture level is absent in the soil. The whole manure turns into wastes in the process. The required operation is of course, labour intensive and costly, but it is necessary to obtain the desired results.

18.10.2 Output marketing problems

It is found that before the beginning of the cultivation of organic crops, their marketability and that too at a premium price over the conventional produce has to be assured. Inability to obtain a premium price, at least during the period required to achieve the productivity levels of the conventional crop will be a setback. It was found that the farmers of organic wheat in Rajasthan got lower prices than those of the conventional wheat. The cost of marketing of both types of products was also same and the buyers of wheat were not prepared to pay higher prices to the organic variety .

18.10.3 Shortage of bio-mass

Many experts and well informed farmers are not sure whether all the nutrients with the required quantities can be made available by the organic materials. Even if this problem can be surmounted, they are of the view that the available organic matter is not simply enough to meet the requirements. The crop residues useful to prepare vermi-compost are removed after harvest from the farms and they are used as fodder and fuel. Even if a portion is left out on the farms termites, insects etc destroy them. Experiments have shown that the crop residues ploughed back into soil increase productivity and a better alternative is conversion into compost. The small and marginal cultivators have difficulties in getting the organic manures compared to the chemical fertilizers, which can be bought easily, of course if they have the financial ability. But they have to either produce the organic manures by utilizing the biomass they have or they have to be collected from the locality with a minimum effort and cost. Increasing pressure of population and the disappearance of the common lands including the wastes and government lands make the task difficult.

18.10.4 Inadequate supporting infrastructure

In spite of the adoption of the NPOP during 2000, the state governments are yet to formulate policies and a credible mechanism to implement them. There are only four agencies for accreditation and their expertise is limited to fruits and vegetables, tea, coffee and spices. The certifying agencies are expensive, the direct access of organic markets are non-existent, the trade channels are yet to be formed and the infrastructure facilities for verification leading to certification of the farms are complicated and limited.

18.10.5 High input costs

The small and marginal farmers in India have been practising a sort of organic farming in the form of the traditional farming system. They use local or own farm renewable resources and carry on the agricultural practices in an ecologically friendly environment. However, now the costs of the organic inputs are higher than those of industrially produced chemical fertilizers and pesticides including other inputs used in the conventional farming system. The groundnut cake, neem seed and cake, vermi-compost, silt, cow dung, other manures, etc. applied as organic manure are increasingly becoming costlier making them unaffordable to the small cultivators.

18.10.6 Marketing problems of organic inputs

Bio-fertilizers and bio-pesticides are yet to become popular in the country. There is a lack of marketing and distribution network for them because the retailers are not interested to deal with these products, as the demand is low. The erratic supplies and the low level of awareness of the cultivators also add to the problem. Higher margins of profit for chemical fertilizers and pesticides for retailing, heavy advertisement campaigns by the manufacturers and dealers are other major problems affecting the markets for organic inputs in India.

18.10.7 Absence of an appropriate agriculture policy

Promotion of organic agriculture both for export and domestic consumption, the requirements of food security for millions of the poor, national self-sufficiency in food production, product and input supplies, etc. are vital issues which will have to be dealt with in an appropriate agriculture policy. These are serious issues the solution for which is hard and consistent efforts along with a national consensus will be essential to go forward. Formulation of an appropriate agriculture policy taking care of these complexities is essential to promote organic agriculture in a big way.

18.10.8 Limited financial support

The developing countries like India have to design a plethora of national and regional standards in attune with those of the developed countries. The adoption and maintenance of such a regulatory framework and its implementation will be costly.

The cost of certification, a major component of which is the periodical inspections carried out by the certifying agencies, which have freedom to fix the timings, type and number of such inspections appears to be burdensome for the small and marginal farmers. Of course, the fees charged by the international agencies working in India before the NPOP were prohibitive and that was a reason for the weak response to organic agriculture even among the large farmers in the country. No financial support as being provided in advanced countries like Germany is available in India. Supports for the marketing of the organic products are also not orthcoming neither from the State nor from the Union governments. Even the financial assistance extended to the conventional farming methods are absent for the promotion of organic farming.

18.10.9 Low yields

In many cases the farmers experience some loss in yields on discarding synthetic inputs on conversion of their farming method from conventional to organic. Restoration of full biological activity in terms of growth of beneficial insect populations, nitrogen fixation from legumes, pest suppression and fertility problems will take some time and the reduction in the yield rates is the result in the interregnum. It may also be possible that it will take years to make organic production possible on the farm.

Small and marginal farmers cannot take the risk of low yields for the initial 2-3 years on the conversion to organic farming. There are no schemes to compensate them during the gestation period. The price premium on the organic products will not be much of help, as they will disappear once significant quantities of organic farm products are made available.

18.10.10 Inability to meet the export demand

The demand for organic products is high in the advanced countries of the west like USA, European Union and Japan. It is reported that the US consumers are ready to pay a premium price of 60 to 100 per cent for the organic products. The upper classes in India are also following this trend as elsewhere in the world. The market survey made by the International Trade Centre (ITC) during 2000 indicates that the demand for organic products is growing rapidly in many of the world markets while the supply is not attuned. India is known in the world organic market as a tea and spices supplier and there is a good potential to export coffee, vegetables, sugar, herbs, nuts, cosmetics etc. In spite of the several initiatives to produce and export organic produces from the country, the aggregate production for export came to only about 42000 tonnes. This also includes the production of organic spices in about 1000 ha under certification.

18.10.11 Vested interests

Hybrid seeds are designed to respond to fertilizers and chemicals. The seed, fertilizer and pesticide industries as also the importers of these inputs to the country have a stake in the conventional farming. Their opposition to organic farming stems from these interests.

18.10.12 Lack of quality standards for bio-manures

The need for fixing standards and quality parameters for bio-fertilizers and bio-manures has arisen with the increasing popularity of organic farming in the country. There are a very large number of brands of organic manures, claiming the high levels of natural nutrients and essential elements. But most farmers are not aware of the pitfalls of using the commercially available bio-manure products. While the concept of organic farming itself lays great stress on the manures produced on the farm and the farmers' household, many of the branded products available in the market may not be really organic.

18.10.13 Improper accounting method

An understanding of the real costs of erosion of soil and human health, the loss of welfare of both humans and other living things and the computation of these costs are necessary to evaluate the benefits of organic farming. These costs will have to be integrated to a plan for the implementation of organic agriculture.

A recent study shows the inappropriateness of the cost and return accounting methods adopted to find out the economics of the organic farming. An economic evaluation of the bad effects of inorganic agriculture and their internalization through environmental taxes is proposed for a market based approach to promote organic farming in India.

18.11 FUTURE PROSPECTS

Indian agriculture should be able not only to maintain but also must strive to increase the production of food grains. It appears that given the availability of organic infrastructure, minimum efforts for conversion due to the low use of chemical farming methods and the limit of the public investment, organic farming can be progressively introduced. The potential areas and crops, which fulfil the above constraints, could be explored and brought under organic agriculture. The rainfed, tribal, north-east and hilly regions of India where the traditional farming is more or less practised could be considered . Agriculture production in these areas is still almost on the traditional eco-friendly lines and making the farmers aware of the methods of organic farming may not be very difficult. A strategy to prevent sudden and substantial yield losses is to convert to organic production in phases to reduce the risks during the initial years. The question of the vast requirement of organic matter to the country's farms in order to switch over to organic agriculture is also answered. Chemical fertilizer is applied only in 30 per cent of the cultivated area, which is irrigated, and the remaining land is under rainfed agriculture with almost no fertilizer application. Also the rainfed area under cultivation accounts for only 40 per cent of the food grain production of the country. The introduction of organic farming in these areas will allay the fears of a sudden sharp decline of food production. Thus the demand for biomass for the production of organic manures can also be controlled in a phased manner. Moreover, the simple technologies with low input use have been developed for dry farming and they can be transferred to the farms for organic cultivation. The resulting increase in productivity and sustainability of production will increasingly contribute to the betterment of the economic condition of the dryland faming community, which is very poor in the country. Several alternatives for supply of organic soil nutrients like vermicomposts and biofertilizers exist. Technologies have been developed to produce large quantities of these nutrients. Crop specific biofertilizers for cereals, millets, pulses and oil seeds are also available. Vermicomposting and bio-fertilizer manufacturing can be undertaken to increase the supply of organic manure to meet the demand.The basic rules and regulations for accreditation and certification of organic products are in place in India. This has become necessary to cultivate the acceptable levels of crop, livestock and human nutrition products and above all to protect both the crops and humans from pests and diseases through the use of bio-chemicals and bio fertilizers. Such a situation is suitable for making the farming community aware of the organic farming methods to make the switch over less troublesome. Currently there is a good demand for organic products in the domestic market/ international market, which is not met out by suppliers. The linkages between the two do not exist which in turn discourage production. The wholesalers/ traders play a major role in the distribution of organic produces as they source from the small farms. Large farmers have access to international buyers, supermarkets and own stalls for distribution. Mumbai, Delhi, Kolkota, Chennai, Bangalore and Hyderbad are the major domestic markets for organic products. Retail chains are trying in India to make organic foods a separate division. Domestic organic standard is finalized and will be enforced very shortly then in India also organic market will be consolidated and organized.

18.12 CONCLUSIONS

A country like India can enjoy a number of benefits from the adoption of organic farming. The price premiums for the products, conservation of the natural resources in terms of improved soil fertility and water quality, prevention of soil erosion, preservation of natural and agro-biodiversity are major benefits. Economic and social benefits like generation of rural employment, lower urban migration, improved household nutrition, local food security and reduced dependence on external inputs will be large gains in the Indian conditions. The protection of environment and the consequent increase in the quality of human life will be other contributions of organic farming. The ill effects of the conventional farming system are felt in India in terms of the unsustainablity of agricultural production, environmental degradation, health and sanitation problems, etc. Organic agriculture is gaining momentum as an alternative method to the modern system. Many countries have been able to convert 2-10 per cent of their cultivated areas into organic farming. The demand for organic products is growing fast (at the rate of 20 per cent per annum in the major developed countries).

It appears that India is slow in the adoption of organic farming. So far, few achievements are:

- Commencement of National Programme for Organic Production (NSOP)
- Approval of 18 certification agencies (government, private, national and international bodies) whose expertise is in wide range of crops.
- Support provided through NHM, MoAgri. , ICAR etc.

Some of the issues, which require attention at the government policy are as under :

1. Substantial financial support is absolutely necessary to promote organic farming. In India, organic farmers do not receive the benefits of government subsidies as they are targeted at the conventional cultivation.
2. To make the organic farming an attractive proposition at support price must be declared.
3. Government has introduced organic farming in the north-east region and in the dry farming areas of India. A programme for organic agriculture in these

areas must be fully supported by the full compensation both in cash and kind to the farmers in the event of the crop failure.

4. Market development for the organic products is a crucial factor to promote export and domestic sales.
5. Supports to the producer and consumer associations to market the organic products.
6. The producer organizations/ traders/ operators must be encouraged to get certification in accordance with the NPOP.
7. Reduction in the costs of certification besides the simplification of the process.
8. A vigorous campaign to highlight the benefits of organic farming against the conventional system is essential to increase the awareness of the farmers and consumers.
9. Package of practices for organic cultivation should be available to farmers for better output.
10. Equivalence of NPOP with organic standards of EU, USA, Canada, South Korea and other countries.

18.13 REFERENCES

Gehlot, Dushyent (2009) *Organic farming*, published from Agrobios, Jodhpur, 1p.

Hirada M. (1975) Minamata disease : a medical report In : Minamata, Smith WE and Smith AM (Eds) Holt Reinhart, New York.

Malik, Mahmood N (2000) *Horticulture*, published from biotech books, Delhi, 10p.

Naik, Rama K (2007) Nutritional value of organic food. In *Organic family*, APA, Udaipur, pp. 276-280.

Padmadevi, K.; Vijaya, T. and Vijaykumar, M. (2008) Organic farming of plantation crops. In : *Innovative Horticulture* (Eds. R. Arun Kumar *et al.*) New India Publishing Agency, New Delhi, 105p.

Chapter – 19

Homa Jaivik Krishi : A Hope for Sustainable Horticulture – Mardi's Approach

R. K. Pathak

19.1 INTRODUCTION

Indiscriminate use of agro-chemicals over the last five decades has adversely affected soil fertility, crop productivity, produce quality and particularly the environment. These degenerative effects of intensive agriculture have compelled to think for alternative and sustainable system of agriculture. As a result, number of alternative systems *viz Biodynamic, Rishi Krishi, Panchagavya Krishi, Natueco Organic Farming, Homa Organic Farming, Natural Farming and Jaivik Krishi* have emerged in different parts of the country. In organic farming production is achieved without use of chemical input (Singh, 2011 and Malik, 2000). In these systems, maximum reliance is placed on self-regulatory agro systems and locally available or farm derived renewable resources. Since the soil, water and environment have been contaminated, hence even organic production techniques once effective are now are not capable of mitigating these issues. This has created number of myths in the mind of policy makers and even large number of farmers.

19.2 MYTHS IN ORGANIC FARMING

- Can yield level be maintained in organic farming?
- Are there enough organic wastes available for conversion of entire area in the country under organic farming?
- Is organically produced food are nutritive as compared with those grown by conventional systems?
- Can pests and diseases be managed organically?

- Is the organic techniques are eco friendly and cost effective?
- Are there enough market for organic produce?

In fact, the Indian Farming Systems had been a sustainable way of life for thousands of years. The ancient system was primarily based on harnessing the energy sources available in our neighborhood. The four sources of energy i.e. energy are Cosmos, Cow, mother Earth and Plants are available in plenty. We at MARDI are of the view that if systematically harnessed these can ensure sustainable production and "Ever Green Revolution" even at this juncture.

The state of Rajasthan is well suited for harnessing the strength of organic production. The main reasons are minimum use of agro-chemicals as compared to other states, sizeable areas dominated by tribal population, presence of enough cattle population, strength of horticultural crops which have export value (seed spices), water shortage and climatic vagaries. Most of these problems can be resolved by adaptation of *Homa Jaivik Krishi* in intensive way. Brief accounts of these have been dealt.

The cosmic energy is harvested by adopting Agriculture Calendar (*Krishi Panchang*), proper habitat development through encouraging plantation, creation of water bodies, storage of rain water and its recycling, integration with cow, caring mother earth for increased humus, her protection and its integration with *Homa* Organic Farming are the major strength of our technology.

19.3 STEPS OF MARDI APPROACH

19.3.1 Habitat development

Proper habitat development is done by encouraging wide range of plants, bushes, water bodies for creation of diversity of flora and fauna in the area through involvement of farmers. Green vegetation helps in capturing of solar energy, release of oxygen and storage of CO_2 as carbon in the wood. They also act as source of biomass; provide congenial conditions for birds, predators and parasites helpful in farming. A large number of plants Biological Nitrogen Fixers (BNF) are potent source for capturing N (78.6%) freely available in the atmosphere. It is interesting to record that rhizosphere soil of certain trees such as *Banyan* (*Ficus bengalensis*), *Khejari* (*Prosopis cineraria*) etc are rich in microbial consortia, while different plants parts have pesticides properties. These are utilized in preparation of bio-enhancers and bio-pesticides. All these are helpful in one or other way in assuring sustainable production.

19.3.2 Agriculture calendar

Life pattern of all living organisms are interwoven into cosmic rhythms. Human life, as well animal and plant life is strongly dependent on the rhythms of the earth and moon. On the basis of such influences, planting calendar is prepared for the year. The agricultural operations such as compost preparation, use, field preparation, seed sowing

/transplanting, harvesting, storage are performed during different timings of the year. It is interesting to record that 10-15 per cent increase in yield, minimization in pest's incidence and better quality can be ensured by simply adopting the calendar (as annexed 2010-11).

19.3.3 Seeds/Plants

Preferably locally adopted varieties are preferred over the hybrids and imported ones. Organically produced seeds and planting materials, their storage by organic techniques without any chemicals are being encouraged. This is almost virgin area, lot of research and development is required.

19.3.4 Cow

The cow is a very special creature. No other animal gives so many products which are useful to human being The five products derived from the cow, and which are very beneficial for good health and in agriculture i.e. *Panchagavya* are: Milk, Curd, ghee, dung and urine. The indigenous breed of cow (*Bos indicus*) with hump has capacity of absorbing cosmic energy its dung is rich in microbial consortia, while urine and butter milk are used for preparation of composts, bio- enhancers and bio- pesticides.

19.3.5 Bio- enhancers

With the help of cow products, number of bio-enhancers and bio-pesticides are prepared and used in organic farming. The main bio- enhancers are BD-500, CPP, *Amrit Pani, Amrit Mati, Jiwamrita, Panchagva*, Biosol, Vermiwash etc. Out of these *Jiwamrita* is cheap and easy to be practised; hence it has been included here.

(i) **Seed and seedlings are very tender in nature:** hence they need protection in early stages of their establishment. Hence, their treatment with organic products will be helpful in better germination/establishment and vigour. *Bijamrita* is easy to use for seed treatment.

It is prepared in mud pot, cement or plastic container. Five kg fresh cow dung is immersed in water by tying in a cotton bag over night. In next morning the dung is washed thoroughly 3 times with water to get essence of dung in water. In these, 5 litres of cow urine is added and the mixture is stirred thoroughly by mixing 100g virgin soil. 50g lime is mixed with one litre of water separately for a night. The lime water is added and the solution is diluted to the tune of 200 litres with plain water. *Bijamrita* is ready for use. It is advisable to prepare fresh *Bijamrita* and use it for treatment of seeds/ seedlings and other plant parts before sowing/planting/transplanting. It enhances germination, protects infection of pathogens and increases tillering and plant vigour.

(ii) ***Jiwamrita*** **-** It is one of the cheapest bio-enhancers, which can be prepared by individual farmers and used regularly to improve the biological properties of the soil and thus the humus content. Ingredients for preparation of *Jiwamrita* are cow dung (10 kg), urine (5-10 l), and virgin soil (1kg). Since it is rich in microbial consortia, hence 2kg jaggery and pulse flour are also incorporated just as a food to the microbes. These are mixed thoroughly in a plastic or cement container filled to 200 l with plain water. The mixture is incubated for 3-4 days by stirring 3 times a day and has to be used within a week. *Jiwamrita* is Two hundred litres of *Jiwamrita* is enough for one-acre area. It can be applied through irrigation water by flow, drip or sprinkler or even by drenching on organic mulches spread over in the field or in the tree basin.

19.3.6 Mulching

It pertains to covering soil, with a thin layer (10-25cm) of waste/crop residue or other similar locally available material. Mulch prevents soil erosion, water loss, weeds growth, minimizes temperature fluctuation, and improve organic carbon content in the soil. Mulches modify the micro-climate, alter the environment of soil microbes, and enhance soil flora and fauna activities.

In perennial trees even its own leaves, pruned shoots etc if properly recycled, can meet 60-80 per cent nutrition requirement. These can be further riched by growing some legumes during rainy season and in the ploughing down them back into the soil.

19.3.7 Tree pasting

Tree pasting with paste prepared with cow dung, urine, clay along with some dung ash has been found very effective in management of gummosis, infestation of pests like stem bores etc. In case, *Agnihotra* is performed in the orchard, *Agnihotra* ash has better response than cow dung ash. If possible, pasting twice (after rains and during spring) in a year should be practised particularly in crops like mango, guava, *aonla*, citrus etc. Besides protection, it encourages better transport of nutrients and water from the ground and photosynthetic materials from foliage to root systems.

19.3.8 *Homa* organic farming

Its concept has been drawn from *Vedas* the most ancient body of knowledge known to mankind. It is a super technology, which simultaneously negates the effects of polluting factors while increasing per unit area of quality production. The knowledge can used beneficially to grow all kinds of crops without chemicals and pesticides. It is known as *Homa Organic farming*. The process of "*Agnihotra* Pyramid Fire" is the basic *Homa*. *Agnihotra* is a must to sustain life now on this planet. In fact, it is gift to humanity from ancient most *Vedic* sciences of bio- energy, medicine, agriculture and climate engineering.

Agnihotra is the basic fire in *Homa* therapy. It is process of purification of the atmosphere through the agency of fire, prepared in a copper pyramid tuned to the biorhythm of sunrise/sunset. It works on the principle that one heals the atmosphere and the healed atmosphere in turns heals the micro-atmosphare. As an impact of *Agnihotra*, tremendous amount of energy are gathered which creates a magnetic field? This neutralizes the negative energies and reinforces positive energies. Pyramid is a generator and fire is turbine. Cow dung, *ghee* and rice then interact to form a composition, which thrust, and provides nutrients for survival, yield and quality production.

It injects nutrients into atmosphere to prevent diseases favours natural predators and improves honeybee activity. As a result healthy micro-flora and fauna is created. This gives rise to a microenvironment or micro-system, which is comparatively less toxic to growth of plants. After creation of microenvironment, creatures like earthworms and microbes thrive better. They eat the soil, digest it again and replenish the soil.

It has been observed that when "*Agnihotra* ash" is added to the normal soil it increases the water soluble phosphate content of the soil and nutrients are absorbed readily by the root hairs of the plants. Absorption of macronutrients like nitrogen, phosphorus and potash because of small cells and active transport improves in *Homa* atmosphere.

A special bio-formulation developed by Gloria company and named as "Gloria Biosol" from Peru is prepared after a series of operations and processes that lead to the biodegradation of organic matter, worm humus, fresh cattle manure and water, until reaching mineral grade. It is powerful bio-food and bio-fertilizer for the plants with high macro and micronutrients.

Gloria Biosol acts like a powerful foliar substance, allowing development of good amount of functional leaves (with a shortening of more than 50% in its vegetative cycle), what assures a good production in the plant.

Because of its high content of enzymes, phyto-hormones, beneficent microorganisms, and other special and useful components the plants and for soil, it contributes to health recovery and the nutrition of the resource soil, if it is directly applied to the agricultural field as foliar spray.

Organic farming is effective in maintaining sustainability, presserving biological property of soil, conserving natural resouces, minimizing environmental degradation, reducing cost of cultivation and over and above in protecting human health (Satyanarayan, 2011)

From the above description, it is apparent that *Homa Jaivik Krishi* which is capable of enriching rhizosphere is suitable option to ensure sustainable agriculture and can mitigate the present day crises in agriculture including climatic disasters. Hence, there is urgent need to initiate systematic research and promote modern tools of transfer of technology (TOT) so that it is promoted on wide range of crops in different agro climatic situations in the country in assertive way.

AGRICULTURE CALENDER 2010-11

Month	Best Dates for Sowing		Sowing Dates as per Moon in Zodiac ***				Agric activities****			
	Moon sat. 180 *	48 Hrs. Before full moon **	Seed & Fruit Eg. Grains, Pulses, Oil Seeds & Fruits	underground crops Eg. Potato, Onion, Garlic, Carrot etc.	Flower Eg. Rose, Marigold, Crysanthemum etc.	Leaves Eg. Salad, Green, Leafy Green, Veg. Grass	Nodes Rahu & Ketu	Apogee	Perijgee	New Moon
April	12	26	5,15,16,24	7,8,17,18,26,27	1,10,11,19,27,28	3,12,13,22,30	6,20	9	25	14
May	9	25	2,12,13,21,22,31	5,6,15,23,24	8,18,25,26	1,10,11,19,27,28	3,17,30	7	20	14
June	6	24	9,10,17,18,26	1,2,11,19,20,28,29,30	4,5,13,21,22	6,7,16,24,25	14,27	3	15	12
July	3,31	24	6,7,14,15,23	8,9,17,26,27	2,10,19,20,28	3,4,5,12,21,22,31	11,24	1,29	13	11
August	27	22	2,3,11,1219,21,29,30,31	5,6,13,14,22,23	8,15,16,24,26	1,9,17,18,27,28	7,20	25	10	10
Suptember	24	21	7,17,26,27	1,2,9,10,18,19,28,29	3,11,12,22,30	5,6,13,14,23,24	4,16	21	8	8
October	21	21	5,14,23,24	15,16,17,25,26	9,10,19,29	3,11,12,20,21,22,30,31	1,13,28	18	6	7
November	18	19	1,2,9,11,19,20,28,29	4,12,13,22,23,30	5,14,16,25	7,8,17,18,26,27	10,24	15	3	6
December	15	19	8,17,18,26	9,10,19,20,28	3,12,22,30,31	6,14,15,23,24	7,21	13	1,25	5
January	12	17	13, 14, 23,	6, 7,15, 16,24, 25	8, 9,19, 26, 27,	1, 2, 11,20, 21,28, 29	3,18,30	22	10	4
February	8	16	1, 9, 10, 11, 18, 28	2, 12, 13,20, 21	4, 5, 6, 15,22, 23	17, 24, 25	14,27	19	7	3
March	8	17	9, 10,18, 19, 27	1, 2, 11, 12,28, 29	5, 14, 22, 23,31	7, 16, 24, 25,	13,26	20	6	4

Works to be done in : * Best the for sowing all crops, Crops Soon tolerate pest and disease infection; ** Better germination due to good moisure content ;

*** Ideal dates for sowing & performing other agric. activities for different sets of crops **** Avoid sowing and other agric. activities on these days, except Apogeeis good for potato sowing;

Prepared by : Shri T.G.K. Menon, Dr. O. N. Solanki

19.4 STRATEGY FOR PROMOTION OF *HOMA JAIVIK KRISHI*

- Intensive research on organic farming systems by networking of research institutions of ICAR institutes and State Agricultural Universities to develop reliable data for convincing policy makers and development of *Homa Jaivik package,*
- Introduction of organic farming curriculum at school, college and at University,
- Creation of organic input production units at the main campus, zonal research stations, KVKs, KGKs for skill upgradation and initial supply of inputs in the initial years,
- Promotion of Organic Farmers Interest Groups (OFIGs), to undertake organic farming activities,
- Encourage to federate these groups through formation of Farmers Federation at district, state level to encourage organic production, value addition and marketing,
- Efforts need to be initiated to divert the Govt. funds to the practicing farmers' through Producer Company Concept (for the farmers, of the farmers and by the farmers),
- Co-ordination with Animal Husbandry department to increase population of local cattle breeds for quality input production and maintenance of soil fertility,
- Reintroduction of mixed farming system by integration with sheep, goat and poultry for optimum utilization of land, family labours and local resourses
- Efforts are required to encourage group to produce their own seeds and plants by adopting organic techniques
- Encourage to plant trees, shrubs, bushes at the farm boundaries and vacant places for congenial organic atmosphere through habitat development
- If possible, arrangement of water body for conservation of rain water and it's sharing as life saving irrigation by fellow farmers
- Try to develop Public Private Partnership (PPP) model in development of *Homa Jaivik* package and its adoption by the farming community
- Provide skill upgradation training and some basic infrastructure for input generation at the farm in organic village
- Develop and provide enough literature which can be understood by the farming community for its implementation
- Try to use modern extension techniques by utilizing mass, electronic media, NGOs, Corporate Sector including local bodies

- Arrange visit of group members to established organic farms in the local area to closely observe and learn secrets of successful organic farmer
- Help them to prepare cropping plan suiting to their need for Group Certification through Participatory Guarantee System (PGS) for certification and marketing
- Establishment of organic hubs which could serve as Information Technology Centers (ITC) and collection centers to facilitate and provide information on value addition and marketing
- Creation of better infrastructure facilities like processing units, vacuum packing units and community cold storages, warehouses etc for Organic produce which can be exported to distant markets
- Improved credit facilitates for organic farming through banks
- Increasing consumer awareness on the values of *Jaivik* products consumption
- Facilitating Contract Farming of *Homa Jaivik Krishi* on large scale for specific commodities by linking through Super markets and Multinational companies
- Establishment of dedicated Organic Resource Centres by pooling all organic projects of Central and State Govt. in order to avoid duplication in approach in villages for promotion of *Homa Jaivik Krishi.*

19.5 REFERENCES

Malik, Mahmood, N (2000) *Horticulture*, Biotech Books, Delhi, p.10.

Satyanarayan (2011) *Unnat Krishi Ke Sidhhant evan jaivik Kheti, RSSC* Jaipur, p. 6

Singh, Jitendra (2011). *Basic Horticulture*, Kalyani Publishers, Ludhiana, p.15.

Chapter – 20

Plant Bio- Enhancers : Potential and Affordable Source for Fertigation

R. K. Pathak

20.1 INTRODUCTION

With rapid increase in area under micro irrigation, now fertigation is getting momentum in many states in the country. At present, most of the soluble fertilizers are imported in the country. Hence, these are very expensive and beyond the reach of common farmers. It is interesting to record that for number of nutrients, which are becoming constraints in most of the Indian soils, soluble formulations have yet to be developed. This calls for some alternative options. As an alternatives liquid manures like cow urine, biogas slurry, or fermented manures etc can be sprinkled on the soil, applied with irrigation water or diluted and sprayed as foliar fertilizers (Frank *et al;* 2005).To our understanding, organic fertilizers /organic liquid manures referred now as bio-enhancers are essential to maintain the activity of microorganisms and other life forms in the soil. These are prepared locally, can resolve number of apprehensions and will be helpful in boosting production and mitigating number of nutritional disorders in most of soils and crops. It is interesting to record that these can be prepared at the farm with some infrastructure facilities and hands on training. Most of these formulations are used through irrigation. These organisms (bacteria and moulds) improve the soil health by solubilzing the complex organic substrates into simple forms and make it available to the plant, resulting in increased productivity. If these are integrated with micro irrigation techniques after proper filtration their efficacy can further be improved.

Cow dung and urine are important component in organic farming. Cow dung is rich source of microbial consortia *Actinomycetes*, which are helpful in control of many plant diseases. A potential species *Streptosporangium psedovulgare* of *Actinomycetes* and four strains of *Bacillus subtalis* have been identified at CISH, Lucknow (Pathak *et.al.*, 2009).These showed anti pathogenic potential against anthracnose gummosis, stem end rot and dieback pathogens. These were observed to attack mycelia cell wall and finally the host cell degenerates completely. Organic liquid manures play a key role in promoting growth and providing immunity to plant system (Sreenivasa *et al.*, 2010).

Cow urine is rich in macro, micronutrients and has disinfectant and prophylactic properties. It purifies the atmosphere and improves the soil fertility. Simple cow urine diluted with water in ratio of 1: 20 and used as foliar spray is not only effective in management of pathogens and insects, but also acts as effective growth promoter for the crops. In plant based organic residue, leaves and other plant parts from leguminous plants such as *Glyricidia, Sesbania, Leucanea leucocephala, Casia tora* etc. are good source as biological nitrogen.

A brief write up for preparation of bio-enhancers and their role in organic, Integrated crop management practices are enumerated below.

20.2 LIQUID BIO- ENHANCERS

Liquid manures, liquid fertilizers, henceforth termed as bio-enhancers are organic preparations, obtained by active fermentation of animal and plant residues over specific duration. These are rich source of microbial consortia, macro and micronutrients and plant growth promoting substances. These are utilized to treat seeds/ seedlings, enrich the soil and induce better plant vigour. Farmers of different states are using many variants of these liquid manures. Brief accounts of these are enumerated below (Frank *et al.*, 2005).

20.2.1 Cow urine

Collect cow urine, mix 1 litre of cow urine with 10 to 15 litres of water and spray the mixture on the plants every two to three weeks during crop growth.

20.2.2 Slurry from biogas

Take 10-15kg slurry from biogas plant, tie it in a piece of cloth and allow if to suspend it in a drum of 100 litres water for 10-15 days, so that the water in the drum turns grey to blackish. Mix slurry solution and water in 1:1 ratio in a spray pump and spray at every 2-3 weeks interval till flowering.

20.2.3 Matka khad

Mix 15kg cow dung + 15 litres cow urine + 250g black jaggery. Keep for 8 days in an earthen pot (*matka*), then dilute in 200 l of water. Two to three times application 3-4 weeks after germination is done.

20.2.4 *Charota* (*Cassia tora*)

A leguminous weed very common during rainy season. Fill a drum with *charota* weed. Fill it with water and cover the drum with a piece of cloth. Let it to ferment for a week. Dilute the liquid ratio 1:1 in with water and spray or sprinkle 3-4 times during crop growth.

20.2.5 *Panchagavya*

It is a special bio-enhancer prepared from five products obtained from cow, i.e. dung, urine, milk, curd and *ghee*. When suitably mixed, incubated and used, these have miraculous effects. The preparation is rich in nutrients, auxins, gibberellins, and microbial fauna and acts as tonic to enrich the soil, induce plant vigour with quality production. Its remarkable effects have been demonstrated in fruits like mango, guava, acid lime, banana, spice-turmeric, flower-jasmine and vegetables such as cucumber, spinach etc. The spray of *Panchagvya* on chillies produces dark green coloured leaves within 10 days and its role has been reported by Sreenivasa *et al;* (2009).

20.3. PREPARATION METHOD

There is need of a wide mouth mud pot, concrete tank or plastic cans. Metal containers should be avoided. In fact use of *Panchagavya* was for the first time standardized by a medical doctor Dr. Natarajan. Later on it was thoroughly studied by the scientist from Tamil Nadu Agricultural University. To prepare it first fresh cow dung and cows *ghee* are put into the container and are added it is thoroughly mixed twice for three days. On the fourth day rest of the ingredients and stir red twice daily for 15 days. The *Panchagavya* stock solution will be ready after 28th day. It should be kept in the shade and covered with a wire mesh or plastic mesh net to prevent houseflies from laying eggs and formation of maggots (worms) in the solution. It has population of useful microbes such as :

Table 1 : Microbial Load in Panchagavya

Fungi	38800/ml
Bacteria	1880000/ml
Lactobacillus	2260000/ml
Total anaerobes	10000/ml
Acid formers	360/ml
Methanogen	250/ml

Physico-chemical studies revealed that *Panchagavya possess almost all macro*, micronutrients and growth hormones (IAA, GA) required for plant growth. Predominance of fermentative microorganisms like yeast and *Lacto bacillus* are *due* to combined effect of low pH milk products, and addition of jaggery/sugarcane juice which acts as substrate for their growth. In general 3-4 (3-4kg/100l) per cent solution of *Panchagavya* has been found effective for most of the crops. This solution can be mixed with irrigation water @ 50 litres per hectare either through drip or flow irrigation. This solution is also used to treat the seeds; seedlings or other plant parts before sowing/planting. Seed treatment with *Panchagavya* before storage and its drying in shade is helpful to prolong storage life.

20.4 PERIODICITY

- Pre flowering phase two sprays at 15 days, depending upon crop duration
- Flowering & pod setting two sprays at 10 days intervals
- Fruit/ pod maturation One during pod maturation

Its remarkable effects have been demonstrated in fruits like mango, guava, acid lime, banana, spice-turmeric, flower-jasmine, medicinal plants such as Coleus, Ashwagandha, and vegetables such as cucumber spinach, okra, radish, and wide range of crops like maize, green gram, and sunflower. It has also shown the antifungal potential activities (Sugha, 2005).

6. Jiwamrita: *Jiwamrita* is prepared by fermenting cow dung, urine, jaggery, pulse flour, virgin soil) by simple facilities created in the village with minimum expenditure. Credit for development of recopies for *Jiwamrita* and its extensive use goes to Palekar (2006), a strong promoter of Natural Farming. Its can be used at 15 to 30 days interval through irrigation water, coupled with mulching (green/ dry, monocot + dicot). *Jiwamrita* is a rich bio-formulation contains consortia of microbes. This formulation is used within 3-7 days of preparation Ingredients for preparation of *Jiwamrita* are indicated as under :

- Cow dung 10 kg
- Cow urine 5-10 litres
- Jaggery 2-kg/ 4 litres sugarcane Juice
- Virgin soil 1kg
- Pulse flour 2 kg (As per availability)
- Water 200 litres
- Container Plastic/earthen/cement

Two hundred litres of *Jiwamrita* is enough for one-acre area. In general two – three times application is recommended. It can be placed on mulches either by drip irrigation or through drenching. It is also effective in quick decomposition of crop residues if applied with irrigation water given for field preparation. With micro irrigation, 3 to 4 times area can be covered with 200 litres of Jiwamrita.

20.5 BD- LIQUID MANURES

It is prepared using different materials i.e. cow dung, urine and leaves of leguminous tree, neem leaves, *Calotropis,* Lantana, *Ipomea,* custard apple leaves, *Vitex, Adothoda,* (Pathak and Ram; 2009) as per their availability in the region. In a tank crushed material of these are mixed with 10kg cow dung and 10litres of cow urine and a little quantity of soil along with 200 litres of water. The material is stirred regularly three times a day and it is ready in three weeks. BD liquid manures are used to promote the vigour and quality production.

20.5.1 Preparation

- A plastic drum with capacity of 250 litres is taken
- After adding 10 litres of cow urine and 10 kg of cow dung, drum is filled up with 150 litres of water
- 20-25 kg leaves are chopped off in pieces and put in the drum
- One set of BD-502-507 is mixed and put inside in folded leaves
- BD-507 is mixed in 2-3 litres of water for 15-20 minutes and poured in the drum
- Liquid is stirred every day in the morning and evening
- Drum is covered with gunnysack
- Preparation gets ready in 2-3 weeks depending upon the temperature and humidity
- One litre of liquid manure is dissolved in 2-3 litres of water and sprayed on the plants either to encourage vigour or to control pests and diseases.

20.5.2 Considerations

- Incorporation of succulent plant which parts are rich in zinc and boron, enrich the preparation.
- *Caliotropis* and *Glyrecidia* are rich source of Nitrogen potash; hence for quality production of fruits and vegetables these should be utilized
- Biodynamic system is almost new to India, but the observations over a decade by us and other organizations have shown encouraging responses.

20.6 *Amritpani*

It is a special bio-formulation rich in plant nutrients and microbial consortia. Ingredients for preparation of Amritpani and its intensive use was advocated by Deshpandey (2003). It is basically used to improve seed germination, soil fertility and plant vigour.

Materials Required

- 10kg cow dung
- 250g cow ghee
- 500g honey

- 200 litres water
- Cement/Plastic drum or Earthen container

PREPARATION

- Thoroughly mix 250g ghee into 10kg of cow dung
- Blend 500g honey into this mixture
- Add 200 litres of water stirring all the time
- Filter it and spray on crop in field
- 200 litre preparation is enough to cover one acre area
- It can also be sprayed on mulch for its quick decomposition
- It can be given in tree basin over the mulching through drip irrigation or by drenching.

20.7 VERMIWASH

Vermiwash is a liquid leachate obtained by excess water after saturation the vermi composting substrate. It is collection of excretory products and mucus recreations of earthworm along with nutrients from the soil organic molecules. In fact vermiwash is enriched bio enhancer prepared from the heavy population of earthworms reared in earthen pots/plastic or cement container of suitable size. The extract contains major and micronutrients, vitamins (such as $B_{12)}$ and hormones (gibberellins, cytokinins) secreted by the earthworms (Zambare *et al.*, 2008). Vermiwash can be sprayed on crops and trees for better growth, yield and quality production. In fact vermiwash is cocktail of enzymes such as proteases, amylases, urease and phosphatase. In recent study, Gupta *et al*; (2011), isolated 53 bacteria and 18 of these isolates were selected as efficient phosphate solublilizers. Among these bacterial isolates, 14 produced indole acetic acid and 8 isolates produced siderophores. Recently microbial study of vermiwash revealed that it contains nitrogen-fixing bacteria like *Azotobacter* sp. *Agrobacterium* sp. *Rhizobium* sp. and some phosphate solubilzing bacteria. The protease in soils helps in seed germination, while amylases help in availability of simple carbon source for enhancement of plant vigour and productivity as well. Soil born micro flora is essential for growth of plants because organic nitrogenous compounds and phosphorus are decomposed and mineralized by nitrogen fixing and phosphate solubilzing bacteria. Vermiwash micro flora contains *Azotobacter, Agrobacterium, and Rhizobium* and Phosphate solubilzing microbes. Presence of these microbes makes available inorganic nitrogen, amino acids and inorganic phosphate to plants through amonification and nitrification processes. Presence of large number of beneficial microorganisms helps in plant growth and protects from a number of infections in the field. Repeated spray of vermiwash has been found effective even in management of thrips and mites in chilies (George *et al.*, 2007). Extract diluted in the water 1:5-10 ratios) can be used as foliar spray for any crop. Its impact is better than spray of urea.

20.8 BIO-DIGESTER EXTRACT

The extract is prepared by fermenting crushed leaves of some plants along with cow dung and urine in a plastic container of suitable size known as bio-digester. In general green leaves of neem, Caliotropis, *Vitex, Adothoda, Ipomea,* custard apple, and agave (5kg each) are mixed in little soil and 200 litres of water are mixed in it. The mixture is stirred thrice a day and ready for use in three weeks.

The microbial load and nutrient status in *Panchagyva, Beejamrita, Jiwamrita* and bio digester extract has been estimated by Sreenivasa *et al.,* (2010). The data in table indicates presence of micro flora especially nitrogen fixers and P-solubilizers in all liquid formulations in addition to both major and micro nutrients. Presence of naturally occurring beneficial microorganisms predominantly bacteria, actinomycetes, yeast, photosynthetic bacteria and certain fungi are the major strength of these bio-enhancers.

The results of the study revealed that the nutrient status and microbial load present in the bio-enhancers may differ with the type and quantity of material used, period of fermentation, environmental conditions, etc. However, the nutrients and micro flora present in bio-enhancers support the improvement in soil fertility and in turn better yield when these are used irrespective of any crop.

Table 1 : Microbial load in different organic liquid manures

Organims	Colony count (cfu/ml)			
	Panchagavya	Beejamrutha	Jeevamrutha	Biodigester
Bacteria	26.1×10^5	15.4×10^5	20.4×10^5	12.9×10^5
Fungi	18.0×10^3	10.5×10^3	13.8×10^3	9.2×10^3
Actionomycetes	4.20×10^3	6.8×10^3	3.6×10^3	3.0×10^3
Phosphate solublising Organisms	5.72×10^3	2.7×10^2	4.5×10^2	1.0×10^2
Free living N_2 - fixers	2.70×10^2	3.1×10^2	5.0×10^2	$21. \times 10^2$

Institute of Organic Farming, UAS Dharwad

Table 2 : Nutrient status of different organic liquid manures

Parameter	Panchagavya	Beejamrutha	Jeevamrutha	Biodigester
pH	6.82	8.2	7.07	7.29
Soluble salt (EC)	1.88 dSm^{-1}	5.5 dSm^{-1}	3.40 dSm^{-1}	1.09 dSm^{-1}
Total Nitrogen	0.10 per cent	4 ppm	770 ppm	255 ppm
Total Phosphorus	175.4 ppm	155.3 ppm	166 rpm	79ppm
Total Potassium	194.1 ppm	252.0 ppm	126 ppm	42 ppm
Total Zinc	1.27 ppm	2.96 ppm	4.29 ppm	0.52 ppm
Total Copper	0.38 ppm	0.52 ppm	1.58 ppm	1.24ppm
Total Iron	29.71 ppm	15.35 ppm	282 ppm	9.60 ppm
Total Manganese	1.84 ppm	3.32 ppm	10.7 ppm	8.30 ppm

20.9 BIO–ENHANCERS WITH *AGNIHOTRA* ASH

Agnihotra is a process of harnessing cosmic energy through science of Pyramidology, biorhythm of nature (sunrise and sunset), sonic energy (science of vibration of specific *mantras*), burning of organic substances, duly amplified and energized through vaporization and its impact is extended in a given area through establishment of Resonance point.

It works on the principle that you heal the atmosphere and healed atmosphere will heal us. *Agnihotra* ash and Biosol are two potent bio- formulations, which are frequently used. This negates the effect of polluting factors while increasing quality production. *Agnihotra* ash in fact is full of energy and can be utilized for many purposes in agriculture and human health. Two bio-enhancers i.e. *Agnihotra* enriched water and Biosol is prepared and used in agriculture.

(i) *Agnihotra* enriched water: This is prepared by incorporating 1kg Agnihotra ash + 1kg cow dung powder + 5litres of cow urine with 200 litres of water. The mixture is stirred 3 times a day and used through drip/ sprinkler or as foliar spray.

(ii) A special bio-formulation has been developed by Gloria and named as "Gloria Biosol" from Peru (Weir, 2009). It is prepared after a series of operations and processes that lead to the biodegradation of organic matter, worm humus, fresh cattle manure and water, until reaching mineral grade. It is powerful bio-food and bio-fertilizer for the plants with high macro and micronutrients. Brief account of *Biosol* is enumerated below.

20.9.1 Materials used (500 litre tanks)

- *Shree Yantra* 1 unit
- Vermicompost 80 kg
- Fresh Cow dung manure 80 kg
- Cow urine 10 litre
- *Agnihotra* Ash 1 kg
- Water 200 litre

20.9.2 Bio-digester for biosol

The bio-digester tank may be of 200- 1,000 litres in volume. It may be cylindrical rigid plastic tank with air valve, liquid outlet and a lid. Care should be taken that after carefully filling the material in proper proportion lid should be sealed to avoid any leakage. The diameter of air valve opening should be one and half inch. The air ball valve should be fixed properly. Outlet for Biosol liquid should be of six inches diameter and oriented so to facilitate easy removal. Lid should be sealed with good quality adhesive and Teflon tape.

10.9.3 Procedure

- Put one Copper *Shree Yantra* at the bottom of the tank facing upwards. (*Shree Yantra* geometrical design is engraved in copper and, according to traditional knowledge, it is a powerful energy attractor)
- Make two hundred litres *Agnihotra* ash water solution (i.e. one kilogram of *Agnihotra* ash in two hundred litres of water). Let it stand for 3 days before use
- Collect 80kg fresh cow dung, 80kg vermicompost and 10 liters cow urine
- Divide the cow dung and vermi-compost into 3 piles each
- Mix one part vermicompost, one part cows dung, about 3 litres of cow urine and about fifty litres of *Agnihotra* ash water solution and stir to slurry. This can be done outside the tank. After stirring thoroughly, pour it into the tank. (Make sure the valves are closed before pouring)
- Repeat the same process with the second portion of the materials while stirring the material continuously and then add the second slurry to the tank
- Finally repeat the process a third time and add the slurry to the tank
- Add the remaining *Agnihotra* ash water solution to the tank and again stir
- In this way filling of 160 kilogram of organic matter + ten litres cow urine + 200 litres of *Agnihotra* Ash solution is done. Make sure that all the material is fully stirred. Seal the lid with adhesive and Teflon tape. Care should be taken that glue should not fall in the tank while sealing
- At an interval of seven days, release the air valve to release the gas formed in the Bio-digester
- In this way for thirty days the biodegradation of organic matter will take place in the presence of *Agnihotra* ash
- After thirty days outlet valve is opened to remove the *Biosol* liquid. It should be filled in airtight plastic cans of suitable sizes. The moist solid material, which remains after removing *Biosol* liquid, should be squeezed and more liquid *Biosol* is collected
- One may get about two hundred litres of Biosol liquid. It should be used for foliar applications with water at a ratio of 1:15 to 1:20 depending on density of plant population
- One should spray *Biosol* liquid on any crop at an interval of 7-10 days in the beginning
- Roughly 50 litres of *Biosol* is required for one-hectare area
- If properly preserved *biosol* in airtight cans it will last longer, say six months

20.10 STRATEGY FOR PROMOTION OF BIO-ENHANCERS

- Bio-enhancers are potential tools to boost the soil fertility and crop vigour. It is effective tool for pest and disease management
- These can be prepared with little support and skill upgradation trainings
- There is need for delineation of nutrient status (macro and micro nutrients), plant growth promoting factors etc for quick adaptation,
- Proper filtration will help use of these bio-enhancers through drip/sprinkler as fertigation
- Its impact on soil fertility, crop productivity need to be delineated
- Comparative evaluation of aforesaid bio-enhancers for their nutritive value and impact will help for their preparation and use
- There is need to work out its contribution in organic production and frequency of their use

20.11 REFERENCES

Deshpande, (2003) *Organic Farming vs Cosmic energy*, Non Violence Rsishi –Krishi, Mrs Pushpa Mohan Deshpande, Khede-Ajra, Kholapur, Maharashtra.

Frank, Eyhorn, Saro,G.R. and Ramakrishnan, M,T (2005), *Organic Cotton Crop Guide*, A manual for practitioners in the tropics, FiBL, Frick, Switzerland p. 27

George, S; Giraddi, R.S. and Patil, (2007) Utility of vermiwash for the management of thrips and mites in chili (*Capsicum annum*) amended by soil organics, *Karnataka, Jour. Agric.Sci;* 20: 657-659.

Natarajan, K, (2003) *Panchagvya- A Manual*, Other India Press, Goa, India, pp:1-23.

Palekar, S, (2006) *The Philosophy of Spiritual Farming* (Zero budget of Natural Farming in 3 volumes) Amrit Subash Palekar, Amravati, Maharashtra.

Pathak, R.K. and R.A.Ram, (2009) *Manual on Jaivik Krishi*, CISH, Lucknow.

Pathak, R.K., Ram, R.A, Garg, N, Kishun, R, Bhriguvanshi, S.R., Hasseb, M and Sharma, S (2009), Critical review of indigenous technologies for organic farming in horticultural crops. *Organic farming Newsletter*, 6 (2) : 3-16.

Sreenivasa, M.N. Nagaraj, N and Bhat, S.N. (2009) Beneficial traits of microbial isolates of organic liquid manures –*First Asian PGPR Congress for Sustainable Agriculture, 21-24 June, 2009*, ANGARU, Hyderabad.

Sreenivasa, M.N. Nagaraj, N and Bhat, S.N., (2010) *Beejamrutha*: A source for beneficial bacteria, *Karnataka J Agril. Sci*, 17(4) : 731-735.

Sreenivasa, M.N. Nagaraj, N and Bhat, S.N. (2010) Organic Liquid Manures: Source for beneficial Microorganisms and Plant Nutrients, *Organic Farming Newsletter*, 6 (4).

Sugha, S.K. (2005) Antifungal potential of *Panchagavya, Pl. Dis. Res;* 20(2) : 156-158.

Weir, M.L. (2009) Preparation and application of Gloria Biosol : Proc. Brainstorming Conference on "Bringing Homa Organic Farming into the mainstream of Indian Agriculture, Organized by Five Fold Path Mission in collaboration with Planning Commission, GOI. New Delhi, pp. 95-96.

Zambare, V.P.; Padul, M.V; Yadav, A.A and Shete, T.B. (2008) Vermiwash: Biochemical and Microbial Approach as Eco friendly soil conditioner, ARPN *J. Agric & Biol. Sci;* 3(4): 1-4

Chapter – 21

DRIS Norms for Precision Farming in Horticultural Crops

B. D. Sharma

21.1 INTRODUCTION

High population pressure has forced us to plough 46 per cent of our country area for food production. The present productivity is 12.36 tonnes/ha of fruits needs to be raised to 40.0 tonnes/ha by using N : P_2O_5: K_2O with other inputs, since recommended ratio is 0.75:0.50:1.00 only.

Since cost of the inputs like fertilizers and manures are very high, hence, efficient nutrient management is essential for sustainable fruit production and protection of the environment from different types of hazards occurring due to misuse of costly fertilizers.

The soils of India and especially of arid and semi-arid regions are impoverished to plant nutrients. Considering economy, energy and environment, it is imperative to use nutrients effectively by adopting the appropriate doses proper, placement and correct timings of application and to sustain the available nutrients in soil at the optimum level.

21.2 NUTRIENT MANAGEMENT OF ARID AND SEMI-ARID FRUIT CROPS

To ensure high economic productivity to sustain the available soil nutrients status at the desired level, correct dose of fertilizers and manures must be applied by use of suitable diagnostic tools designated to avoid nutrients shortage and excesses. Several diagnostic methods have been developed to arrive at need based manuring schedule.

21.3 LEAF ANALYSIS

Leaf is the centre of physiological activities in plants where translocated minerals are known to involve in vital reactions such as photophosphorylation, synthesis of chlorophyll, amino acids, nucleic acid, proteins etc. Leaf analysis methodology consists of leaf sampling technique, sample preparation, analysis and making diagnosis and nutrient recommendations using leaf nutrient guide. A large number of fruit plants, environmental and procedural factor vitiate leaf nutrient concentration, however, a careful worked out sampling technique for index tissue will make a sound foundation of leaf analysis programme. Modern methods of interpretation of leaf analysis data such as DRIS and Boundary Line Concept can be adopted to diagnose the growth/yield limiting nutrients and to recommend the optimum use of nutrients.

Importance of Leaf Analysis

1. To understand the internal function of the nutrient in the plant
2. To confirm toxicity/deficiency
3. To identify mineral imbalance
4. To ascertain whether applied nutrients have their assimilation the plant system
5. To identify whether hidden hunger is there or not

21.4 DIAGNOSIS AND RECOMMENDATION INTEGRATED SYSTEM (DRIS)

Diagnosis and recommendation integrated system (DRIS) represents a holistic approach to the mineral nutrition of crops and has an impact on the integrated set of norms representing calibrations of plant tissues, soil composition, environmental parameters and farming practices as the functions of the yield of a crop (Beaufils, 1973). Once such norms are developed, it is possible to make a diagnosis of the conditions of the crop thereby isolating the factors, which are likely to be responsible for limiting the growth and production. The most important advantages of DRIS approach are its ability to make a diagnosis at any stage of crop development and to list the nutrient in the order of importance, which are responsible for limiting the yield. Diagnosis and Recommendation Integrated System has been developed to fulfil the diagnosis and predictive use of the leaf analysis. Optimizing these factors create conditions, which are likely to increase the chances of obtaining higher yields and quality.

21.4.1 Development of DRIS norms

In contrast to the classical field experiment approach for the leaf analysis, the DRIS employs survey technique, where a large number of randomly distributed sites throughout the area are selected. At each site, the samples of leaf and soil are taken for the analysis and the details of the applied manures and fertilizers are recorded.

Experiments conducted at one place employing various levels of 1, 2 or 3 plant nutrients can also be used to develop DRIS indices or its verification. The procedures for development of DRIS norms are as under :

- Define the parameters to be improved and the factors likely to affect them
- Collect all the reliable data available from the fields and experimental plots
- Study the relationship between the yield and environmental or external factors such as soil composition, weather conditions etc. so that the most favourable condition may be determined in a particular case. This would help in taking the suitable corrective measures on such interrelationship
- Study the relationship between the yield and the internal parameters such as petioles composition using following steps:

- Each internal plant parameter is expressed in as many as possible forms e.g. N / DM, DM / N, N x DM, N / P, P / N, N x P etc.
- The whole population is divided into a number or subgroups based on the economic optimum yield of the area/zone
- The means of the each sub-population are calculated for the various forms of expressions
- If necessary, class interval limits between the average and the outstanding yields are readjusted, so that the means of below average and average yielding population remain comparable
- Chi-square test is performed to know that population confirms a normal distribution
- The variance ratios between the yields of sub-populations for all the forms of expressions are calculated together with the co-efficient of variation
- The forms of expressions, are selected and related to one another in expression with common nutrient
- For the calculation of DRIS index, Bhargava and Chadha (1988), Bhargava and Raghupathi (1993) and Bhargava (1999), worked out formulae for petiole/ leaf N, P, K, Ca, Mg, S, Fe, Mn, Zn, Cu and yield for grapes, mango and acid lime. The formulae developed under one set of condition and for one crop are given below:

N = 1/10[f(N/P)-f(K/N)+f(N/Ca)+f(N/Mg)+f(N/S)-f(Fe/N)+f(N/Mn)+f(N/Zn) (Cu/N)+f(N/Y)] where $f(N/P = \left[\frac{N/P}{n/p} - 1\right]\left[\frac{1000}{CV}\right]$

and $= \left[1 - \frac{n/p)}{N/P}\right]\left[\frac{1000}{CV}\right]$

N/P = the actual value of the ratio of N and P in the plant under diagnosis. The n/p = the value of the norm (which is the mean value of high yielding orchards) and C.V.= co-efficient of variation for population of high yielding orchards.

P=1/10[(-f(N/P)+f(P/K)-f(Ca/P)+f(P/Mg)-f(S/P)-f(Fe/P)+f(P/Mn)+f(P/Zn) f(Cu/P)+f(P/Y)]

K = 1/10[f (K/N)-f(P/K)-f(Ca/K)-f(Mg/K)-f(S/K)-f(Fe/K)+f(K/Mn)+f(K/Zn)-f(Cu/K)+f(K/Y)]

Ca = 1/10[(-f(N/Ca)+f(Ca/P)+f(Ca/K)+f(Ca/Mg)+f(Ca/S)-f(Fe/Ca)+f(Ca/Mn)+f(Ca/Zn)-f(Cu/Ca)+f(Ca/Y)]

Mg=1/10[-f(N/Mg)-f(P/Mg)+f(Mg/K)-f(Ca/Mg)-f(S/Mg)-f(Fe/Mg)+f(Mg/Mn)+f(Mg/Zn)-f(Cu/Mg)+f (Mg/Y)]

S = 1/10[(-f (N/S)+f(S/P)+f(S/K)-f(Ca/S)+f(S/Mg)-f(Fe/S)+f(S/Mn)-f(Zn/S)-f(Cu/S)+f(S/Y)]

Fe = 1/10[(f (Fe/N)+f(Fe/P)+f(Fe/K)+f(Fe/Ca)+f(Fe/Mg)+f(Fe/s)-f(Mn/Fe)-f(Zn/Fe)-f(Cu/Fe)+f(Fe/Y)]

Mn=1/10[(-f(N/Mn)-f(P/Mn)-f(K/Mn)-f(Ca/Mn)-f(Mg/Mn)-f(S/Mn)+f(Mn/Fe)-f(Zn/Mn)-f(Cu/Mn)+f(Mn/Y)]

Zn =1/10[(-f (N/Zn)-f(P/Zn)-f(K/Zn)-f(Ca/Zn)-f(Mg/Zn)+f(Zn/S)+f(Zn/Fe)+f(Zn/Mn)-f(Cu/Zn)+f(Zn/Y)]

Cu=1/10[(-f(Cu/N)f(Cu/P)+f(Cu/K)+f(Cu/Ca)+f(Cu/Mg)+f(Cu/S)+f(Cu/Fe)+f(Cu/Mn)+f(Cu/Zn)+f(Cu/Y)]

Y = 1/10[(-f (N/Y)-f(P/Y)-f(K/Y)-f(Ca/Y)-f(Mg/Y)-f(S/Y)-f(Fe/Y)-f(Mn/Y)-f(Zn/Y)-f(Cu/Y)]

Please remember that the composition of the formulae will depend the situation and there is no fixed formula to be used under all crop/site/situations. The investigator can make an attempt to workout his/her formulae for the crop and use it.

21.4.2 Diagnosis

Initially standard values or norms were determined from DRIS. These standard values were then used to develop indices by following the equations given above. The value of each ratio function is added to one index sub-total and subtracted from another, prior to averaging. Therefore, all indices are balanced around zero. The higher negative index shows that the corresponding nutrient is relatively deficient. Alternatively, a large positive index indicates the nutrient is excessive in quantity.

21.4.3 Recommendations

Based on the diagnosis made, recommendation is proposed. Some examples are given below based on the diagnosis made in grapes, mango, banana and acid lime.

21.4.3.1 Ber

Banana is grown (in Alfisols of southern Karnataka) by applying the recommended dose of primary nutrients i.e., 200g N, 300 g K_2O/plant in 4 splits along with 50 g P_2O_5 as base dose. However, full yield is not obtained unless Mg, Zn and B are applied at the rate of 25 to 50g magnesium sulphate, 10 to 20g zinc sulphate and 2.0g borax/plant through soil.

21.4.3.2 Acid lime

Sulphur is the most deficient nutrient and application of elemental S super phosphate (11 per cent S) supplies S. The next nutrient needed to boost yield is N followed Mg and then P.

The diagnosis made on four fruit crops, brings out the need for applying secondary and micronutrients along with primary ones. It is necessary to provide a base of organic manure, which creates a situation for efficient utilization of applied nutrients.

Table 1 : Nutrient diagnosis using DRIS indices

Crop												**Yield**
Pineapple	Nutrients found deficient and low				Nutrients found High and excess							
Nutrient	DM	Zn	Mg	Mn	Ca	K	S	P	Cu	N	Fe	
Status	22.24	18	0.26	30	1.26	1.34	0.12	0.52	14	1.07	118	22.24
DRIS Indices	-282	-269	-97	-45	-18	5	64	67	147	158	270	
MANGO	Nutrients found deficient and low				Nutrients found High and excess							
Nutrient	DM	Zn	Mg	S	N	Cu	Fe	Ca	K	Mn	P	
Status	4.60	10	0.11	60	1.00	16	810	1.26	1.42	480	49	4.60
DRIS Indices	-187	-134	-129	-108	-51	-14	-9	23	60	204	344	
Ber	Nutrients found deficient and low				Nutrients found High and excess							
Nutrient	DM	K	Mg	N	Zn	P	S	Ca	Cu	Mn	Fe	
Status	15.00	1.50	0.17	1.10	20	110	0.24	1.50	10	200	180	5.00
DRIS Indices	-285	-260	-180	-35	-10	5	18	47	110	195	215	
ACID LIME	Nutrients found deficient and low				Nutrients found High and excess							
Nutrient	DM	S	Mg	P	N	K	Cu	Mn	Zn	Ca	Fe	
Status	3.10	0.22	0.22	0.11	1.8	1.33	13	50	19	4.13	489	3.10
DRIS Indices	-708	-272	-231	-137	-83	-22	28	65	166	189	1007	

21.5 DEVELOPING NUTRIENT NORM FROM DRIS

Close examination of DRIS clearly indicates that the system calculate DRIS indices for various nutrients taking into consideration the status of other nutrients and factors influencing growth, yield and or quality. System also lists the order of nutrients influencing crop performance. However, it does not provide the methodology for developing nutrient norms for various nutrients. Beaufils (1971) stated that mean + 4/3 SD form the range for status of a nutrient / nutrient ratio, which is considered as balanced. This range cover 50% of the data forms, the range provides norms for sufficient status of a nutrient. Further, it is stated that mean – 4/3 SD to mean – 8/3 SD as well as mean + 4/3 SD to mean + 8/3 SD gave situation of unbalanced nutrition. These two ranges, which cover 16% of the data in each zone, gave us status for 'hidden hunger' and status for 'high' or 'more than required' and the value below mean – 8/3 SD was considered as deficient and covers 9% of the data in the normal distribution. Value above mean + 8/3 SD was taken as excessive or toxic and covers about 9% of the data. Nutrient norms for nutrient and potential yield relationship was developed as outlined above and presented.

Table 2 : Leaf nutrient Standards for *Alphonso* mango grown in coastal region of Maharashtra

Nutrient	Unit	Status				
		Deficient	Low	Optimum	High	Excessive
N	g/100g	< 0.31	0.32-0.77	(1.21)0.78 – 1.65	1.66-2.09	> 2.10
P	mg/100g	< 16	16-22	(28)23 – 33	34 - 40	> 40
K	g/100g	< 0.27	0.27–0.76	(1.25)0.77 – 1.73	1.74 – 2.22	> 2.22
Ca	g/100g	< 0.31	0.31–0.75	(1.20)0.76 – 1.63	1.64 – 2.07	> 2.07
Mg	g/100g	< 0.21	0.21-0.39	(0.52)0.40 – 0.65	0.65 - 0.87	> 0.87
S	mg/100g	< 20	20-34	(83)35 – 131	132 - 179	> 180
Fe	ugg^{-1}	< 504	504-656	(809)657 – 961	962 - 1114	> 1115
Mn	ugg^{-1}	< 10	10-13	(210)14 – 408	408 - 605	> 605
Zn	ugg^{-1}	< 2.4	2.4–7.7	(13.0)7.8 – 18.3	18.3 – 23.7	> 50
Cu	ugg^{-1}	< 12.5	12.5–14.2	(16.0)14.3 – 17.8	17.9 – 19.6	> 50
Yield	tha^{-1}	< 4.2	4.2–5.3	(6.3)5.3 – 7.4	7.5 – 8.5	> 8.5
	$kgha^{-1}$	< 4,200	4,200–5,300	(6350)5,301–7,400	7,500–8,500	> 8,500

Table 3 : Leaf nutrient Standards for *Totapuri mango*

Nutrient	Unit	Status				
		Deficient	Low	Optimum	High	Excessive
Nitrogen	g/100g	< 0.47	0.47 - 0.83	(1.18) 0.84 – 1.53	1.54 – 1.88	> 2.00
Phosphorous	mg/100g	< 26	26 - 63	(105) 64 – 147	148 - 186	> 200
Potassium	g/100g	< 0.22	0.22 – 0.51	(0.81) 0.52 – 1.10	1.11 – 1.40	> 2.00
Calcium	g/100g	< 1.33	1.33 – 1.96	(2.58) 1.97 – 3.20	3.33 – 3.80	> 3.80
Magnesium	g/100g	< 0.21	0.21 - 0.39	(0.52) 0.40 – 0.65	0.66 - 0.87	> 0.90
Sulphur	mg/100g	< 112	112 - 146	(181) 147 – 215	216 - 249	> 250
Iron	ugg^{-1}	< 28	28 - 47	(67) 48 – 86	87 - 105	> 105
Manganese	ugg^{-1}	< 20	20 - 57	(116) 58 – 174	175 - 232	> 235
Zinc	ugg^{-1}	< 5	5 – 24	(29) 25 – 33	34 – 43	> 100
Copper	ugg^{-1}	< 0.47	0.47 – 3.00	(5.50) 3.01 – 8.00	8.01–10.60	> 50
Yield	tha^{-1}	< 9.20	9.2 – 10.40	(11.60) 10.50 – 12.70	12.80–14.00	> 14.00
	$kgha^{-1}$	< 9,200	9,200–10,400	11.600 10,501–12,700	12,800–14,000	> 14,000

Table 4: Leaf nutrient Standards for *Rajapuri* mango grown in tribal regions of Southern Gujarat

Nutrient	Unit	Status				
		Deficient	Low	Optimum	High	Excessive
Nitrogen	g/100g	< 0.35	0.36 - 0.88	(1.41)0.89 - 1.93	1.94-2.45	> 3.00
Phosphorous	mg/100g	< 39	40 - 63	(87)64 – 111	112-135	> 200
Potassium	g/100g	< 0.51	0.52 - 1.01	(1.51)1.02 - 2.01	2.02-2.52	> 3.00
Calcium	g/100g	< 0.32	0.33 - 1.10	(1.89)1.11 - 2.67	2.68-3.45	> 4.00
Magnesium	g/100g	< 0.22	0.23 - 0.28	(0.34)0.29 - 0.40	0.41-0.46	> 0.60
Sulphur	mg/100g	< 75	76 - 100	(140)107 – 169	170-200	> 400
Iron	ugg^{-1}	< 12	13 - 57	(101)58 – 146	147-190	> 500
Manganese	ugg^{-1}	< 5	5 - 28	(114)29 – 199	200-285	> 500
Zinc	ugg^{-1}	< 5	6 - 10	(16)11 – 24	25-30	> 100
Copper	ugg^{-1}	< 10	10 - 15	(19)16 – 24	25-30	> 100
Yield	tha^{-1}	< 1.00	1.0 - 2.0	4.752.0 – 7.5	7.5-11.8	> 25.0
	$kgha^{-1}$	<1000	1000–2000	(4750)2001–7500	7501-11800	> 25000

Table 5 : Leaf nutrient norms for Acid Lime

Nutrient	Unit	Status				
		Deficient	Low	Optimum	High	Excessive / Toxic
Nitrogen	%	< 1.24	1.24 – 1.52	(1.81) 1.52 – 2.10	2.11 – 2.39	> 2.40
Phosphorus	mg/100g	< 70	71 - 90	(120) 91 – 150	151 – 180	> 200
Potassium	%	< 0.60	0.60 – 0.96	(1.31) 0.97 – 1.66	1.67 – 2.01	> 2.10
Calcium	%	< 2.85	2.86 – 3.04	(3.24) 3.05 – 3.43	3.44 – 3.63	> 3.65
Magnesium	%	< 0.22	0.22 – 0.24	(0.28) 0.25 – 0.30	0.31 – 0.32	> 0.35
Sulphur	%	< 0.21	0.21 – 0.24	(0.27) 0.25 – 0.29	0.30 – 0.31	> 0.35
Iron	ppm	< 77	77 - 116	(156) 117 – 195	196 - 234	> 250
Manganese	ppm	< 10	10 - 21	(43) 22 – 64	65 - 85	> 100
Zinc	ppm	< 3.2	3.2 – 9.5	(13.9) 9.6 – 18.2	18.3 – 22.5	> 50
Copper	ppm	< 5.6	5.6 – 8.6	(11.75) 8.7 – 14.8	14.9 – 17.8	> 50
Yield	t/ha	< 13.7	13.7 – 15.6	(17.55) 15.7 – 19.4	19.5 – 21.4	> 21.5

Table 6 : Leaf Nutrient Norms for Pomegranate

Nutrient	Unit	Observed			Nutrient Norms		
		Range	Deficient	Hidden Hunger	Optimum	High	Excessive
Nitrogen	%	0.40-2.20	< 0.54	0.54–0.90	0.91 – 1.66	1.67 – 2.04	> 2.10
Phosphorous	%	0.08–0.33	< 0.09	0.09–0.11	0.12 – 0.18	0.19 – 0.21	> 0.25
Potassium	%	0.20–2.05	< 0.20	0.20–0.60	0.61 – 1.59	1.60 – 2.26	> 2.26
Calcium	%	0.60–2.40	< 0.13	0.13–0.76	0.77 – 2.02	2.03 – 2.65	> 2.65
Magnesium	%	0.15–0.49	< 0.03	0.03–0.15	0.16 – 0.42	0.43 – 0.55	> 0.55
Sulphur	%	0.04–0.70	< 0.10	0.10–0.15	0.16 – 0.26	0.26 – 0.42	> 0.45
Iron	ppm	25 – 297	< 34	34 – 70	71 – 214	215 – 286	> 300
Manganese	ppm	14 – 99	< 15	15 – 28	29 – 89	90 – 119	> 120
Zinc	ppm	07 – 44	< 08	08 – 13	14 – 72	73 – 94	> 100
Copper	ppm	21 – 86	< 07	07 – 28	29 – 72	73 – 94	> 100

Multivariate Nutrient Diagnosis : The premise underlying Uni-variate (CVA= Critical Value Approach), bi-variate (DRIS = Diagnosis & Recommendation Integrated System) diagnostic systems were re-examined with regard to compositional data analysis (CND), CDA recognizes a structure of dependence among plant nutrients, the 'Bounded Sum Constraints' to one (the whole composition equals 100%) and removes the curvature problem carried by crude components and by dual ratios or log ratios when treated in isolation. Liberalization by "Row-centered - Logrationary" of nutrient fractions shows great potential for carrying multi-variate diagnosis and principal component analysis on nutrient data. Compositional Nutrient Diagnosis (CND) is supported by the theory of CDA. CND is a multi-variate expansion of CVA and DRIS and is fully compatible with PCA. CND takes all possible nutrients interactions into account. CND nutrient indices are composed of two separate functions based on nutrient balances (as defined by nutrient geometric means), of individual and target specimens. These functions indicate that nutrient in-sufficiency can be corrected by adding either a single nutrient or taking advantage of multiple nutrient interactions to improve nutrient balance as a whole.

Mineral composition of plant tissues, expressed as concentration or relative (ratio) values, forms the basic numerical information for diagnosing nutrient status of crops. CVA uses crude components expressed as proportions of interpretation of nutrients deficiency, sufficiency, dry matter (DM) or as concentrations in the petiole or are excess was affected to a large extent by nutrient interactions and plant age. DRIS was found to improve diagnosis as compared to CVA by taking into account duel interactions expressed as nutrient ratios and further transformed into DRIS indices. However CVA and DRIS could carry distorted or redundant efficiency used in the logarithmic transformation to improve data distribution of nutrient ratios from population data to generate DRIS norms. When dry matter values (DM) are identical over all nutrients components, i.e., in individual plant times, dual log ratioing can provide a single correction factor for a nutrient X_1 and nutrient X_K :

$$\log (X_1 / X_K) = \mathrm{Log}\,(X_1) - \mathrm{Log}\,(X_K)$$

A reasoned application of principal component analysis (PCA) could lead to a greater understanding of the effects of fertilization treatments on leaf composition compared to univariate, dual ratio, or linear combinations of original variates. Using the PCA approach to compositional data analyses, compositional nutrient diagnosis (CND) has been developed, CND requires less computational efforts than DRIS by computing row – centered logratios from original concentration values instead of using various procedures of function and index calculations.

An improved knowledge of nutrient balance in horticultural crops is necessary to comprehend the nature of nutrient interactions the plant and to facilitate the interpretation of analytical results.

Why CND and PCA?

The nutrient expression for diagnosis with CVA is a percentage relative to DM. Thus, for any nutrient N, P, K, Mg, or Zn no adjustment to other nutrients is provided.

DRIS expands CVA by considering further adjustment of a given nutrient to another nutrient i.e. by dual ratioing nutrient values expressed as percentage. Nutrient percentages and dual ratios for independent samples are first compared to DRIS norms using one of two equations to compute DRIS functions. The calculations of DRIS indices from DRIS functions could adjust or row – centre the nutrient under diagnosis relative to all other nutrients and DM. The nutrients form should appear preferably as numerator or denominator in the DRIS functions to be added up. A generalized nutrient imbalance index (NIl) can be calculated by summingup DRIS indices irrespective of sign. Until now, the DRIS indices have not been projected into a multi-dimensional space using PCA.

Multi-normal data distribution is a pre-requisite for conducting multivariate analyses. Nutrient data are relative and typically compositional, and, as such, is not directly amenable to standard statistical linear modeling due to the "Bounded – Sum constraint to 100 per cent (the DM content) of plant components. Before conducting PCA, DM should be split into nutrient proportion and a filling value between 100 per cent and the known proportions.

Compositional Nutrient Diagnosis (CND): The multivariate technique, namely, compositional nutrient diagnosis has been recently used for developing norms and interpretation of leaf analysis data. It is presented below in brief:

The CND in plant tissues can be described by the following simplex S^D computed to 100%.

S^D= [(N, P, K,…R): N > 0, P > 0, K > 0,…, R >); N + P + K +…+ R = 100%].1

Where 100% is the dry matter content (i.e. the invariable sum of all the components or the full relative composition of the diagnostic tissues) N, P, K are the nutrient concentrations, and R is the filling value between 100% and sum of the nutrient concentrations. The value of R is, thus, composed of undetermined components and experimental error. It is required to linearize the compositional data.

Correcting nutrient concentrations by geometric mean, G of all the D components including R, alleviates the bounded sum computed to 100% of compositional data.

$G = [N \times P \times K \times \ldots \times R]^{1/D}$ 2

Then, generating row centered log ratios V_N to V_{Zn} as follows

$V_N = \ln(N/G), \ldots V_{Zn} = \ln(Zn/G)$ 3

Expressions, such as N/G,…Zn/G are multinutrient ratios rather than dual ratios, such as in DRIS, because each nutrient is divided by geometric mean of all the components (the determined nutrients and the filling values). The row centered log ratios are linearized (distorted) estimates of the original components that are fully compatible with PCA.

A PCA can be performed on row centered, log ratioed nutrient values characterizing the high yielding sub-population. The correlation matrix in PCA generates linear combination of standardized, row centered, log ratios as follows:

$PCj = a_{ij} (V_N - V_N^*) / SD^*_N + \ldots + a (D-1)_j (V_{Zn} - V_{Zn}^*) / SD^*_{Zn}$ 4

Where PCj is the principal component and a_{ij} to a (D – 1); (D – 1)] components excluding R as specified above are PC score coefficients of standardized row centered log ratios. V^*_N to V^*_{Zn} and SD^*_N to SD^*_{Zn} are the CND norms (indicated by asterisks), i.e. mean and standard deviation of each row centered log ratio in the high yielding sup-population. The standardized variables $(V_N - V_N^*)/SD^*_N$ to $(V_{Zn} - V^*_{Zn})/SD^*_{Zn}$ are the CND nutrient indices which are CND analogs of the DRIS nutrient indices

$I_N (V_N - V_N^*) / SD^*_N \ \ldots I_{Zn} = (V_{Zn} - V_{Zn}^*)/SD^*_{Zn}$ 5

Independent values for V_N to V_{Zn} can be introduced in the equation for diagnostic purpose.

The PC loadings are correlation coefficients between the PC and each row centered log ratio. Only the loadings exceeding the selection criteria were used to designate the PCs as contrast between row centered log ratios. The selection criterion (SC) was computed as follows;

$SC = 0.5 / (\text{igen Value})^{0.5}$ 6

Since CND produces a system of PC axes perpendicular to one another, the generalized distance between independent and target specimen is the Euclidian distance 'd' across the P-dimensional space defined as follow :

$\text{'d'} = (EPC_j^{2})^{0.50}$ 7

The numbers of PCs were determined using a thumb rule that the PCs selected were those that explain at least 100/p per cent of the total variance, where p is the number of variables.

21.5.1 CDN Nutrient norms

The CND nutrient norms for *Anab-e-Shahi*, an important cultivar of grapes grown in India is shown in Table 6. The V values for all the nutrients except that of K was higher during bud differentiation stage (BDS) compared to flowering stage (FS) indicating the greater requirement of nutrients during bud initiation and differentiation. The reconstituted values indicate that the concentration value corresponding to V_N were approximately 1.19 and 0.91% N at BDS and FS respectively, indicating optimum requirement of N during BDS and consequent mobilization of N from vines to inflorescence. The concentration value corresponding to V_P at BDS was 0.81% P. Large differences in CND norm were particularly noticed with Ca and Mg. The V_{Ca} was nearly lowered by two-fold during flowering from BDS, while V_{Mg} values showed a two-fold difference between the two stages. The V_{Zn} value showed wide variation between BDS and FS, the reconstituted value being 78 and 33 ppm for BDS and FS respectively.

21.5.2 PCA on pomegranate

An attempt was made to develop diagnostic norms for evaluation of yield limiting nutrients in low yielding pomegranate orchards. The leaf nutrient status was interpreted using DRIS and CND. Principal Component Analysis (PCA) extracted the correlation structure among the nutrients. The departure of DRIS indices from their CND counterparts was relatively small and a highly significant positive correlation was obtained between DRIS and CND indices. Three principal components explained 59.5% of the variation in the high yielding population and the designated PCs were (N+S+Zn+Fe-Mn-), (N+P+Ca+Zn-), and ((K-S+). Pomegranate is mainly grown on marginal soils with low fertility and hence more than two or three nutrients were found associated in limiting yields. However, N and Zn were most commonly yield-limiting nutrients.

Table 7 : Parameters of principal component analysis in Pomegranate

Composition Nutrient diagnosis variate	N+S+ZnFe-Mn-	N+P+Ca-Zn-	S+K-
	PC Loading		
TNR	1	2	3
V_N	0.506*	0.357*	0.092
V_P	0.186	0.768*	0.029
V_K	0.307	0.261	-0.819*
V_{Ca}	0.206	-0.677*	-0.138
V_{Mg}	0.290	0.162	0.113
V_S	0.513*	-0.083	0.707*
V_{Fe}	-0.836*	0.187	0.231
V_{Mn}	-0.818*	-0.207	-0.096
V_{Zn}	0.596*	-0.468*	-0.091
	Eigen values (Latent roots)		
	2.497	1.573	1.284
	Per cent total variance explained		
	27.74	17.47	14.26

*Significant loading

21.6 CALCULATING NUTRIENT REQUIREMENT FOR ARID ZONE FRUIT CROPS

There are several methods of working out nutrient requirement for the crop. These are **1**. Studies on yield and quality response of the crop based on applied nutrients through organic and inorganic forms. **2**. Nutrient uptake studies, where nutrient removal by the crop is calculated with a targeted yield level. Removed nutrients are added to the soil to replenish the loss from growing environment. If we add the quantity of nutrients removed by 25 tonnes of the crop from an hectare of land, the addition will be less than replenishment, as studies on grapes have clearly indicted that nutrient use efficiency of each nutrient varies and much less than 100 per cent. Thus replenishment needs careful study and considerations of nutrients contributed from vine reserve, soil, irrigation water, organic manure and finally the chemical fertilizers. **3.** Growers have certain practices of application of manures and fertilizers and on computing doses of various nutrients.

Nutrient use efficiency of applied N varies from 20 to 40 per cent i.e. if you apply 100 kg N [218 kg urea or 625 kg ammonium sulphate fertilizer] is applied only 20 to 40kg of N [equivalent to 44 to 88 kg urea or 125 to 250 kg ammonium sulphate] will enter into grapevines and rest will be either utilized by weeds or get lost in ground water [beyond the reach of grape roots] as $NO_3 - N$ is lost into the atmosphere as N_2 gas by the process of de-nitrification [where $NO_3 - N$ will get converted to NH_3 or N_2 gas and escape into the atmospheric gases. Surface applied urea or ammonium sulphate also loose some N as NH_3 gas in alkaline soils.

Similarly, P use efficiency varies from 5 to 20% i.e. if a grower applies 80 kg P_2O_5 [equivalent to 500 kg single super phosphate or 175 kg di-ammonium phosphate to soil only 8 to 16 kg P_2O_5 [50 to 100 kg single super phosphate or 18 to 35 kg DAP] will be utilized by the grapevines. Rest of the P will get immobilized in the soil to insoluble form like $Ca(HPO_4)_2$ [citrate soluble] or insoluble form in soils with alkaline reaction and rich in $CaCO_3$, or taken away by weeds. Phosphorous is present as water soluble form [$H_2PO_4^-$], citrate soluble form [$CaHPO_4$] and water insoluble form [$Ca_3(PO_4)_2$]. Phosphorous is normally applied through well decomposed organic manure like FYM or through cow dung slurry and under such situations, the use efficiency may go up to 15 or 20% of the applied P.

The availability of K is relatively high and the use efficiency may vary from 40 to 60 per cent. When fertilizers containing K are added to soil, there is release of fixed soil K and that is why there is higher use efficiency of added K.

Recommendations jointly made by Mahatma Phule Krishi Vishwa Vidyalaya Rahuri, Maharashtra, National Research Centre for Grapes Pune and Department of Horticulture. Government of Maharashtra is 900-500-700 kg for N, P_2O_5 & K_2O per hectare per year with a population of 2,222 vines per ha., whereas the recommendation was 660 – 880 – 660 kg N, P_2O_5 and K_2O per hectare per year.

Table 8 : Nutrient Use Efficiency

Primary Nutrient	Uptake [kg / hectare]	Nutrient Use efficiency
N	250 kg per hectare	20 to 40 per cent of applied N
P_2O_5	114 kg [50 kg P] kg per hectare	5 to 20 per cent of applied P_2O_5
K_2O	296 kg [250 kg K] kg per hectare	40 to 60 per cent of applied K_2O
Ca	250 kg per hectare	30 to 40 per cent of Foliar applied Ca
Mg	62 kg per hectare	10 to 20 per cent of Soil applied Mg
		40 to 50 per cent of foliar applied Mg
S	2.48 kg per hectare	20 to 40 per cent of applied S
Fe	248 g per hectare	5 per cent of soil applied Fe30 to
		50 per cent foliar applied Fe
Mn	346 g per hectare	5 - 10 per cent of soil applied Mn
		30 to 50 per cent of foliar applied Mn
Zn	180 g per hectare	5 per cent of soil applied Zn25 to
		50 per cent foliar applied Zn
Cu	—	—@
B	124 g per hectare	10 to 20 per cent of soil applied B
		50 to 60 per cent of foliar applied B
Mo	2.50 g per hectare	50 to 60 per cent of foliar applied Mo
Cl	148 kg per hectare	—$

Note : Efficiency of Ca and Mg will be higher in acidic soils and relatively low in soils having alkaline reaction

SSP, SOP and micronutrient fertilizers will supply Sulphur

There is no further need of S as nutrient. Sulphur can be applied as amendment.

Computing Doses of Nutrients for One Year: Considering nutrients removed, the suggested doses will be given in the Table.

Table 9 : Proportion of Nutrients supplied by soil and fertilizers based on soil test value in one year to one hectare area

Nutrient	Recommended dose proposed	Input as		Nutrient supplied by soil [0-30 cm layer]	Fertility of soil [kg/ha] irrigation	Nutrient added through reserve	Nutrient provided to plant	Total Nutrient supplied
		Organic Manures	Chemical fertilizers					
N	334–506kg	200-270kg	136–235kg	75-124kg	336	12.5-86.0kg	37–50kg	420–766kg
P_2O_5	370–830kg	62-114kg	310–716kg	6.2-25.0kg	100	25-37kg	12.5–125kg	414-1015kg
K_2O	380–630kg	173-248kg	210-385kg	100-600kg	862	15-25kg	45-60kg	510-1335kg
Ca	30.15-40.50g	30–40kg	500-1000g	200-248kg	16677	124-248kg	Nil	351–535kg
Mg	20.0-41.0kg	11.0–21.2kg	10-20kg	12.5-37.0kg	4043	60-125kg	12.5-25.0kg	108–226kg
S	No recommend -ation for S as nutrient*	108–163g	161-270kg	Up to 25kg	220	20–25kg	0.012 -0.050 kg	117.40–195.02kg
Fe	60-105kg	50–74.6kg	11.11-30.00kg	Up to 25kg	7.37	20-25kg	12.5-50g	37.66–82.30kg
Mn	5.60–10.50kg	0.44-0.54kg	5.14-10.13kg	Up to kg	25.00	1.48-2.50kg	25-500g	7.082–136.745 kg
Zn	9.30–24.20kg	4.0–4.1kg	5.28-20.11kg	Up to 10kg	9.40	125-15.0kg	25-50g	21.695–64.196 kg
Cu	—@	25-100kg	—	1-10kg	14.35	—	2.5-25g	—
B	2.65–2.75kg	2.500–2.500g	161-247g	Up to 500g	—	2.50-50.0g	12.5–25.0g	2.671–4.052kg
Mo	305–500g	52.0–128.0g	252–380g	Up to 40g	—	2.50-50.0g	12.5-25.0g	318.75–682.00g
Cl	—	12.5–25.0kg	Up to 250kg	10-20kg	—	61-500kg	12.5-62.0g	—
Na	—	— —	—	1350	61-124 kg	—	—	

Note : With all the care, precautions and sincerity, it has been very difficult to decontaminate plant tissue samples from Cu sticking on the surface, as plant protection

Chemicals Efficiency of Ca and Mg will be higher in acidic soils and relatively low in soils having alkaline reaction.

Although Cl is an essential nutrient, total supply made to grapevines toxicity.

No recommendations of sulphur as nutrients. Sulphur can be used as fungicides to control fungal diseases.

21.7 REFERENCES

Annapurna, K., Bojappa, K.M. and Bhargava B.S. (1988). Leaf sampling guide for sapota (*Manilkara achras* M. Fosberg) cv. Cricket Ball. *Crop Res.*, 1: 69-75.

Awasthi, R.P., Dhuria, H.S. and Karkara, B.K. (1979) Nutritional Status of apple orchards in H.P. *Indian J. Hort.*36: 57-63.

Azospirilum on growth and development of banana cv. Poovan (AAB). *South Indian Hort.*, 36(1) : 1-4.

Bates, T. F. (1971) Factors affecting nutrient concentrations in plants and their evaluation. A Review, *Soil Sci.* 112: 116-130.

Beaufils, E. R. (1957) Research for rational exploitation of *Hevea* using a physiological diagnosis based on the mineral analysis of various parts of the plants. *Fertilite,*. 3: 27-38.

Beaufils, E. R. (1971) Phsiological diagnosis- A guide for improving maize production based on priciples developed for rubber trees. *Fertil. Soc. S. Afr. J.*, 1: 1-28.

Beaufils, E. R. (1973) Diagnosis and recommendation integrated system (DRIS). *Soil Sci. Bull. No. 1,* University of Natal, S. Africa.

Bhambal, S.B. (1987). *Effect of foliar application of micro-nutrients on growth, yield, fruit quality and leaf nutrient status of pomegranate (Punica granatum L.) cv. Ghanesh.* M.Sc. (Ag.) thesis. MPAU, Rahuri.

Bhargava, B.S. (1999). Leaf analysis for diagnosing nutrients need in fruit crops. *Indian Horticulture,* ICAR,43 (4) : 6-8.

Bhargava, B.S. and Chadha, K.L. (1993). Leaf nutrient guide for fruit crops. *Advances in Hort.*, Vol. 2. pp.973-1029. Chadha, K.L. and O.P. Pareek (Eds). Maharotra Publishing House, New Delhi.

Bhargava, B.S. and Raghupathi, H.B. (1993). *Analysis of plant materials for macro and micro nutrient Bull.* IIHR pp.49-82

Bhargava, B.S., Reddy, Y.T.N., Kohli, R.R. and Srinivasan, V.R. (1989). Use of boundary line concept in working out petiole nutrient norms in papaya (*Carica papaya* L.) cv. Coorg Honey Dew. *Prog. Hort.*, 21 : 56-61.

Bouma, D. (1983) Diagnosis of mineral deficiencies using plant tests. In: A. Lauchli and Bieleski, R. L. (eds) *Encyclopedia of plant physiology.* Vol. 15A. *Inorganic plant nutrition.* Springer- Verlag, Berlin, pp. 120-148.

Chadha, K.L. and Reddy, B.M.C. (1993). Pineapple nutrition. *Advances in Hort.*, Vol. 2. pp.899-905. Chadha, K.L. and O.P. Pareek (Eds). Maharotra Publishing House, New Delhi.

Chopra, S.L. and kanwar, J.S. (1986). *Analytical Agricultural Chemistry.* pp.250-252. Kalyani Publishers, Ludhiana.

Chundawat, B.S. (1997). *Nutritional Management in Fruit Crops.* Agrotech Publishing Academy, Udiapur.

Erickson, T.; Kelling, K. A. and Shulte, E. E. (1982) Predicting alfa nutrient needs through DRIS. *Proc. Wisconsin Fertil. Aglime. & Pest Management Conference,* 21: 233-246.

Ghose, B., Bose, T.K. and Mitra, S.K. (1986). Chemical induction of flowering and control of fruit drop in litchi (*Litchi chinesis* Sonn.) *Proc.*, 22nd *Int. Hort. Cong.*, California. Abst., 1189.

Iyenger, B. R. V. (1993) Fertilizer management. In: *Advances in Horticulture* Vol. 2 pp. 1055-1072. Chaddha, K. L. and Pareek, O. P. (eds). Malhotra Publishing House, New Delhi.

Khanduja, S.D. and Garg, Y.K. (1984). Macro-nutrient element composition of leaves from *Jujuba* (*Zizyphus mauritiana* Lamk.) tree in north India. *India J. Hort.*,41 : 22-24.

Ramaswamy, N. (1971). *Studies on the effect of nitrogen on the growth and development of 'Robusta' (Musa cavendishi* L.). M.Sc. (Ag.) thesis. Annamalai University, Annamalainagar.

Reuther, W., Batchelor, H.J. and Webber, H.J. (1962). *Citrus Industry.* Vol. II. Univ. of California Press, Berkeley, California.

Samra, J.S., Thakur R.S. and Chadha, K.L. (1978). Evaluation of existing limits of leaf nutrient standards in mango. *Scientia Hort.*, 8: 349-355.

Sharma, B. D., Dhandar, D. Gand Vashishtha, B. B. (2002) Response of ber to drip fertigation in aridisols of western Rajasthan. In. *Nat. Sem. On sustainable management of water resources for enhanced agricultural production,* BSKVV, Dapoli.

Shende, D.G. (1977). *Effect of N, P and K on growth, yield and quality of pomegranate (Punica granatum* L.) M.Sc. (Ag.) thesis. MPAU, Rahuri.

Singh, N.P. (1981). Studies on guava nutriation with special referance to leaf analysis. *National Symp.Trop.Subtrop. Fruit Crop.* Bangalore, pp. 75.

Tandon, H.L.S. (1987). *Fertilizer recommendations for Horticultural Crops in India.* pp.16. F.D.C.O. New Delhi.

Wavhal, K.N., Rane, D.A.; Kamble, K.B. and choudhary, S.M. (1985). Nutritional studies in pomegranate. *Third National Workshop on Arid Zone Fruit Research.* MPAU,Rahuri.

Chapter – 22

New Dimensions in Post-harvest Handling, Processing and Value Addition of Horticultural Produce

D. S. Khurdiya

22.1 INTRODUCTION

The life-styles, economic conditions and food habits have a great effect on food marketing. There is an increasing demand for fresh fruits and vegetables and their processed products. Fruits and vegetables are processed into a variety of products and it is required to be picked at its right stage. For making juice, fruits should be picked at ripe and soft stage, for canning purpose it should be picked at fully mature but firm just before ripe stage. Fruits and vegetables are passed through a long channel before their consumption; which may lead to a number of undesirable changes in their composition. Thus effective post harvest management of fruits and vegetables is essential to provide their acceptable quantity and quality worth consumption. In this regard handling, processing and preservation and value additions form important loci.

22.2 HANDLING OF FRUITS AND VEGETABLES

Post harvest management is defined as methods and techniques applied to increase the shelf life and retain quality of either fresh or processed horticultural produce. Another consideration is to reduce the post harvest losses, which occur in quality and quantity after harvesting till consumption.

Handling is a process through which horticulture produce from the field is transferred to the consumer. A large number of aspects are involved in the process such as harvesting, assembling, packaging, storage, transportation and marketing.

22.3 HARVESTING

Harvesting is a unit operation of detaching fruit or vegetable or edible part or other economic part from the source plant. Mostly horticultural produce are highly perishable in nature due to high moisture content, may be unacceptable for consumption if not handled properly during harvesting and after harvest. Thus, harvesting is an important operation in horticultural crop production. Some considerations for harvesting are :

i) Harvesting with improper method result in damage of produce by bruising which may be caused by compression, impact or vibration.

ii) Factors like delicacy of crop, maturity criteria, time and method of harvesting, mode of packaging and transportation, the speed during and after harvest, economy of operations to fulfil the market requirement, should be taken into considerations.

iii) Maturity indices are important for deciding when a given commodity should be harvested to provide some marketing flexibility and to ensure the attainment of edible quality to the consumer.

Harvesting Methods: (I) Hand harvesting: (II) Mechanical harvesting (a) Mechanical assistance and (b) Harvesting machines.

Harvesting practices should cause as little mechanical damage to produce as possible. Gentle digging, picking and handling will help reduce crop losses. For some crops, a natural break point forms at the junction of the stem and the stalk when produce is mature. Harvesters should grasp the produce firmly but gently and pull upward. Wearing cotton gloves, trimming finger nails and removing jewellery such as rings and bracelets can help reduce mechanical damage during harvest. Mechanical harvesting devices employ direct contact methods such as combing, cutting, pulling, snapping, twisting, stripping and compacting (O' brien *et al.*, 1983).

22.3.1 Harvesting containers

Picking bags, baskets and buckets of different sizes and shapes are used for harvesting. Recently, use of plastic crates has gained immense popularity since these are durable, reusable, and easy to clean. When empty, they can be nested to save space in storage or transport.

22.3.2 Harvesting tools

Some fruits need to be clipped or cut from the parent plant. A lot of advancement has been made in designing of harvesting tools such as clippers, knives, shears etc. Using a cutting tool attached to a long pole can aid harvesting from taller trees such as mango, avocado etc. Different types of pole mounted cutting shears with or without catching bags have been developed in recent years with slight modifications suiting to local needs.

22.4 PACKAGING

Packaging of fresh fruits and vegetables reduces the wastage of commodities by protecting them from mechanical damage, pilferage, dirt, moisture loss and other undesirable physiological changes and pathological deterioration during the course of storage, transformation and subsequent marketing. The best quality of produce should be packed for uniform and quality packed commodities. It cannot improve the quality of produce but maintain as it protects against hazards of journey. Packaging serves as an efficient handling unit to carry produce from field to consumer (Roy *et al.*, 2000).

Striking developments have been in the field of packaging of horticultural produce and the gunny bags, grasses and stem leaves used so far for packaging are now replaced by a variety of containers such as wooden boxes, baskets woven from bamboo or twigs, Hessian sack/jute bags, plastic punnets and corrugated fibre board (CFB) boxes. Due to problems of poor dimensional stability and stacking strength with basket and sack bags and heaviness and adverse effects of deforestation on ecological balance by using wooden boxes, some alternative ecofriendly packaging materials have been designed to replace them.

22.5 NEW INNOVATIONS IN PACKAGING OF FRUITS AND VEGETABLES

Designs and material used for packaging may be different under various condition viz. vacuum packaging, active packaging technologies, controlled atmosphere packaging, modified atmosphere packaging, or gas flush packaging, ripening chambers etc. and shipping container.

22.5.1 Pallet boxes and shipping container

Loading and unloading is one of the very important steps in the post-harvest packaging of fruits, vegetables and root crops. Post-harvest losses can be considerably reduced by using pallets. All the subsequent handling and transportation operation become very easy once the boxes are placed on pallets. Pallets literally form the base on which most fresh produce is delivered to the consumer.

22.5.2 Palletisation/ Unitization

Palletization – A pallet is a platform mark to hold one or more boxes, bags, cartons etc in a group. In advance countries almost all horticultural packages are unitized for handling. The packages must be designed for secure palletisation. The package must withstand expected stresses in the stacking column. If the packages are cross-stacked, their vents must be located on both ends and sides to allow air circulation through all packages on the pallet. The package dimensions must be compatible with pallet dimensions.

Advantages of palletisation include

a) Savings in handling costs averaging 40 to 45%

b) Reduced handling time

c) Fewer personnel accidents by substituting mechanical or manual handling and lifting

d) More efficient utilization of storage space

e) Reduced product damage

f) Uniform stacking of produce.

22.5.3 Biodegradable plastics

Plastics used in food packaging at present are non-biodegradable. The overflowing landfills cannot absorb the large quantity of plastic trash created in many countries. To reduce the waste, the best solution is a combination of biodegradable plastics, plastics recycling and composting. Biodegradable plastics are biodisintegratable plastics, which disintegrate into polymer chains and break down into water and carbon dioxide.

The first generation biodegradable plastics are made of polymers containing about 6 per cent starch (normally corn starch), which biodegrades and a pro-oxidant, which enhances the reaction of the polymer with oxygen in the air. New biodegradable plastics contains up to 97 per cent starch.

Other than biodegradable plastics, chemical and photodegradable plastics have also been developed. They are chemical plastic, for which a natural catalyst like metal salt is added and the photodegradable plastic for which a light-sensitive chemical is added to or sprayed on the plastic to make it disintegrate in ultraviolet light.

22.5.4 Special packaging

22.5.4.1 Edible films

Natural edible films extend shelf life, although for shortest time periods than synthetic non-edible packaging materials. Examples of edible films include casings, such as in sausage, and edible waxes, such as those applied to fruits and vegetables, the waxes function to improve or maintain appearance, prevent mold, and contain moisture while still allowing respiration. Food may be coated with a thin layer of polysaccha-rides such as cellulose, pectin, starch and vegetable gums, or proteins, such as casein and gelatin. Cut and dried, fruit pieces are often sprayed with an edible film prior to their inclusion into items such as breakfast cereal.

22.5.4.2 Foil

These may be used in snack bags (chips, etc.) or as a laminate in aseptic packaging. It is used as a wrapping for dry, refrigerator, or freezer storage. It provides a moisture-proof and vapour-proof barrier,

22.5.4.3 Laminates

These are multilayer of foil, paper, or plastics which may be utilized selectively according to the specific food packaging need, In combination, the various laminates may provide more strength and barrier protection than the individual laminate material. Laminates provide barriers useful in controlling O_2, water vapour, and light transmission, and they provide good burst strength. The laminates may resist pinholes and flex cracking. Retort pouches are examples or laminates used in packaging and contain polyester film, aluminum foil, and polypropylene.

22.5.4.4 Flexible packaging

It is available for packaging use in the foodservice industry and is finding more applications at the retail level. It keeps food products fresh by providing flavour and aroma barriers, which keep outside doors out and flavours in. It is used for fresh fruits and vegetables and matches respiration rate as closely as possible.

Manufacturers are offering more food products in flexible packaging and find that "cost savings and environmental concerns are two of the driving forces behind the switch to flexible packaging".

22.6 STORAGE

Storage of fruits and vegetables is very much essential, because of their highly perishable nature. Their storage can be extended by various treatments applied to them after harvest. The most important of these is temperature management. Orderly marketing of these perishable commodities requires some storage facility in order to balance day to day fluctuation between product harvest and their sale.

A proper storage system reduces physiological activity of horticultural commodities by maintaining lowest safe temperature that will not cause chilling or freezing injury. It also reduces growth of spoilage causing microorganisms. With some commodities, the storage facility may also be used to apply special treatments like curing of tuber and other fruits and vegetables. For example potatoes and sweet potatoes are held at high temperature and high relative humidity for few hours/days to cure wounds sustained during harvest, drying of scaly leaves of onion and garlic, and degreening of oranges and lemons and application of ethrol for uniform ripening before marketing or shipment.

22.6.1 Low cost storage

Clamps

Pits

Cellar

Night time cooling

Well water

Naturally formed ice

High altitude cooling

Underground storage

Ventilated storage/onion storage structure - Ventilation is essential

Evaporative cooling/Pusa zero energy cool chamber

22.6.2 Ventilated storages

A ventilated store may be used for onions, garlic, yams and sweet potato.

22.6.3 Improved pusa zero-energy cool chamber

A low cost zero-energy cool chamber, developed at IARI, New Delhi, recently is based on the principle of evaporating cooling (Roy and Khurdiya, 1982). Raw material readily available is used for installing the short-term double-walled storage for fresh vegetables and fruits. The inside temperature of the chamber maintained is between 17-18°C (lower than outside) with relative humidity of about 90 per cent during the peak summer months and also throughout the year. Such chambers are quite suitable for storage of fresh vegetables for short duration. Moreover, these are cheap, economical and within the reach of an average vegetable grower.

22.6.4 High cost storage

22.6.4.1 Refrigerated/Cool Storage

One of the first successful cold storages for apples was built in 1856 by Reverend Benjamin Nyce in Illinois, where ice harvested from the ponds was used to keep the fruits cold (Taylor, 1900). The first installation of mechanical refrigeration in the U.S. was made in Boston in 1881 (Anderson, 1953). Today refrigerated storage is a well-established technology widely used for storing horticultural crops all over the world and is undoubtly the most effective method of prolonging the storage shelf life and reducing post-harvest losses by arresting metabolic breakdown and fungal deterioration of the horticultural commodity. Its application is limited only by cost and benefit considerations. Earlier ammonia gas was used as refrigerant but recently fluorocarbons, such as Freon- 12 and 22 have replaced ammonia (Patchen, 1971). Essentially, all crops can benefit by being stored at a suitable low temperature, which extends the storage life and preserves quality.

22.6.4.2 Controlled Atmosphere (CA)

The fresh horticultural produce consumes oxygen for respiration and releases carbon dioxide and ethylene. The ethylene further enhances ripening. The major benefit of CA storage is derived from the reduction in respiration as a result of low O_2 content. High CO_2 competes for ethylene receptor sites (Burg and Burg, 1967). In CA storage the levels of CO_2, O_2 and N_2 in the storage room are monitored. In CA, oxygen is reduced from 21 per cent to 2-5 per cent and CO_2 is increased from 0.03 to 1-5 per cent. These results slow down of physiological activity of fruits and vegetables such as rate of respiration, ethylene production and other bio-chemical reactions. The reduced rate of these activities actually decreases the ripening and senescence process considerably. CA storage combined with refrigeration reduces respiration and delays yellowing and quality changes. However, the tolerance of individual varieties of horticultural crops needs to be considered. Commercial application of CA storage is limited to only a few crops, like apples and pears being the most popular ones. All types of controlled atmosphere storage require frequent gas analysis to determine when aeration is required to add oxygen or to remove carbon dioxide

22.6.4.3 Modified Atmosphere (MA)

While in modified atmosphere (MA), the respiration of the commodity is allowed to reach a desirable low level of O_2 and high level of CO_2 inside a closed chamber, container, plastic bag or plastic tent and the gases are maintained at those levels. Both the systems are best used with refrigerated storage involving manipulation of CO_2, O_2 and N_2. Other gases such as ethylene, acetylene and propylene are also considered. MA differs from CA only in how precisely the addition of pressures is modified. For prevention of accumulation of too much CO_2 and decrease of too much O_2, a few holes can be made. For maintaining a desired atmosphere within a package, absorbers and adsorbers of O_2, CO_2, ethylene and water vapours provide additional benefits (Moleyar and Narsimham, 1994).

Danish cabbage has been reported to be kept better for four to five months at 0ÚC in gas mixtures with 2.5 to 5% oxygen and carbon dioxide than it has in air. Mature green tomatoes may keep green for five to six weeks at 13ÚC in an atmosphere at 3% oxygen with 97% nitrogen. After removal to air at 18ÚC, these tomatoes ripened normally with acceptable flavour. Mushroom in refrigerated storage has been kept better for short periods in atmosphere with 5 to 10% carbon dioxide than air.

22.6.4.4 Ice bank cooler

A recent development in refrigeration is the Ice Bank Cooling system with positive ventilation. The system positively directs ice cold air through the boxes or crates containing fresh fruits and vegetables. In this way quicker cooling is possible and large amount of heat can be removed in a relatively short period of time. In this system, a bank of ice is built up with the help of a small refrigeration plant. The ice is accumulated on extended surface plate, which is suspended in a tank of water. The water surrounding the ice is pumped to the top of a cooling tower and falls through the tower passing over the extended surface, giving a large surface area of heat exchange. The air leaving the tower maintains high relative humidity. Once initial cooling is complete the store maintains a temperature of 0.5 - 0.8°C and relative humidity of 98 per cent. In view of the continuous humid cool air circulated through the stored commodities, this system of storage is found very useful for storage of fresh horticultural produce. This will be preceding step for cool storage of fruits and vegetables.

22.6.4.5 Hypobaric storage

Fruits and vegetables can be stored under less than atmospheric pressure (Burg and Burg, 1966). This technology is called hypobaric storage. Fruits were held at about 0.2 to 0.5 atmospheric pressure at 59 to 75°F (15 to 24°C) and humidified air was passed through the chamber in order to maintain humidity and to remove ethylene and other volatile compounds given off by the fruits and vegetables. Hypobaric storage is a form of controlled atmosphere storage in which the produce is stored in partial vacuum. In hypobaric storage system, the commodity is placed in a vacuum tight and refrigerated container and air is evacuated by a vacuum pump to the desired low pressure. This is a very expensive method and used in case of high value commodities.

22.6.4.6 Cold chain infrastructure

Provision of lower safe temperature to the commodities immediately after its harvest till it reaches the consumer is called cold chain. The various steps are involved in cold chain are:

1. Pre- cooling
2. Cold storage at field site
3. Refrigerated transportation
4. Cold storage at destination/marketing centre
    - Setting up of cold rooms in regulated markets
    - Establishing cold storage facilities at the airports/ports
    - Building multipurpose cold storages

22.7 TRANSPORTATION

Transportation refers to carrying a produce from one place to another. Rapid and reliable transport of fresh fruits and vegetables is of critical importance to successful marketing. The delay in transport not only affects shelf life and the quality of fresh produce, but also reduces its value in the market. Transport is an important link in the handling, storage and distribution of fruits and vegetables.

The transport of produce starts from the field to the assembly points or collection centres. From these places, bulk transport is undertaken by retailers, middlemen, wholesalers, processors, exporters and importers at packing stations, storage plants and shipping points. Several modes of transport are used to move fresh fruits and vegetables from farm to final consumer. They include head load, bullock carts, bicycle, rickshaw, autorikshaw, lorry, boat, ship and aeroplane.

Factors affecting transportation

Some factors which affect transportation are as under :

- Product handling
- In-transit temperature and humidity
- Air circulation and ethylene removal
- Compression/abrasion during movement

Poor handling of produce during loading and unloading of transport vehicle results in high post-harvest losses. Each commodity must be transported under optimum conditions of temperature and humidity. Packing and mode of transport selected should be such that it causes no or minimum vibration and bruising injury to the product. The produce is transported through shipping containers, bulk packs, consumer packages and stacking containers.

22.8 FOOD PARK

Food Parks are comprehensive industrial estates for food-processing units where the industries would have the provision of common facilities like cold storage, ripening room, cold chain, effluent treatment plant, ware housing, power connection, water facilities, sewerage etc. The creation of these common infrastructures would benefit the individual units particularly the small and medium scale units, because these are expensive to be set up by any single individual unit. Therefore a common park with all the infrastructures would help in the growth of this industry.

The main objective of setting up the food park is to attract several units in the food processing industry into a common area under the industry cluster concept so as to ensure low operating cost, more flexibility of operations of the units in the park and reduce post-harvest losses. This would also help to establish better linkages between farmers and processors, which would finally result in better realization to the farmers as well as lower cost to the consumers. It is promoted with an idea of targeting small- and medium-sized food-processing units that will process the agricultural/horticulture produce of surrounding areas. We are now losing about 37 per cent of our agricultural produce due to lack of proper processing and storage facilities. Food parks can play a vital role in eliminating this huge wastage."

The food parks can act as a bridge between farmers and the industries. We can facilitate a firm buyback agreement between a farmer and a food-processing unit located in a park. This will address two problems. One, the food-processing unit is assured of good quality agriculture/ horticulture input and the farmer is assured of sale and the price.

The Department of Food Processing Industries and Horticulture of Indian states is encouraging and facilitating private sector, joint venture, and PSU initiatives for establishing Food Parks at different places of their states in order to provide common facilities to the entrepreneurs for setting up food processing units. Ministry of Food Processing Industries has approved about 57 food parks in the country (Anon., 2011).

22.9 PACKINGHOUSE OPERATIONS

Packinghouse operations can be as simple as moving produce from a field lug into a shipping container, or may include a variety of handling practices, from cleaning, waxing, sizing, and quality grading to colour sorting. The provision of shade during the packing operations is extremely important. Shade can be created using palm leaf fronds, a plastic mesh or canvas sheet hung from temporary poles, or via a permanent roofed structure. When deciding upon where to locate a packinghouse, access to the field and market point, adequate space for vehicles to enter and leave the packinghouse and ease of access to labour all will be considerations.

In the simplest packinghouse, produce is delivered in picking containers, immediately after harvest, directly to the packers. The packers then sort, grade, size and pack the produce directly into appropriate transport containers. In this case, each worker must be knowledgeable regarding produce defects, grade and size requirements, and packing methods. As the size and complexity of the packinghouse increases, more operations and workers trained in specific tasks might be added. Some additional activities may include degreening, curing, washing, bunching, chemical treatments and pre-cooling. The sequence of activities varies with different crops. These operations are essential preparatory steps for transportation and subsequent marketing.

22.10 MINIMALLY PROCESSED FRUITS AND VEGETABLES

Minimally processed horticultural products are prepared and handled to maintain fresh natural while providing convenience to the user. Other terms used to refer minimally processed products are "lightly processed", "partially processed", "fresh processed", and "prepared". The purpose of MP foods is to deliver to the consumer a like-fresh fruit or vegetable product with an extended shelf life and at the same time ensure food safety and maintain sound nutritional and sensory quality. The increased health consciousnesses of consumers, diminished food preparation times, and increased purchasing power have recently combined to increase the demand for minimally processed fruits and vegetables. Restaurants and eating outlets opened by developed countries (MNC) in India also demands about the minimally processed fruits and vegetables. The MP fruits and vegetables are usually living respiring tissues, but respiration is greatly increased by cutting, slicing, low temperature, heat treatments and preservatives. MP fruits and vegetables require cold-chain refrigerated or slightly higher temperatures depending on the commodity or mixtures of commodities. MP fruits and vegetables are very different than frozen or dehydrated or thermally processed foods. Minimally processed fruits and vegetables require special treatments/preparatory operations, preservation, packaging, storage and marketing.

MP fruits and vegetables are defined as those prepared by minimally or lightly processed operations such as washing, sorting, peeling, slicing, shredding, juicing etc. given a partial but not end-point preservation treatment including use of minimal heat, a preservative or radiation. But that do not affect the fresh-like quality of the fruit or vegetable.

However, because the procedure is essentially wounded, the shelf-life is often greatly diminished. The preservation or hurdle treatment may include pH control, antioxidants, chlorinated water dips or combination of these or other treatments. It is important to take advantage of the synergies of all preservation treatments (Wiley, 1994).

Minimally processed fruits and vegetables include peeled and sliced potatoes; shredded lettuce and cabbage; washed and trimmed spinach; chilled peach, mango, melon, and other fruit slices; vegetable snack, such as carrot and celery sticks, and cauliflower and broccoli florets; packaged mixed salads; cleaned and diced onions; peeled and cored pineapple; fresh sauces; peeled citrus fruits; and microwaveable fresh vegetable trays.

Most MP food raw materials production is seasonal, usually in far away places from consumption areas, and concentrated at certain region where yield and quantity can be optimized. The raw material remains a living entity and highly perishable, bulky, price and quantity-variable commodity. An example of fresh sliced strawberry (Rosen and Kader, 1989) is cited below:

Fresh sliced strawberry processing

1. Cultivar selection - Parjaro
2. Harvest - Hand or mechanical at full bright red or pink
3. Precooling - Rapid forced air or hydro precooling at 7° C (8min to 2hr.)
4. Field processing - Wash to remove sand, separate berry
5. Transportation - To packing house with dry ice (0 ° C)
6. Sorting and grading - Defectives and colour sorting, small or large
7. Processing - Slicing - quarter, water washing – to remove exudates, dip in $CaCl_2$ solution and spin-drying
8. Packaging - CA-12% CO_2, 2%O_2: 95% RH, 0°C , 500g, 1kg, 5kg
9. Storage at Warehouse - 95%RH, 0°C, Standard Pallets, Max.7 Days.
10. Storage and Display at Retail - Display Cabinet 1 day at 18-20 ° C

22.11 PROCESSING AND PRESERVATIONS

Various Processing and Preservation Techniques Used For Fruits and Vegetables

S.L. No.	Processing	Preservation
1.	Canning/Bottling/Retort processing	Heat application - Blanching, Pasteurization and Sterilization
2.	Drying and Dehydration	Water removal
3.	Freezing and Dehydro-freezing	Low temperature preservation
4.	Irradiation	Dosing with U.V. or Ionizing Radiation
5.	Pickling	Use of salt, vinegar, oils and spices etc.
6.	Fermentation	Alcoholic, acetic Acid and lactic acid fermentation
7.	Evaporation or Concentration (concentrate, puree and paste)	Water removal; Use of heat and low temperature
8.	Concentrated pulp, spreads etc (jam, jelly, marmalades, preserve, candy, etc.)	Use of high sugar and acids
9.	Aseptic processing	HTST, (High temp. and short time) UHT (Ultra - high temp)
10.	Sauces and ketchups	Use of chemical preservatives
11.	Fruits juices and beverages	Use of heat application and /or chemical preservatives
12.	Carbonation	Anaerobic environment
13.	Ultra-High Pressure	Inactivation microorganisms and Enzymes
14.	Microwave heating	Inactivation microorganisms and Enzymes
15.	Ohmic/Electric current	Inactivation microorganisms and Enzymes
16.	Light pulses	Inactivation microorganisms and Enzymes
17.	Pulsed electric field processing	Inactivation microorganisms and Enzymes
18.	Super Critical Extraction Technology	Extraction of essential oils
19.	By Other Means	Hurdle technology

22.11.1 Aseptic processing and packaging

In order to destroy any ***Clostridium botulinum*** spores and extend the shelf life of low-acid foods, aseptic processing and packaging may be utilized. Independent aseptic processing is not a new technology. In fact, it has been used commercially in Europe and Japan for many years. In India, a considerable increase in the production of aseptically packed products has been made as the best method of preservation in the last decade. Aseptic processing and packaging is a remarkable technological advancement over traditional retort processing. It is the process by which microorganisms are prevented from entering into the package during and after packaging. Aseptic processing and packaging can be defined, the process in which a commercially sterile, cooled product is packed, into pasteurized containers in a sterile environment at an ambient temperature. The packaged product could be stored without the need for refrigeration or preservatives for periods up to one year.

22.11.2 The aseptic process

The aseptic process is a technique for processing liquid food products by exposing them to brief intense heating. Sterility is the key word in the process and involves the same principles as discussed with respect to sterilization. Since very high temperatures are employed (125 °C to 145 °C), the process is also referred to as Ultra High Temperature (UHT). The holding time of the product at high temperature is very short (3 to 15 seconds) and therefore, thermal stress on the product is less while ensuring safety (Nelson, 2010). This rapid heating and cooling process also reduces the energy use and nutrient losses associated with conventional sterilization. Processing at high temperature usually requires that the process be based on enzyme inactivation rather than microbial destruction. Liquids such as creamers, milk, or juices may be packed in this manner. Triple or multiple packs of flavoured milk and juice, with attached straws, are available on grocery shelves. The market leaders of aseptic packages have introduced easy-open, easy-pour features into their canons. The plastic devices are injection moulded and adhered to the package tops.

22.11.3 Ohmic heating

In ohmic heating, an electric current is passed through a food material which then heats as a result of its inherent electrical resistance. The process involves the passage of low frequency alternating current (50 or 60 Hz.) through the product (Lima, 2008). The electrical energy is transformed into thermal energy. The extent of heating depends on the uniformity of the electrical conduction throughout the product and its residence time in the heater. A typical use of ohmic heaters is in an aseptic processing system.

22.11.4 Microwave processing

Microwave processing refers to the use of electromagnetic waves of certain frequencies to generate heat in a material through two mechanisms-dielectric and ionic. Microwave heating for pasteurization and sterilization are preferred to conventional heating because this requires less time and reduced nutrient losses, particularly of thermolabile components (Brewer, 2005).

22.11.5 Light pulses

This method uses intermittent flashes of white light (20% UV, 50% visible and 30% infra-red) with an intensity claimed to be 20,000 times that of the sun at the earth's surface. One to twenty flashes per second are typical pulse rates which lead to significant surface reductions in microorganisms when used on meat, fish and bakery products. This technique is ideal for surface decontamination of packaging materials and works best on smooth, dust-free surfaces.

22.11.6 Pulsed Electric Fields (PEF)

This process involves applying repeated short pulses of a high-voltage electric field (10 – 50kV/cm) to a pumpable fluid flowing between two electrodes. It does not use electricity to generate heat, but instead it inactivates microorganisms by disrupting the walls and membranes of cells exposed to the high-voltage pulses. PEF is mostly used in refrigerated or ambient products and because it is applied for just one second or less, it does not result in heating of the product. It is for this reason that it has nutritional advantages over more traditional thermal processes which degrade heat-sensitive nutrients (Vega- Mercado *et al.*, 1999).

22.11.7 Irradiation

Ionizing irradiation has huge potential in area of post harvest management of fresh horticultural produce. It can be applied by exposing the crops to radioisotopes (gamma rays or X-rays) and from machine which produce a high-energy electron beam. The ionizing radiation has numerous applications in post harvest handling of fresh fruits and vegetables (Kader, 1985) as:

- Inhibiting sprouting in tubers, bulbs and root crops.
- Delaying the processes of ripening and senescence in certain climacteric fruits and strawberries (irradiating dose 1.5 to 3.0 KGy).
- Controlling the growth of fungal pathogen which causes decay of stored fruits by combination procedures involving irradiation and hot water dip treatments.
- Disinfestations of insects which cause spoilage in fresh fruits, dry fruits and nuts as well as quarantine restrictions in their international trade.
- Delaying stem elongation and cap opening in button mushrooms.

22.11.8 Carbonation

Aerobic microorganisms such as moulds and yeasts require oxygen for their growth. They become inactive in the presence of carbon dioxide. In carbonated drinks, the oxygen of the air is displaced with carbon dioxide. The absence of air and the presence of carbon dioxide in them help to prevent the growth of mould and yeast. Further, the keeping quality of carbonated fruit beverages is enhanced by adding about 0.005 per cent of sodium benzoate and or light heat processing.

Carbonation is the process of mixing carbon dioxide gas under pressure with water or beverages. The water or beverages are mixed with certain additives. The drink when serves gives off gas in fine bubbles and has the characteristic pungent taste of desired beverages. It adds to the life of the product and contributes in some way to its tang. Carbonated beverages are generally bottled with carbon dioxide content varying from 1 to 8 gram per litre. It provides anaerobic condition and reduces the oxidation of ascorbic acid.

22.11.9 Ultra-high pressure

At a pressure of 4000-9000 atmospheres enzyme and bacteria are inactivated but taste and flavour remain unaffected. Besides preservation, it improves rheological and functional properties of food and is environmental friendly causing less affluent problem. High pressure with 4000-9000 bars extends refrigerated produce shelf life from two months to six months at 4°C.

22.11.10 Super critical extraction technology

Supercritical fluid extraction technology is an alternative method for new or improved applications in food processing. It has the potential to be an environmentally friendly "green" processing technique, and in some cases, replacing the traditional organic solvent based extraction techniques with more benign solvents like supercritical carbon dioxide (CO_2), ethanol and water.

22.11.11 Hurdle technology

Hurdle technology (HT) concept was introduced by Leistner is another paradigm shift in preserving fruit and vegetables, offering a viable alternative to canning and dehydration. The HT processing is a combination of processing technique in which several technological approaches and unit operations used as freezing, refrigeration, dehydration, irradiation, pulse electric fields, high pressure processing, ultra-sonication, use of chemical preservatives, pH and water activity controls are used in unison in a synergistic fashion to bring about desired microbiological safety and sensory acceptability. Convenience, safety, microbiological stability and fresh-like appearance are qualities of HT products.

22.12 VALUE ADDITIONS TO FRUITS AND VEGETABLES

Value Addition is a process through which value is added to fresh fruits and vegetables and / or their processed products for better quality and fetching better returns.

1. Unit packing of primarily/minimally processed vegetables such as cauliflower, peas, leafy vegetables, etc. can be so that inedible parts are removed before being sent on to the metropolitan city markets. Similarly, bananas should be transported in hands as is done in other countries. This process will reduce the transportation and handling cost of inedible parts and provide convenience food.

2. Each packing station should have the facility for processing. Physically damaged fruit and vegetables that are without infection can be made into durable and value added processed products. It could considerably reduce the price of main product as a result more processed food products will come within the reach of common people.

3. About 25-30% of total production are not fit for fresh marketing (undersized, oversized and malformed/deformed and physically damaged fruits and

vegetables but microbiologically free) are called culled produce. It could be utilized, if processed into various value added by products.

4. Factory generate large amount of valuable waste such as peel, stones and other inedible plant parts that are generally rejected as garbage. If they are gainfully utilsed at the proper time they can become value added products such as pectin, starch, colours/ pigments, essential oils, sugars, vinegar, alcohol, and many compounds, and also reduce the price of the primary processed products.
5. Premature fruit drop due vagaries of climate such as dust or hailstorm a substantial quantity of fruits and vegetables are lost before they attain proper maturity stage. They can be processed into value added products such as pulp, pickle, chutney, dried powder etc.
6. During peak harvesting season an abundant quantity of quality fruits and vegetables resulting in market glut. It can not be marketed in the fresh form. Its result in huge spoilage and wastage. This wastage can be reduced if they are timely processed into different value added products and it can be used throughout the year and can be transported to distant market. They are not only palatable but also nutritious and give economic gain, besides acts as price stabilizer.
7. Value addition also includes peeling, slicing, cutting into pieces, processing and packaging. All such activities increase value of the products e.g. Minimally Processed Product.
8. Some indigenous fruits that are underutilized are tropical/subtropical in nature and grow even under adverse agro-climatic conditions. They possess therapeutic/medicinal and nutritive value and have excellent flavour and very attractive colour. e.g. *bael* fruit and *Kokum* are not acceptable as a fresh fruit but can be used in beverage industry or for culinary purpose in the dried form. Similarly *aonla* as a fresh fruit is not liked because of its strong astringent taste. All these fruits have a great potentiality to processing into a value added fruit products of commercial importance.
9. Waxing and surface coating treatment improves shelf-life thus add value to the fresh produce.
10. Degreening in citrus fruits also add value to the fruits.

22.13 REFERENCES

Anderson, O.E. Jr. (1953). Refrigeration in America. Princeton University Press, Princeton, U.S.

Anonymous (2011). Statement of food parks approved for financial assistance. MOFPI, GOI official website, http://mofpi.nic.in.

Brewer, M. (2005). Microwave processing, nutritional and sensory quality. In: Schubert, H. And Regier M. (eds.). The *Microwave Processing of Foods*, CRC Press, New York, p 335.

Burg, S.P. and Burg, E.A. (1966). Fruit storage at subatmospheric pressure. *Sci.*,153: 314.

Burg, S.P. and Burg, E.A. (1967). Molecular requirements for the biological activity of ethylene. *Pl. Physiol.* 42: 144- 152.

Kader, A.A. (1985).Potential applications of ionizing radiation in post- harvest handling of fresh fruits and vegetables. *Food Technol.* 40 (6): 117-121.

Lima, M. (2008). Food preservation aspects of Ohmic heating. In: Rahman, M.S. (ed.) *Handbook of Food Preservation.* Second Edition. CRC Press, New York.

Moleyar, V. And Narsimham (1994). Modified atmosphere packaging of vegetables- an appraisal. *J Food Sci. Technol.* 31 (4): 267.

Nelson, P.E. (2010). *Principles of Aseptic Processing and Packaging.* Perdue University Press, West Lafayette, Indiana.

O' brien, Cargill, M.B.F.and Fridely, R.B. (1983). *Principles and Practices for harvesting and handling of fruits and nuts.* AVI Pub. Co. INC. Westport, Connecticut, p. 636.

Patchen, G.O. (1971). Storage for apples and pears. U.S. Deptt. Agri. Mktg. Res. Rept.: 924.

Rosen, J.C. and Kader, A.A.(1989). Post-harvest physiology and quality maintenance of sliced pear and strawberry fruits. *J. Food Sci.* 54 (3): 656- 659.

Roy, S.K. and Khurdiya, D.S. (1982). Keep vegetables fresh in summers. *Indian Hort.* 27: 5.

Roy, S.K., Pal, R.K. and Sen, Nita (2000) Packaging technology for fruits, vegetables and their products. In : *Post harvest technology of fruits and vegetables.* Indus Publishing Company.New Delhi.

Taylor, W.A. (1900). The influence of refrigeration on the fruit industry. U.S. Deptt. Agri. Year Book. 561-580 pp.

Vega- Mercado, H., Gongora- Nieto, M.M., Barbosa- Canovas, G.V. and Swanson, B.G. (1999). Non-thermal preservation of liquid foods using pulsed electric fields. In: Rahman, M.S. (ed.) *Handbook of Food Preservation.* Second Edition. CRC Press, New York.

Wiley, R.C. (1994). Minimally processed refrigerated fruits and vegetables. Chapman and Hall, New York.

Chapter – 23

Post Harvest Handling and Processing of Mandarin – Prospects and Constraints

S. K. Jain, L. K. Dashora and Jitendra Singh

23.1 INTRODUCTION

Mandarin oranges are the most popular among the citrus fruits and alone occupy more than 50% share under widely divergent citrus crops grown across the globe. In India, oranges have a place of prominence among fruit kind after mango and banana. It's delicious, relishing, nutritious and handy fruits offer best mouth feel especially during summer when other fruits fail to quench thirst of throat. There is a need of scientific handling so as to accentuate trade and to exploit market potential of mandarin fruit.

Country wide, orange covers an area of 5.02 lakh hectares and the production is of the order of 43.96 lakh tonnes accounting for 12.03% and 6.40 % share under total acreage and production respectively in the fruits (Anon., 2010). Orange's production in developed countries is projected to grow at an annualized rate of 0.6 per cent and that in developing country at 0.8 per cent.

In Rajasthan, 'Nagpur' mandarin and 'Kinnow' fruits occupy a place of prominence among various citrus crops grown. Cultivation of 'Nagpur' mandarin is mainly concentrated to Jhalawar district where it covers about 6040 ha area producing 87586 tons fruits, annually. After the implementation of National Horticulture Mission the area is further increasing steadily in the district.

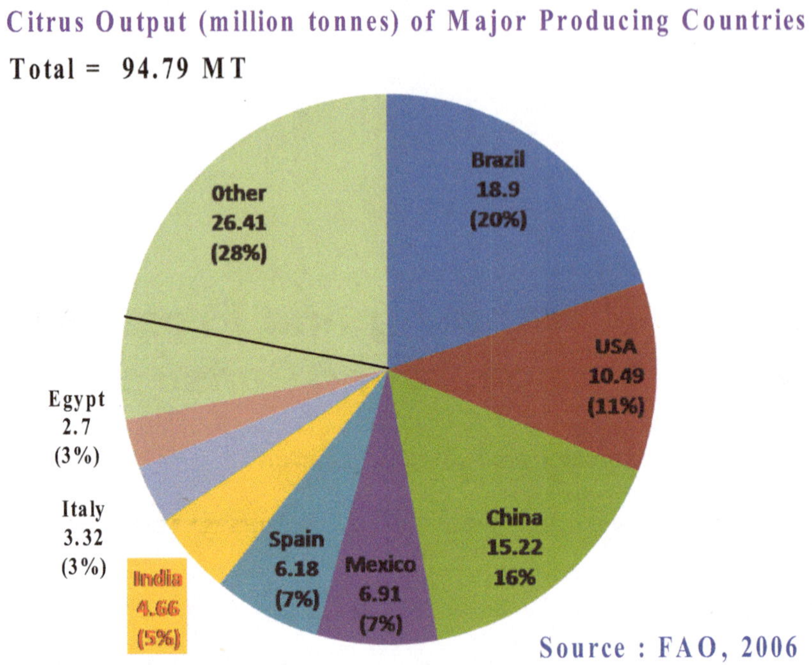

23.2 PRESENT STATUS

In our state, growers, being small land holders follow traditional system of orcharding. Fruits are grown without reference to its purpose of utilization. Most of the orchards are surrendered to the pre- harvest contractors who rarely show their concern for the health of the trees.

Harvesting is done manually by the unskilled labour which follows traditional method of plucking, causing button hole injury to the fruits. These fruits are liable to rot quickly during subsequent storage and transportation. After harvesting the fruit is packed in locally available packaging material such as gunny bags or baskets or otherwise most of the times load loosely in to the truck causing damage to the fruits during transportation. At farm level the practice of proper post harvest handling such as sorting, grading, precooling or packaging remains neglected and thus the produce fetches poor price in the market.

The marketing channel prevailing is by and large not amenable to the growers as a whole. Different marketing channels prevailing in the distribution system of mandarin oranges are presented in Figure1.

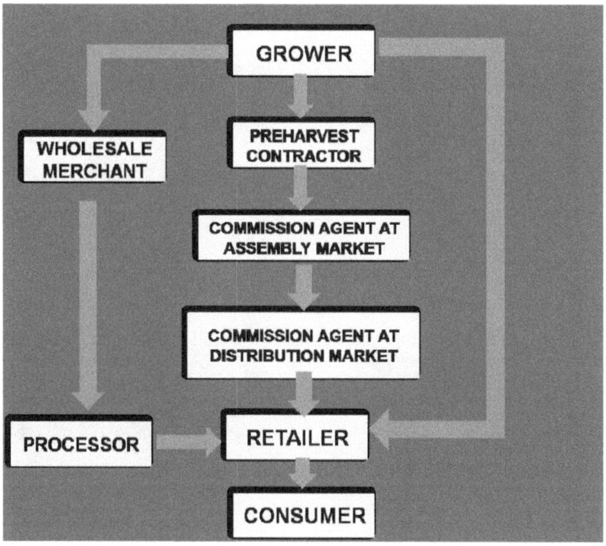

Fig. 1: Existing marketing channels in the distribution of mandarin oranges

The poor production technology, insect- pest and disease incidences, lack of proper post harvest handling infrastructure and long marketing channels cause enormous proportion being wasted. The results of a post harvest survey done on mandarin (table 1) show that approximately 25- 45 per cent of production goes waste annually. The condition becomes more pathetic during peak harvesting season when there is glut in the market and due to non availability of proper storage and processing facilities the farmers are compelled to sale their produce at throw away prices. It should be noted here that in India, less than 1 per cent of total fruit production is being processed as against 80 per cent in Malaysia and 78 per cent in USA.

Table 1: Estimation of post harvest losses in mandarin at various stages of distribution

S.No.	Level of distribution	Per cent loss	Causes
1.	During sorting at farm	6	Culls, harvesting injury, insect damage
2.	At whole sale market	8	Cull, insect damaged, pressed, rotted and sunburn injury
3.	At retail level	10	Rotting , pressing and insect damage
4.	Storage	20	Normal recommended storage temperature for mandarins = 6-7°C but Nagpur mandarins are generally stored in cold storage meant for apple and potato (1-4° C) causing chilling injury, stem end rot and *Penicillium* rots.
	Total	**44**	

Losses due to improper handling

Button hole injury due to wrong harvesting method

Peel injury due to piercing of long stalk of neighboring fruit in the package

Post Harvest Decay due to poor orchard sanitation

Green mold *(Penicillium digitatum)*

Blue mold *(Penicillium italicum)*

Stem-end rot *(Diplodia natalensis)*

23.3 POST HARVEST HANDLING OF MANDARIN

23.3.1 Harvesting

Harvesting of fruits should be done when they reach the maturity. The development of rind colour is best indication of harvesting orange fruits. But, for all practical purposes, grower's perception works well as they develop experience of adjudging maturity simply by visual inspection of fruits. The maturity indices for 'Nagpur' mandarin are given in Table 2 (Saraswathy *et al.*, 2008). In conventional practices, mandarin fruits are harvested by snap method using ladder. In this method, the jerk exerted causes button hole injury to the fruits. Such injury has been reported to affect 8-9% fruits in a single harvest. Use of clippers (Figure 2) for harvesting fruits reduces such injury. But care should be taken that not more than 2 mm stalk should be left otherwise it will puncture other fruits inside the package (Satyanarayana and Reddy, 1994).

Table 2 : Maturity indices of Nagpur mandarin grown in central India (*Mrig bahar* crop)

S.No.	Parameter	Indices
1.	TSS: acid ratio	14
2.	TSS	12 – 14° Brix
3.	Peel colour	Green- yellow
4.	Juice content	35-50 %
5.	Days from anthesis	240

Frequency of harvesting is also important for getting quality fruits from orange orchards. Sometimes, constrained by labour, input, agricultural operations etc, the growers harvest entire crop at a time. In this process, poor quality fruits are also harvested along with good quality fruits. This practice needs to be discontinued. Fruits should be plucked time to time as and when they reach harvestable stage. Such type of staggered harvest also help in regulating marketing of the produce and getting plus price.

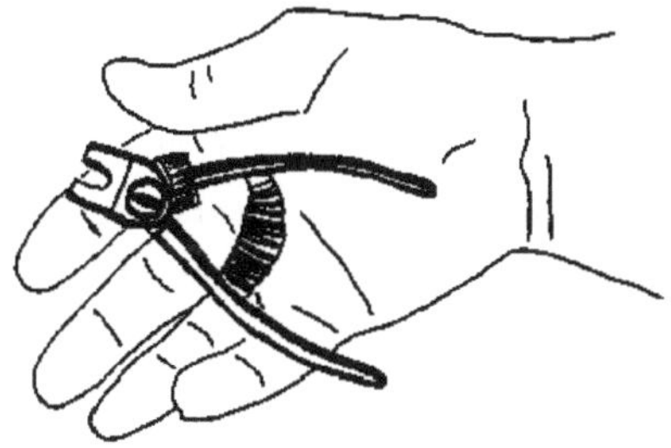

Fig.2 : Clipper for harvesting citrus fruits

23.3.2 Degreening

In oranges, fruits grown in central and south India, fail to develop attractive peel colour, although they reach maturity. The colour break in the rind of citrus fruits takes place, when the fruits are exposed to a minimum temperature below 4°C. The deep orange colour of Nagpur mandarin pulp is due to Cryptoxanthin while the peel contains violaxanthin, cryptoxanthin and neoxanthin (Subramanyam and Cama, 1965). Development of these carotenoids is favoured by cool night temperature, cool day temperature and cool soil temperature. When night remains warm at the time of fruit maturity, chlorophyll synthesis continues on and unmasking of carotenoids doesn't take place. This type of situation renders greenness to oranges although the fruits are ripe.

Ethylene application induces the accumulation of carotenoids such as Cyptoxanthin, β- citraurine and violaxantin in flavedo (Stewart and Wheaton, 1972). In India, ethylene treatment was standardized for the 'Nagpur' mandarin for the first time during early 1990's, using compressed gas applied through the trickle method in specially designed degreening chambers. For 1 tonne of fruits, an ethylene concentration of 5±2 ppm by trickle method (continuous flow) was found effective (Ladaniya, 2000). Fruits with 50 per cent surface green need 48 h for degreening at 27- 29°C and 90-95 per cent relative humidity with four air changes per hour. Degreening of mandarin fruits can also be achieved by 'ethephon dip' (2500- 5000 ppm) method but it develops yellow colour rather than orange colour (Kohli and Ladaniya, 1988).

23.3.3 Pack house operations

Pack house has become an integral part of the citrus industry, worldwide. The sequential operations in a modern pack house include- (i) receiving the fruits, (ii) degreening, (iii) sorting, (iv) washing, (v) waxing and drying, (vi) grading/ sizing (vii) packing and palletization, and (viii) storage or truck loading.

The operation 'degreening' has been described earlier. Remaining important operations are being dealt here :

23.3.4 Sorting

Sorting is done visually on the basis of surface blemishes and colour and to remove the rotten and small sized fruits. Sorting is mostly done manually by experienced labours, on a roller conveyor belt by trained personnel. The removed fruits/ cull are continuously removed from the area. Supervisors have to ensure proper sorting to avoid good fruits being removed along with cull or any cull being retained with good ones. In developed countries high speed electronic graders that recognizes blemishes using photo- optical grading system, have been introduced.

23.3.5 Washing

Washing is done by sprays of wash water containing detergent and chlorine while the fruits move on rollers, covered with nylon brushes to remove dirt and natural wax from the fruit surface. Washing is followed by rinsing with fresh/ plain water. After rinsing, fruits are run on sponge rollers to remove excess water so that fruits are just damp at the time of wax coating.

23.3.6 Waxing and surface drying

Oranges have natural wax coating over the peel but in course of washing, it is removed. As a result, loss of water is exacerbated and the fruits soon lose its appearance and consequent marketing is badly affected. Wax coating is an important operation in the citrus packing houses. Waxing is done with the main objectives of: (1) providing gloss and luster to the fruits, (2) reducing moisture loss during handling and (3) acting as carrier for fungicide or growth regulators such as 2,4- D. Effective protective coating reduces weight loss by 30- 40 per cent (Johnson, 1991). It is convenient to add fungicides to the wax emulsion such as Sodium ortho- phenyl Phenol (SOPP0, however separate use of fungicide give better results commercially (Gutter, 1981). Different types of waxes – edible (bee wax, carnuba wax, candelila wax, sugarcane wax) and non- edible (paraffin wax, shellac based wax etc.) are available in the market. These coatings must be approved 'Food Grade' by any competent authority. A layer of wax is formed over the surface by finely spraying metered amount of wax solution through nozzle mounted over the conveyor rollers. The applied coating is uniformly applied with brushes on subsequent rollers.

23.3.7 Grading/Sizing

Grading of fruits according to dimension (sizing) is important operation before packing of fruits to ensure uniform quality within the package. The mechanical graders used for sizing citrus fruits particularly mandarin oranges are roller sizers. In this case individual fruit is dropped on the divergent rollers, which are positioned with increasing spacing. As the fruit moves from beginning to the end of the roller, small fruit falls on to the pull- out belt below the roller while larger fruit moves forward along the rollers until appropriate spacing between the rollers allow them to drop. These rollers can be adjusted to suit the different sizes of citrus fruits. Graders with a capacity to grade as many as 5000 fruits/ h with 92- 97 per cent accuracy are available in the market.

Normally, oranges have an average diameter of 5.5 to 8.5 cm and within the packing box, there shouldn't be much deviation in size exceeding the range of 0.6 to 0.7 cm. Of course, grading standard varies according to the destination, for Indian markets, oranges are graded into three grades (Table 3).

Table 3 : Grades for Nagpur mandarin

Grade	Size	Diameter of fruits (cm)
1	Large	6.71 to 7.30
2	Medium	6.22 to 6.70
3	Small	5.50 to 6.10

In Bhawani mandi-the largest mandi in Jhalawar, Rajasthan, grading is done on the basis of number of fruits packed in an average size crates. The grading standards in vogue are: 141, 171, 191, 205, 225 and 240. The grading denotes the number of fruits packed in a particular crate. In general, crate containing such type of graded fruit holds 25-26 kg fruit weight.

23.3.8 Packing and labelling

Generally, packing practices followed by growers are very meagre. At the grower's level, hardly standard packaging is practised. Corrugated fibre board boxes being permeable to gases are highly suitable for packing oranges. An ideal packaging material should have sufficient mechanical strength to protect the content during handling and transportation. The base material of packing shouldn't contain any chemical toxic to product or human health. The weight of package must meet handling and marketing requirements in terms of shape, size and weight. The package should add identity to its content. It should be additive to retail presentation and over and above, safe to dispose off.

Mandarin fruits must be packed systematically in layers using 3-ply CFB separators which act as cushioning material between the layers. For mandarins, fully telescopic CFB boxes should be used owing to better strength. A vented polythene liner/ bag is very useful moisture barrier to minimize weight loss in fruits. However, using plastic liners in cartons can be risky at temperatures of 15- 20°C and above since pathogens can grow and cause decay. The produce should have attractive labeling. It should proffer complete detail regarding packed material. For oranges, the label should contain information with regards to date of harvest, date of packing, name of grower, name of packer, grade of produce, weight of fruits etc.

23.3.9 Storage

Storage is the most important operation during marketing of fruits. The main aspects of storage are protection from decay and slowing down the rate of moisture loss. 'Nagpur' mandarin grown in central parts of India develop chilling injury even at 3.5- 5.0°C temperature after 30 days, while fruits grown in northern India can be stored at 2.2-3.9°C without chilling injury. Injury increases with extension of storage up to 90 days (Ladaniya and Sonkar, 1996). A temperature of 5.2- 6.8°C and 90- 95 % relative humidity were optimum conditions for storage of 'Nagpur' mandarins (Ladaniya, 2008).

Citrus fruit is non- climacteric, and fruits are harvested when fully mature/ripe. Therefore there is no vital process that needs to be controlled after harvest to extend the storage life. In general, citrus fruits are difficult to store in controlled atmosphere except for lime, due to off flavour development. Very high doses of CO_2 can however be used as pretreatment or a lower level throughout the storage could be used to minimize chilling injury (Vakis *et al.*, 1970). The proposed optimum conditions for the CA storage of Satsuma mandarins are 8- 12 per cent O_2, 0-2 per cent CO_2 at 1-4°C and 83- 90 per cent relative humidity (Kubo and Heginuma, 1980).

23.4 EXPORT DESTINATION

Oranges find regular place in export trade. European countries are major export destination. However, oranges are also exported to middle-east countries-UAE, Kuwait, Saudi Arabia, Qatar, Bahrain, Oman; Nepal, Sri Lanka, Singapore and Bangladesh. The requirements for export are rigorous and the fruits intended to export, must meet the specific requirements suiting to a particular country.

Table 2 : Parameters for exporting mandarin oranges to European countries

Quality	Minimum standard
Colour	Light reddish, bright yellow to full orange colour or nearly 1/3 to 2/3 portion should be of orange colour
Size	Minimum diameter 4.5 cm preferable
Rind	Medium thick but compact, loose skinned fruits not referred
Seed	Seedless or very less seeded
Get up	Flat or spherical at both the ends
Structure	Having smooth or bright skin
Juice	38-40%
TSS	10-12%
Flavour	Typical aroma, Characteristic of variety
Sweetness	Ratio of sweetness to sourness should be 5:1 to 7:1
Defect	Fruit should be spotless

The export consignment should have no residue level beyond permissible limit. For oranges, the residue limit of carbendazim shouldn't exceed the limit of 5ppm and that of monocrotophos and quinalphos to 0.2 and 0.25ppm respectively.

23.4.1 Measures for increasing export

- Use the latest production technology for getting fruits of high quality.
- Nurseries for preparing orange budlings of required quality with most suitable rootstock must be encouraged. There must be reliability of getting high quality planting materials.
- Integrated pest management technology should be adopted.
- Railways should give 'B' grade priority standards for transport of fruits like oranges.
- Air cargo service should be available at the close distance from the main producing areas.
- Air transport cost should be lowered and may even be subsidized to encourage export and to earn foreign exchange.
- Packing cost should be reduced while maintaining packing of international standards.
- Encouragement for export of processed fruit products should be given. Basic facilities for this should be created.
- Survey of the foreign markets for the quality standards prescribed by them for import of fruits and fruit products may be carried out. The preferences of the foreign buyers must be ascertained.

- Facilities for cooling chambers, cold storage and transport through controlled vans (for temperature and humidity) must be available.
- At the exporting air port, enough space for temporary storage till the fruit packages are loaded in the cargo should be available.

23.5 VALUE ADDITION THROUGH PROCESSING AND WASTE UTILIZATION

A number of value added products could be made from the juice and waste of Nagpur mandarin.

23.5.1 Juice

Extraction of juice from mandarin fruits is somewhat complicated since it turns bitter due to the presence of limonoids which are located in the albedo and juice sacs. Juice from mandarin fruits can suitably be extracted by screw type juice extractor with less bitterness.

23.5.2 Single strength canned juice

Mandarin juice is canned with or without sweetener or consumed in its original form. The sweetened juice contains 85 per cent juice, TSS not less than 16 per cent and acidity 0.5 per cent. The juice is pasteurized at 88-90°C, filled hot in tin containers, hermetically sealed and cooled. The juice after deaeration can also be packed in tetra packs following aseptic processing and packaging.

23.5.3 Ready to serve (RTS) beverages

Fresh or concentrated mandarin juice can also be utilized to prepare RTS which contains 10 per cent each juice and TSS and 0.3 per cent acidity. This is bottled or canned.

23.5.4 Squash

Squash is a beverage prepared from the juice which according to FPO specifications should contain 25 per cent juice, at least 40 per cent TSS and 1 per cent acidity with or without added permitted colour. Squash should be preserved by Potassium meta bisulphite (SO_2 350 ppm) and filled in sterilized glass bottles.

23.5.5 Marmalade

Marmalade is a jelly in which shredded pieces of citrus peel are suspended.

23.5.6 Canned segments

The segments of Nagpur mandarin are canned. Segments are lye peeled (NaOH 1-2 per cent for 20–30 seconds) rinsed and dipped in HCL (0.5- 1 per cent) for 20-30 seconds to remove traces of alkali. Segments are then canned in 50°Brix syrup and cans are exhausted and processed for 20 minutes at 85-90°C.

23.5.7 By products

Mandarin contains 25-30 per cent peel, 15-20 per cent pomace and 5-7 per cent seeds. Therefore, waste from mandarin processing industry can efficiently be utilized to produce byproducts such as peel oil, pectin, cattle feed, flavonoids etc.

23.6 CONSTRAINTS

The proper post harvest handling, marketing and processing of mandarin fruit in India and particularly in Rajasthan is constrained by the following factors:

i. Perishable nature and poor quality of fruits at harvest

ii. Cull (under sized and deformed) fruits due to non adoption of latest production technology.

iii. Damage to fruits due to insect pests – Fruit sucking moth, Black fly, Thrips etc.

iv. Rotting off of fruits due to pre and post harvest infections – *Penicillium digitatum* (Green mould), *Penicillium itallicum* (blue mould), *Diplodia natalencis* (Stem end rot), *Alternaria* rot, *Phytophthora* rots etc.

v. Moisture loss and consequent shrivelling due to high ambient temperature and lack of pre cooling, waxing, packaging and cold storage facilities.

vi. Chilling injury - The recommended storage temperature for mandarin is 6°C. But mostly, mandarin fruit is stored in cold storage meant for apple and potato i.e. at a temperature of 1-4°C. This causes appearance of symptoms of chilling injury.

vii. Peel disorders – 'Sun burn' and 'Peel breakdown' due to exposure to sun and ageing, respectively.

viii. Harvesting injury due to use of unscientific and traditional system of harvesting.

Normally, mandarin has a fairly long storage life of (15 to 25 days under ambient condition and upto 3 months in cold storage) but due to problems mentioned underneath the shelf life is reduced to a great extent :

i. Non- availability of proper infrastructure for post harvest handling of fresh fruits (i.e. grading, packaging etc.) at production centers.

ii. Lack of sufficient cold storage facilities at production site as well as at destination markets.

iii. Lack of cool chain and modern transportation facilities e.g. refer vans.

iv. Lack of organized whole sale markets in the production area.

v. Involvement of long market channel.

vi. Lack of any laws, regulations, incentives or disincentives for marketing the crop (e.g., price supports or controls, banned pesticides or residue limits).

vii. Lack of services available to producers and marketers (for example: credit, inputs, technical advice, subsidies etc.).

viii. Lack of regulated and functional producer or marketer organizations.

ix. Lack of market information regarding current price and demand.

x. Non availability of sufficient infrastructure for processing (e.g. industries, market etc.) for fruits.

xi. Non availability of sufficient institutional support.

xii. Lack of proper coordination between different line departments.

xiii. Lack of innovative research in the field e.g. development of varieties suitable for processing (High TSS, freedom from bitterness etc.), processing technology etc.

xiv. Lack of Government policies to attract investors in the sector.

xv. Lack of Export Zone in the area.

23.7 FUTURE STRATEGIES

i. Creation of awareness among growers, extension workers and market functionaries regarding proper post harvest handling practices through trainings.

ii. Creation of infrastructure facilities for post harvest handling of mandarin, such as pre cooling, waxing, grading, packing, transportation and storage.

iii. Creation of strong marketing system.

iv. Creation of market information system.

v. Provision for minimum support price.

vi. Encouraging contract farming system.

vii. Establishment of Agri Export Centre near to the production area.

viii. Government intervention in creating a platform for increasing private investment in the region for establishment of processing units and by product industries suiting to the crop spectrum and load.

ix. Research initiation pertaining to production technology, quality, post harvest handling and processing (especially removal of bitter principle from mandarin juice).

23.8 REFERENCES

Anonymous (2010). *Indian Horticulture Database, 2010.* NHB, Ministry of Agriculture, Gurgaon, India.

Gutter, Y. (1981). Investigation on new post- harvest fungicide in Israel. Proc. *Int. Soc. Citric.*, Japan, Vol. 2, pp. 810- 811.

Johnson, T.M. (1991). Citrus post- harvest technology to control losses. *Proc. Int. Citrus Symp.*, China, pp. 704- 708.

Kohli, R.R. and Ladaniya, M.S. (1988). Degreening of 'Nagpur' mandarin. Paper presented in 6th workshop of AICRP on Post- Harvest Tech. of Hort. Crops., KVK, Dapoli.

Kubo, N. and Heginuma, S. (1980). Effect of storage conditions on the quality and some of the constituents of Satsuma mandarins. *J. Japanese Soc. Hort. Sci.* 49: 260- 268.

Ladaniya, M.S. (2000). Response of Nagpur mandarin to degreening by ethylene trickle method, mechanical wax coating and storage at ambient condition. *J. Fd. Sci. Technol.* 37: 103- 110.

Ladaniya, M.S. (2008).*Citrus fruits: Biology, Technology and Evaluation*, Academic Press, Elsevier Inc., USA, p 558.

Ladaniya, M.S. and Sonkar, R.K. (1996). Influence of temperature and fruit maturity on 'Nagpur' mandarin (*C. reticulata* Blanco) for storage. *Indian J. Agric. Sci.* 66, 109- 13.

Saraswathy, S., Preethi, T.L., Balasubramanyan, S., Suresh, J., Revathy, N and Natarajan, S. (2008). *Post-harvest management of Horticultural crops*, Agrobios (India), Jodhpur.

Satyanarayana, G. and Ramasubba Reddy, M. (1994). *Citrus cultivation and Protection*. APAU and Department of Horticulture, Andhra Pradesh, Wiley Eastern Ltd., New Delhi, p 66.

Stewart, I, and Wheaton, T.A. (1972). Carotenoids in citrus: Their accumulation induced by ethylene. *J. Agric. Fd. Chem.* 20: 448- 449.

Subramanyam, C. and Cama, H.R. (1965). Carotenoids in 'Nagpur' orange pulp and peel. *Indian J. Chem.* 3: 463-464.

Vakis, N., Grierson, W. and Soule, J. (1970). Chilling injury in tropical and subtropical fruits III. The role of CO_2 in suppressing chilling injury symptoms of grapefruit and avacado. *Proc. Trop. Region Am. Soc. Hort. Sci.* 14: 89- 100.

Chapter – 24

Recent Trends in Value Chain Management of Mandarin for Export

Dinesh Kumar and Lallan Ram

24.1 INTRODUCTION

Citrus fruits occupy third position after mango and banana in the country and ranks 6th in the world citrus production. The productivity, however, is only about 40 percent of that in Spain, Brazil, USA or Japan (8.4 t/ha) only. Of the total production, mandarin contributes more than that of sweet orange in which major chunk of 57.72 per cent was shared by Maharashtra, followed by 19.81 per cent by Madhya Pradesh state. At present, the Brazil is the largest producer of citrus, followed by USA. There are number of issues which need to be addressed to make fruit quality comparable to the international standards. Post harvest losses of fruits are also enormous mainly due to unscientific method of harvesting, lack of utilization of post-harvest handling techniques due to which 25 - 30 per cent losses occur in developing countries in comparison to 5 - 10 per cent in developed countries. Synergistic effect of pre-harvest practices and post harvest handling is the key to get the fruit of high quality and enhanced shelf life.

Recent trends are emerging as use of less chemicals and Good Agricultural Practices (GAP) during cultivation. Maintenance of quality standards and regulations up to retail market level is of utmost importance. Whereas, in marketing sector, now-a-days participation of private players on large scale in retail marketing of produce is noticed.

Value chains have become the centre of considerable academic and practical interests. Part of the reason for the rapidly growing interest in value chains in the citrus sector has been the result of changes in global retailing. In a large number of countries, supermarkets chains now dominate food retailing. Supermarkets have increasingly integrated their operations involving many aspects of the post farm gate operations.

Value chain management is one of these. Some identified issues relates to supply chain bottlenecks resulting in farmers experiencing gluts of commodities, processors not being able to procure sufficient raw materials for their plants, retailers not getting sufficient products to meet the demand of consumers, and exporters are unable to meet foreign consumer requirements. Lack of innovation that affects the entire chain, an example being the introduction of plastic crates for a fruit processor involves a vast number of actors in the chain to coordinate their efforts.

In simple terms a value chain is a series of participants along the entire marketing spectrum who collaborate to satisfy market demands for specific products to get better return, which includes planting material to harvesting, processing and marketing.

In India, less than 2 per cent of total fruit production is processed as against 80 per cent in Brazil and USA. This shows a very big scope of fruit processing in the country. The major bottleneck in mandarin juice processing is development of bitter taste due to presence of bitter compounds like limonin and naringin, Promotion of small scale processing units will create better employment opportunities for the rural masses resulting in more income generation and minimizing the post harvest losses and value addition through efficient processing and better marketing at domestic and international levels.

24.2 FRUIT MATURITY STANDARD AND HARVESTING

Fruit has to be harvested according to maturity criteria for optimum taste and shelf life. The maturity indices for some important citrus cvs. such as Nagpur mandarin, acid lime and sweet orange 'Mosambi' have been already developed by this Centre. Thus for Nagpur mandarin indices for *Ambia* (spring blossom) and *Mrig* (Monsoon blossom) crop are set as TSS content 10 per cent, TSS/acidity ratio 14 and juice content 40%. At least $1/3^{rd}$ of fruit surface should have developed colour with desired taste and firmness. *Ambia* crop takes 270 days while *Mrig* crop takes 230-240 days to mature. Harvesting should be done properly. Mandarin should be harvested by clipping or twisting not by pulling.

24.3 VALUE CHAIN MANAGEMENT : SOME ASPECTS

24.3.1 Spoilage control

Fruit is treated and packed as quickly as possible except unavoidable holding after its receipts in packing house. The Imazalil and Thiabendazole are the principal fungicides used in citrus post-harvest decay control at present. Many countries have banned the use of benomyl and carbendazim on harvest fruit although they can be used 15-30 days before harvest. The residue limits of these fungicides have been set and fruit should not have residues of these compounds more than MRL.

24.3.2 Dipping or drenching of fruit

Low temperature and high humidity delays button abscission. Field sprays of 2,4-D increased the storage life of citrus by delaying the abscission and senescence of fruit button. 2,4-D (100 to 500 ppm) applied in water-wax formulation to lemon before storage delays button deterioration, rate of coloration, and water loss by fruit during storage.

24.3.3 Mechanized handling and grading

The mechanized sorting, washing, wax coating, and size grading has been standardized for "Nagpur" mandarin. In case of 'Nagpur' mandarin much needs to be done as the marketing is still in the hand of traders on conventional line. This technology coupled with orchard management (timely fungicide sprays, sanitation), packaging and transportation would bring down losses from 25% to about 5%.

24.3.4 Ambient storage

Nagpur mandarin fruit can be stored for 3 weeks at 30±5° with mechanical wax coating and packing in vented polythene liner with 5-6 hrs after harvest without curing. However, edible coating of Carnuba wax has also been found suitable for extending the shelf life of Nagpur mandarin fruits (Kumar and Ram, 2011).

24.3.5 Packing house machinery

The development in packing house machinery witnessed a sea change in last 50 years and fully automatic citrus packing houses handling up to 300 tonne fruits every day are common in industrially developed countries. In India, growth has taken place slowly in this area since last decade with the result a complete auto packaging line has been established in *Varud Tehsil* of Amravati district which is having 5 tonne per hour capacity with eight grading systems based on computer operations.

24.3.6 Sizing and sorting

With developments in electronic sorting machine (vision technology), now industry has fully automatic packinghouses. Hand sorters have been found to reject fairly large quality of good fruits. Now, with the help of electronic sorting consistently accurate sorting is ensured and rejection is reduced upto 8 per cent or even less.

24.3.7 Sizing on weight basis

In place packing arrangement, fruit is packed in layers in the box. If height or diameter is not uniform and the fruits are not of same shape, the packed fruits do not appear uniform.

24.3.8 Container for packing

Corrugated cartons made of fibreboards are universally used in international and domestic trades all over the world. In Australia, 18kg (30lit volume) and 15kg (24lit) cartons are most common. Telescopic cartons are preferred. Similar is the case in most Mediterranean and South American countries. The cartons of 18-20kg capacity are common for packing different types of citrus fruits in USA.

24.3.9 Corrugated fibreboard cartons (50 x 30 x 30 cm size) for "Nagpur" mandarin

The corrugated boxes (Universal and Telescopic type) are suitable for distant transport and have sufficient strength to withstand storage condition at 90-95 per cent RH.

24.3.10 Corrugated boxes (universal RSC type and telescopic type) from cotton plant stalks (CPS)

These boxes have been made from kraft paper that is made of cotton plant stalks in the collaborative studies between CIRCOT, Mumbai and NRCC, Nagpur. This kraft paper is cheaper than commercially available kraft paper.

24.3.11 Plastic crates

These crates with ventilation from all 5 sides (top open) can be used several times to reduce packing cost in local marketing. The HDPE crates are very sturdy and stacked high in the truck usually used in India. These crates are returnable and can be used multi-trips.

24.3.12 Transport and palletization

Bulk shipments of citrus (loose fruit without packing) in 10-15 tonnes capacity trucks/container are also being made in domestic long distance markets to get advantage of freight cost with proper cushioning to fruit.

24.3.13 Advances in storage technology

Precision in refrigerated storage temperature and related humidity monitoring and ventilation for fresh air movement in periodic manner is considered necessary.

24.3.14 Pre-cooling

It is pre-requisite in export. Fruit has to be pre-cooled before it is placed in refer container. Pre-cooling of one-half tonne capacity has been developed for forced-air cooling of "Nagpur" mandarin packed in vented corrugated boxes. The stacking pattern and ventilation of box that can give efficient cooling have also been worked out. Pre-cooling at 6-7°C with 90-95 % RH reduces losses in fruit. These facilities are now available at export centres in central India and Punjab.

24.3.15 Refrigerated storage

Optimum treatments and storage conditions have been standardized for"Nagpur" mandarin. Optimum storage condition is 6-7°C temperature and 90-95 per cent RH. The temperature at 3-4°C results in chilling injury.

24.3.16 Evaporative and cooling storage

Walk-in type evaporation cool chamber (working on active evaporative cooling principle) with low cost (about rupees ten thousand) has been designed and developed at NRCC, Nagpur. The inside dimension of the chamber are 255 cm (L) × 180 cm (W) × 210 cm (H), In it, 1.5 tone mandarin, lime and sweet oranges fruits can be stored up to 21,30 and 40 days, respectively.

24.3.17 Grade standards

A: Minimum quality requirements

Grades:	1.	Extra class	-	Fruit of superior quality
	2.	Class I	-	Fruit of good quality
	3.	Class II	-	Fruit satisfying minimum requirement but may not satisfy quality requirement or "Extra class" and Class I

24.3.18 Tolerances

Extra class:5 per cent by number or weight of fruit meeting quality of Grade-I

Class I: 10 per cent by number or weight of fruit, meeting quality of Grade-II

Class II: minimum requirement 10 per cent by number or weight not meeting quality of this class but meeting

Size tolerances: minimum size requirement 10 per cent to the size immediately below or above but not less than

24.3.19 Minimum size requirements

Fruit	Minimum size (mm)
Mandarins	
Nagpur, Coorg, Khasi	45
Satsuma	45
Clementine	35
Kinnow	50
Grapefruit	70

The packaging in layers in the box is mandatory for "Extra class" fruit and optional for class I & II. The content of each package must be uniform and contain only fruit as marked on the box. The uniformity of variety, quality, degree of ripeness, size and development is must. The size is determined by maximum diameter at equatorial section of the fruit. For oranges size code 'O' is given to fruit having diameter 92-110mm and the code '1' to fruit having 87-100mm diameter. Fruit with 53-60mm size should be given size code 13. The size codes to be mentioned on the box of Satsuma's, Clematises and other hybrids are as follows:- size code '-XXX' for fruit with diameter 78mm and size code '2'for 58-69mm diameter. The size codes '3' and '4' are used for fruit having diameter 54-64and 50-60mm, respectively (Anon, 2007).

24.4 CITRUS MARKETING IN INDIA

Considerable improvements are needed in marketing process, auctioning and distribution. Due to faulty handling and long marketing chain loss is high which are 25-30% depending on the season, the marketing channel and the way commodity handled (Muhammad, *et al.*, 2005).

Storage of citrus fruits: In general, mandarin fruits have storage of life of 6-8 weeks at 5-6°C if handled properly.

24.5 CITRUS EXPORT

Mandarin fruits export as per the available statistics is roughly to a tune of 79 thousand tonnes which contributes a very meager in foreign exchange earnings. Presently massive export is done to Nepal, Bangladesh, Pakistan and Maldives. Export of fruits in suitably packed containers to developed countries is demanding and challenging, to get very remunerative prices. APEDA is coming forward to promote the export of mandarins in reefer containers with the time tested technologies.

Export potential of Mandarins from India

	2009-2010	
Country	**Qty**	**Value**
Pakistan	24,600.00	239,186.00
Nepal	45,000.00	214,500.00
Bangladesh	8,113.00	77,471.00
Singapore	1,852.00	65,640.00
Maldives	303	10,519.00
Total	79,868.00	607,316.00

Source: Anon.. (2009-10)

24.6 DEVELOPMENT OF VALUE ADDED PRODUCTS

The mature fruits of Nagpur mandarin having orange colour surface and recommended TSS and TSS/ acid ratio both for *Ambia* and *Mrig* fruits is required to produce value added quality products.

24.6.1 Juice

The extraction of juice from Mandarin is complicated, since it turns bitter after extraction. The bitterness is caused by limonoids which are located in the albedo portion and the juice sacs. Juice from Nagpur mandarin is extracted either by the application of pressure to the whole fruit or by reaming or pressing the halves of crosswise cut fruits on a suitable rosette. Various types of equipments have been developed for fruit juice extraction. Screw type of juice extractor was found better to obtain the juice with minimal bitterness.

24.6.2 Juice blending

The blending of Nagpur mandarin juice with Mosambi juice was adjudged to be the best over Nagpur mandarin juice alone with regard to Vit. C, viscosity, acidity, colour and bitterness acceptability in sensory property upto 60 days after storage. The other combination of Nagpur mandarin juice with Grapes and Aonla juice in proportion 30G: 40N: 30A: were recorded to be the better than that of alone upto 120 days storage at ambient temperature (Bhardwaj and Mukherjee, 2011).

24.6.3 Ready-to-serve-beverage (RTS)

The RTS has minimum 10 per cent juice, not less than 10 per cent TSS and 0.5 per cent acidity as anhydrous citric acid (w/w). Product is prepared from fresh mandarin juice, permitted food colours and flavouring are added. The sweetened RTS contains not more than 70ppm of sulphur dioxide preservative.

24.6.4 Squash

In preparation of squash, the syrup is prepared from sugar in which citric acid and water is mixed along with juice. Orange oil and colour (sunset yellow/orange red) is added to improve flavour and appearance, respectively. Preservative such as potassium metabisulphite (SO_2 350ppm) or sodium benzoate is added and squash is then filled in sterilized bottles and closed with crown cork/ screw cap. Low methyl pectin or propylene glycol ester of organic acid is added for cloud stability.

24.6.5 Juice concentrate

Screened, deaerated and pasteurized, juice is evaporated under reduced pressure for several fold to have increased brix in comparison to single strength natural juice. It is screened through 24 or 32 mesh screens to avoid gelation during storage by reducing the pulp content (Ram, 2008).

24.6.6 Honey based Nagpur mandarin juice

The 10 per cent honey based Nagpur mandarin health juice prepared were evaluated for storage life under ambient and refrigerated condition for limonoids and quality which indicated the best quality in respect of bitterness, TSS, acidity, pH, colour, viscosity and carotenoids content under both ambient and refrigerated storage. No microbial population until120 days under storage were recorded (Ram, 2009).

24.6.7 Marmalade

Marmalade is a jelly in which the of Santra peels are embedded in suspension. The shreds of orange albedo are added to the jelly prepared from the orange pulp. This product is prepared as jelly marmalade or jam marmalade.

24.6.8 Peel candy

Candied peel is mainly used in bakery products. The peel of ripe fruits is tenderized in hot water, then kept in 3^0c brix syrup for 48 hrs. The brix of syrup is raised by 10^0C every day adding sugar and boiling until brix reaches 60^0C. Then citric acid is added @ 1g/kg peel and brix strength is raised to 75^0C. Then peel is kept in syrup for 2-3 weeks. Syrup is drained and peel is dried.

24.6.9 Peel oil

The oil can be extracted by the whole fruit or from separated peel. In India, peel is separated manually since mandarins are easy peelers; the separated peel is treated with lime to increase turgor pressure so that in roller oil extractor oil glands are ruptured (Anon. 2009).

24.6.10 Peel powder

The peel and pomace is dried under control conditions in tray dryer at 60-65^0C for 72 hours to maintain the 6-8% moisture level and was ground, packed in poly bags and stored for further use in cosmetic industries

24.6.11 Peel pectin

Citrus peel from which oil has been removed and the pomace from the juice extractor and finisher form the raw material for commercial extraction of pectin. The mandarins yield less pectin (13.30% on dry weight basis) of lower grade i.e.150-200 grade as compared to lime and lemon.

24.7 REFERENCES

Anonymous (2007). Commission Regulation E.C. No. 1580/2007. *Official J. of the European* Union No. L 350 of 31.12.2007.

Anonymous (2009). Handbook on citrus fruits cultivation and oil extraction, Asia Pacific Business Press Inc. 2009.

Bhardwaj R. L. and Mukherjee. (2011). Effects of fruit juice blending ratios on Kinnow juice preservation at ambient storage condition. *African J. food Sci.* Vol. 5 (5), pp- 281-286.

Kumar Dinesh and Ram Lallan (2011). Sensory and Physico-chemical Quantity Evaluation of Nagpur mandarin (*C. reticulate* Blanco) *Mrig* fruit coated with edible Carnuba Wax. National seminar on Citrus Biodiversity for Livelihood and Nutritional Security, National Research Centre for Citrus, Nagpur, Maharashtra, 4th -5th October, 2010, pp- 433.

Muhammad Sharif, Umar Farooq and Waqar Malik. (2005). Citrus marketing in Punjab: Constraints and potential for improvement, *The Pakistan Development Review*. 2005.

Ram, Lallan (2008). Freeze concentration of Citrus juices and their storage quality study. 3rd Indian Horticulture Congress, Bhubaneshwar, Nov. 06-09, 2008, pp. 373-374.

Ram, Lallan (2009). Honey Blended Nagpur mandarin juice: A source of nutraceutical products paper presented in the 9th Agricultural Science Congress on Technological and Institutional Innovation for Enhancing Agricultural income during June 22-24, at Sher-e-Kashmir University of Agricultural Sciences and Technology, Kashmir, Srinagar, p. 89.

Chapter – 25

Recent Trends in Post-harvest Treatments of Fruits

Ram Asrey

25.1 INTRODUCTION

With the natures gift, consciously or unconsciously human select the best among the lot based on sensual judgment. Warnings about the dangers of the excessive use of the chemicals for the human health and biodiversity have been well reported over the years. The world health organization estimates the occurrence of some more than 25 million cases of excessive chemical use slow poisoning globally each year (Anon, 2010). In developing countries, sulphur and chlorine based formulations are often used for post-harvest treatments of horticultural produce. Recently, some bio-based and other safe post-harvest treatment materials have been formulated for loss reduction and shelf-life extension of fresh fruits. In this article, the recent and safe materials/agents for postharvest treatments are discussed.

25.2 MATERIALS FOR POST–HARVEST TREATMENTS

25.2.1 Polyamines

Polyamines (PAs) are low molecular weight small aliphatic amines that are ubiquitous in living organisms and have been implicated in a wide range of biological processes, including plant growth, development and response to stress (Smith, 1985). In plants, they have been implicated in a wide range of biological processes, including growth, development and abiotic stress responses. The most common polyamines are Putrescine (PUT), Spermidine (SPD) and Spermine (SPM) found in every plant cell in titers ranging from approximately micromolar to millimolar, together with the enzymes regulating their metabolism and depend greatly on environmental conditions, especially stress.

Polyamines can be isolated commercially from plant as well as microbial sources. Among plant sources they can be isolated from leaves and stems of corn (*Zea mays* L.), cucumber (*Cucumis sativus* L.), oat (*Avena sativa* L.) and radish (*Raphanus sativus* L.). They can also be isolated from *Saccharomyces cerevisiae* and *Candida utilis*.

Polyamines have been proposed to be a new category of plant growth regulator and are proposed to be involved in large spectrum of physiological and biological processes of fruits. Polyamines in their free forms have been reported as anti-senescent agent, from both endogenous and exogenous applications (Valero *et al.*, 1998). The main effects of PAs in post harvest management of fruits are- inhibit biosynthesis of ethylene, increase fruit firmness, reduce respiration rate, reduce chilling injury, retard colour changes and reduce mechanical damage.

To investigate the role of putrescine in ethylene biosynthesis in plum, Khan *et al.*, (2007) treated the fruits by dipping in aqueous solution containing different concentrations of putrescine (1mM, 2mM) and Tween 20 (0.01 per cent) as a surfactant. Following PUT treatment fruits were stored at 0 ± 1^0C and 90 ± 5 per cent RH for 6 weeks. It appeared that PUT treatment reduced ethylene production during storage in 'Angelino' plum fruits. There was an inverse relationship between concentration of PUT and ethylene production rate. The reduction in ethylene production with PUT treatment could be attributed to competitive biosynthesis mechanism between ethylene and polyamine.

The effect of polyamine on maintaining fruit firmness can be attributed to their cross-linkage to the carboxyl group of the pectic substances in the cell wall, resulting in rigidification thus blocking the access of degrading enzymes and reducing rate of softening (Valero *et al.,* 2002). Post-harvest treatment of fruits with polyamine was found effective in increasing fruit firmness in apple, plum mango, apricot, lemon etc. However, the efficiency was generally greater for those molecules with higher number of available cations, that is $SPM^{+4} > SPD^{+3} > PUT^{+2}$.

Malik *et al.*, (2005) worked on the effect of pre-storage application on improving fruit quality of mango ('Kensington Pride'). Fruits were dipped for 6 minutes in aqueous solution containing different concentrations of various polyamines including putrescine (1mM), spermidine (0.5mM) and spermine (0.01mM) with 0.01 per cent Tween 20 as surfactant. Treated fruits were stored at $13^0 \pm 1^0$C and 85 ± 5 per cent RH for 4 weeks. Following storage, fruits were allowed to ripen at a temperature of 22 ± 1^0C. Fruit firmness was significantly reduced in untreated fruits, whereas, PA treated fruits exhibited a significantly higher firmness up to 28 days after storage and spermine was found more effective.

It has been reported that under chilling condition, changes in cell membrane lipid from a liquid-crystalline to solid-gel state are induced in plant tissue which leads to increase in membrane permeability as wall as electrolyte leakage of ions (Gomez-Galindo *et al.*, 2004). Polyamines when applied exogenously they seem to induce cold acclimation, which would lead to maintenance of membrane fluidity at low temperatures and could be responsible for reducing electrolyte leakage and skin browning. These polyamines mainly protect the membrane lipid from being converted in physical state and as it has antioxidant property. It prevents lipid peroxidation and thus severity of chilling injury. Mirdeghan *et al*., (2007) studied the effect of pre-storage application of polyamines by pressure or immersion method on reducing chilling injury in pomegranate fruit. Fruits were treated with 1mM putrescine, 1mM spermidine or with distilled water which served as control. Treatments were performed by low pressure infiltration method by applying pressure 0.05 bars for 4 as well as immersion method with polyamine and Tween 20 (2g/l) as surfactant. Following treatments they were stored at 2^0C and 90% RH. Before analysis of the samples they were exposed to 20^0C temperature for 3 days. The results found that control fruits reached the highest percentage of skin browning without significant differences between them. However, the application of putrescine of spermidine either by pressure or by immersion method led to significant reduction in skin browning.

25.2.2 1-Methylcyclopropene

1-methylcyclopropene (1-MCP) is a synthetic cyclic olefin that inhibits ethylene by blocking access to the ethylene-binding receptor (Sisler *et al*., 1997). The application of ethylene action inhibitors lowers the endogenous levels of most maturation associated genes. Successful application of 1-methylcyclopropene to permit extended storage requires prior assessment of the appropriate concentration range and storage conditions for each type of produce at particular maturity (Martinez *et al*., 2002).

25.2.2.1Effects of 1-MCP on fruits

Avocado treated with 1-MCP for 24 hours showed significantly less weight loss and retained more green colour than control fruit at the full-ripe stage .Dou *et al.* (2005) found that 1-MCP treated at the concentration of 50-500 μg l^{-1} effectively controlled the blue mold rot and post-harvest pitting in citrus. Treated fruit of pears with 1-methylcyclopropene were more than 75 N firmer than control fruit after a 6 day storage period. Lime (citrus) treated with 1-methylcyclopropane at low concentrations (250 to 500 l l^{-1}) had effectively suppressed endogenous ethylene production and the effectively retarded yellowing at ambient temperature. 1-MCP bound at the ethylene receptors in pear fruit is still capable of inhibiting cell wall-degrading enzymes such as PG secreted by pathogens and thus prevent pathogenesis. The 1-MCP treatment of 'd'Anjou' pear fruit effectively inhibited ethylene production after cold storage followed by ripening at 20 °C.

The usage of 1-MCP suppresses the ethylene response pathway by permanently binding to a sufficient number of ethylene receptors. Combination of 1-MCP and controlled atmospheric storage did not affect the total soluble solids content and titratable acidity in the apple fruit. Vitamin C content in 1-MCP-treated fruits of guava was significantly higher than in non-treated fruits (Singh and Pal, 2008). Post harvest application of 1-MCP significantly delayed and suppressed the climacteric ethylene production of plum with reduction in the activities of ethylene biosynthesis enzymes and fruit softening enzymes in the skin as well as pulp tissues . The onset of ethylene production and the rise in the respiration rate was delayed and suppressed in 1-MCP-treated fruits of papaya (Manenoi *et al.*, 2007).

Application of 1-MCP delayed the onset of the ethylene climacteric in immature green and mature green fruits of goldenberry in a dose-dependent fashion, and transiently decreased ethylene production in yellow and orange fruit (Gutierrez *et al.*, 2008). Softening and respiration rates were found decreased in response to 1-MCP treatment (1 μL L"1 for 6 h at 14 °C) of fresh-cut banana slices (after processing). 1-MCP has been reported to delay ripening of avocado, custard apple, mango and papaya. In apricot treated with 1-MCP for 4 hours at 20 °C and subsequently stored at 0 or 20 °C, its onset of ethylene production was delayed and respiration rate was reduced. Banana ripening was delayed when exposed to 0.01–1 μl l"1 1-MCP for 24 hours and increasing concentrations of 1-MCP were found generally more effective for longer periods of time .

25.2.3 Ozone

Ozone is the most effective natural bactericide of all the disinfecting agents. Activated oxygen is considered to be the best available technology and a much better alternative for water purification. Ozone is now being used in food processing or storage of perishables as an anti-microbial agent or as a food processing aid. Ozone is the safe and natural purification and disinfecting agents. It is strong and ideal, germicide, sanitizer, sterilizer and vermicide, anti-microbial, bactericide, fungicide and deodorizer, detoxifying agent .

25.2.3.1 Effects on fruits

Ozone oxidizes the metabolic products and neutralizes the odours generated during the ripening stage in storage of fruits. This helps preserve and almost double the shelf life on fresh produce. It also enhances the taste by retaining the original flavour of the products. Ozone enhances the taste of most perishables by oxidizing pesticides and by neutralizing ammonia and ethylene gases produced by ripening or decay. The reduction of ethylene gas increases the shelf life and reduces shrinkage. It changes the chemicals complex molecular structure back to its safe and original basic elements. Its use does not leave any toxic by-products or residues, does not affect healthy cells or alter its chemistry, and is non-carcinogenic.

The shelf life of perishable foods varies when they are subjected to O_3 treatment. Factors like age, crispness, quality, humidity, temperature, the condition on receiving, and the reduction of pathogens during ozonation will determine the extended shelf life of each different type. Positive effect will show at low and constant levels between 0.05 PPM and 0.1 PPM and it allows workers to enter the storage area and carry out their work comfortably (Perez *et al.*, 1999). Ozone should be constantly consumed and absorbed during the oxidation process. The effectiveness is influenced (lowered) due to the presence of steam or 100 per cent humidity levels. The microorganisms have to be in a certain condition of swelling in order to be attacked. When the humidity level is below 50 per cent, the efficiency of it slows as a bacterial medium (Castillo *et al.*, 2003). Growers and processors can use ozonated water to wash fruits replacing chlorine. Applications of it can be applied safely in process water for hydro-cooler systems, bin, dump and dip tanks, spray wash systems, waste-water, processing and storage areas at an affordable cost. Making full use of ozonated water will increase production (Mehta, 2003).

25.2.4 Edible coatings and films

Edible coatings are thin layers of edible material applied to the product surface in addition to or as a replacement for natural protective waxy coatings and provide a barrier to moisture, oxygen and solute movement for the food. They are applied directly on the food surface by dipping, spraying or brushing to create a modified atmosphere. An ideal coating is defined as one that can extend storage life of fresh fruit without causing anaerobiosis and reduces decay without affecting the quality of the produce. Previously, edible coatings have been used to reduce water loss, but recent developments of formulated edible coatings with a wider range of permeability characteristics has extended their potential in using in handling the fresh produce. The effect of coatings on fruits depends greatly on temperature, alkalinity, thickness and type of coating, and the variety and condition of the produce.

Edible polymer film is defined as a thin layer of edible material formed on a product surface as a coating or placed (pre-formed) on or between food components. Several types of edible films have been applied successfully for preservation of fresh horticultural produce (Baldwin *et al.*, 1995). Fruit based films provide enhanced nutrition for food products, while increasing their marketing value. Edible and biodegradable films must meet a number of special functional requirements, for example, moisture barrier, solute or gas barrier, water/lipid solubility, colour and appearance, mechanical and rheological characteristics, non-toxicity, etc. These properties depend on the type of material used, its formation and application. The benefit of using selective films seems to be the reduction of water loss, which is one of the most important factors in the deterioration of highly perishable commodities. The films provide protection against moisture loss and maintain an attractive appearance of the produce.

25.2.5 Apple wraps

Apple based wraps are made from apple puree with various concentrations of fatty acids, fatty alcohols, beeswax and vegetable oil and have a colour of apple sauce. These wraps are excellent oxygen barriers, particularly at low to moderate relative humidity, but are not very good moisture barriers unless lipids are added. Wrapping apple based films formed around apple pieces significantly reduced moisture loss and browning in cut apples, increased the intensity of apple flavour, and maintained the texture during a 12-day storage period at 5°C (McHugh *et al.*, 2000).

25.2.6 Nature seal (NS)

NS, a cellulose-based edible coating, has been used (in combination with antimicrobials, plasticizers, antioxidants, etc.) to coat fresh-cut apples. This coating significantly reduced weight loss of apples more than those treated with water solutions and were not objectionable in taste during several weeks of storage.

25.2.7 Chitosan coatings

Chitosan, a by-product from crustacean shell wastes, is a high molecular weight cationic polysaccharide, normally obtained by the alkaline deacetylation of chitin (complex sugar substance) and refers to as a range of polymers that, unlike chitin, are soluble in dilute organic acids. Chitosan-based coatings are effective in prolonging the shelf life and improving quality of fruits, by delaying ripening, reducing respiration rate, reducing desiccation, regulating gas exchange, decreasing transpiration losses, modifying the internal atmosphere, maintaining the quality of harvested fruits, retaining fruit firmness, freshness, weight loss, titratable acidity, soluble carbohydrates vitamin C, and by reducing mould growth. Chitosan has beneficial effects on titratable acidity, ripening and vitamin C content, firmness and reduced decay (inhibited spore germination, germ tube elongation, and radial growth of *B. cinerea* and *Rhizopus* species in the culture) of strawberries and raspberries. Chitosan coated berries, and peaches were better in quality when compared to control. The application of chitosan coating delayed changes in contents of anthocyanin, flavonoid, and total phenolics, reduced weight loss and browning of litchi fruit, improved storability, and delayed the increase in polyphenolase activity in litchi fruit. The coating partially inhibited PPO activity of longan fruit, which is associated with peel discoloration (Jiang *et al.,* 2001).

25.2.8 Corn-zein coatings

Zein is a natural corn protein produced from corn gluten meal and is insoluble in water, but soluble in aqueous alcohol, glycols and glycol esters. It has good film- forming, binding and adhesive properties. Its water vapour permeability, however, is about 800 times higher than a typical shrink-wrapping film. The degree of colour change was mainly dependent on the thickness of coating. Increased thickness was associated with an increased carbon dioxide level and a decreased oxygen level. Acidity, overall flavour and off-flavour attributes were not affected by coating.

25.2.9 Wax coatings

Wax, the first edible coating known, is the most effective coating to block moisture migration. There are a number of waxes used but the most effective one is paraffin wax, followed by bees wax. The resistance is related to their compositions. Paraffin wax consists of a mixture of long-chain saturated hydrocarbons while bees wax comprises a mixture of hydrophobic, long chain ester compounds, long chain hydrocarbons and long chain fatty acids. The absence of polar groups in paraffin and low levels in bees wax account for their resistance to moisture transport. An increase in ethanol, acetaldehyde, total soluble solids content and a decrease in total solids and titratable acidity was observed for waxed mandarins during storage relative to unwaxed fruits stored in film lined boxes. Storing waxed mandarins at room temperature can lead to anaerobic respiration with higher levels of ethanol and acetaldehyde. Oranges coated with commercial solvent type wax had less weight loss than those with comparable amounts of water wax or polythene coatings. There was an increase in storage life and a decrease in weight loss when mangoes were coated with a wax emulsion in water compared to mineral oil coated and control samples.

25.2.10 Whey protein coatings

The mechanical properties of whey protein films adequately provide durability when used as coatings on food products or films separating layers of homogeneous foods. Adding glycerol and sorbitol reduces internal hydrogen bonding in films, thereby increasing film flexibility. Incorporating whey protein edible films in food product development can results in reduction of food losses due to spoilage and extension of shelf life.

25.2.11 Casein coatings

Research conducted in different countries of the world showed that caesin- lipid coatings provide protection for fruits from moisture loss and oxidative browning. Calcium caesinate and whey protein solutions efficiently delayed browning of apple by acting as oxygen barriers. Respiration rate of Red Delicious apples increased for both caesinate coated and uncoated apples. This indicated that caesin coated formulations did not modify fruit respiration rate.

25.2.12 Pro-long

Pro-long is a mixture of sucrose fatty acid esters, sodium carboxy methyl cellulose and mono-and diglycerides. The mode of action of Pro-long involves the creation of a selectively permeable barrier creating internal atmospheres which preserve the fruit by reducing water loss and chilling injury characteristics, which might be utilized both in the storage of fruit and for the maintenance of quality during the marketing period. Treatment with 0.75% Pro-long significantly increased the storage life of mangoes, retarded ripening and reduced weight loss and chlorophyll loss, without adversely affecting the sensory quality of the limes. Apples treated with 1.25% sucrose ester formulation were stored in air at 3.5°C for up to 5 months. When applied after storage the coating reduced yellowing, loss of firmness and markedly increased internal carbon dioxide levels during a 21-day simulated marketing period. Coating bananas with Pro-long reduced weight loss, oxygen uptake, and ethylene release and chlorophyll loss and modified their internal atmosphere by reducing the permeability of the fruit peel to gases (Banks *et al.*, 2004)

25.2.13 Semperfresh

Semperfresh, a food- grade coating used to retard moisture loss, ripening and spoilage of fruit is a mixture of sucrose esters with high proportion of short chain unsaturated fatty acid esters, sodium salts of CMC and mixed mono and diglycerides. Semperfresh is an improved formulation of earlier SPE (sucrose polyester). The major difference is improved dispersion due to incorporation of higher proportion of short chain USFA esters. These fruit coatings were found to be significantly effective (at both 12°C and 23±2°C) in retention of reducing sugars, delaying changes in firmness, titratable acidity, pH, soluble solids, sugars, ascorbic acid and lycopene synthesis. Semperfresh significantly reduced water loss and internal carbon dioxide in several fruits and reduced colour changes, retained acid, increased shelf life and maintained the keeping quality of apples. The coating increased titratable acidity, firmness and green colour and decreased weight loss, total soluble solids and pH in mangoes when compared with the non-coated fruits. Ascorbic acid decreased in all stored fruit, but the decrease was slower in coated fruit.

Table 1 : Specific coating applications for different fruits

Coating material	Fruits
Prolong	Banana
Semperfresh	Banana
Semperfresh with organic acid	Banana
Ban-seel	Banana and plantains
Tal prolong, Semperfresh and apple wax	Apple
Nutri-save	Golden delicious apple
Semperfresh	Granny smith apple
Brilloshine	Apple, avocado, melons and citrus fruits
Nu-coatflo, Brilloshine and Citrashine	Citrus fruits
Semperfresh	Guava
Palm oil	Guava
Vapor gard	Mango
Chitosan	Strawberry and raspberry
N,O-Carboxymethyl chitosan	Fruits

25.2.14 Essential oils

Use of essential oil for the control of postharvest pathogens, while maintaining fruit quality, has been alternative to chemical fungicides (Plooy *et al*, 2009). This approach eliminates the need for synthetic fungicides, thereby complying with consumer preferences, organic requirements and reducing environmental pollution. Essential oils (EOs) are volatile, natural, complex compounds characterized by a strong odour and are formed by aromatic plants in different parts (flowers, buds, seeds, leaves, twigs, bark, herbs, wood, fruits and roots) as secondary metabolites. These are registered food grade materials and have the potential to be applied as an alternative treatment to control post harvest decay of fruits. An important characteristic of EOs and their components is their hydrophobicity, which enables them to partition in the lipids of the bacterial cell membrane and mitochondria, disturbing the structures and rendering them more permeable. As a result, damage of membrane proteins and depletion of proton motive force takes place. Leakage of ions and other cell contents can then occur which after attaining extensive loss of cell contents or of critical molecules and ions lead to death of bacterial cells. A number of essential oil components has been identified as effective antibacterial, in several fruits like apple, peach, sweet cherry, strawberry, grape, citrus, mango etc (Regnier *et al.*, 2008)

Table 2 : Essential oils used for the control of postharvest diseases of fruit

Essential oil	**Major component**	**Bacterial spp.**	**Fruit crop**
Clove oil	Eugenol	*P. expansum, M. fructigena, B. cinerea, P. vagabunda*	Apple, Grape
Mint oil	Menthol	*P. italicum, B. cinerea, R. stolonifer*	Orange, Strawberry
Thyme oil	Thymol, carvacrol	*B. cinerea, R. stolonifer, M. fructicola*	Starwberry, grape, sweet cherry
Cinnamon oil	Cinnamal -dehyde	Natural flora	Kiwifruit
Lemongrass oil	Citral	*C. gloeosporioides, P. expansum, B. cinerea, R. stolonifer*	Mango, peach

25.2.15 Biocontrol agents

Biocontrol strategy has made much progress during the last decade. At present, several biocontrol treatments have been approved for commercial applications, and now research is focused on improving the bioefficacy of the antagonist. One of the approaches has been the selection of combinations of antagonists, which may work more effectively (Barkai-Golan, 2001). It is a very challenging work, as microorganisms have differential growth habits, requirements for nutrition and cultural conditions. Naturally occurring microorganisms, which are found to be adhered on the fruits and vegetables surface have been shown potential to protect the fresh produce against postharvest disease causing pathogens. During last decade several products viz Serenade (*Bacillus subtilis* based), Messenger (*Erwinia amylovora* based), Biosave (*Pseudomonas syringae* strain 10 LP *)*, Aspire (*Candida oleophila* strain 1-18), AQ-10 bio-fungicide (*Ampelomyces quisqualis)* have been isolated and registered in the United States and Germany (Zhao et al., 2007, Fravel, 2005).Use of some safe bioactive compounds have been proved beneficial in bringing down the physiological activities of fruits during transportation, storage and minimizing the overall qualitative and quantitative losses .

Table 3 : Microbial antagonists used for the control of postharvest diseases of fruit

Antagonists	**Disease (pathogen)**	**Fruit crop**
Bacillus subtilis	Brown rot (*Lasiodiplodia theobromae*), Gray mold (*Botrytis cinerea*), Green mold (*Penicillium digitatum*), Stem end rot (*Botryodiplodia theobromae*), Alternaria rot (*Alternaria alternata*)	Apricot, Strawberry, Citrus, Avocado, Cherry, Litchi
Bacillus licheniformis	Anthracnose (*Colletotrichum gloeosporioides*), stem end rot (*Dothiorella gregaria*)	Mango
Trichoderma viride	Green mold (*Penicillium digitatum*), Stem-end rot (*Botryodiplodia theobromae*), Gray mold (*Botrytis cinerea*)	Citrus, Mango, Strawberry
Pseudomonas syringae	Blue mold (*Penicillium expansum*), Green and blue mold (*Penicillium digitatum* and *P. italicum*), Gray mold (*Botrytis cinerea*), Brown rot (*Monilinia laxa*)	Apple, Citrus, Peach
Candida oleophila	Penicillium rots (*Penicillium digitatum and Penicillium italicum*), Crown rot (*Colletotrichum musae*), Anthracnose (*Colletotrichum gloeosporioides*)	Citrus, Banana, Papaya
Debaryomyces hansenii	Green and blue mold (*Penicillium digitatum* and *Penicillium italicum*), Blue mold (*Penicillium italicum*), Rhizopus rot (*Rhizopus stolonifer*), Sour rot (*Geotrichum candidum*)	Citrus, Peach

25.2.16 Irradiation

Irradiation can be applied by exposing the fruit to ionizing radiations from radioisotopes (normally in the form of gamma-rays but X-rays can also be used) and from machines, which produces a high-energy electron beam. The FDA has approved two types of radiation sources for the treatment of foods: gamma rays produced by the natural decay of radioactive isotopes of cobalt-60 or cesium-137, x-rays with a maximum energy of five million electron volts (MeV), and electrons with a maximum energy of 10 MeV. The rays are directed onto the fruit being irradiated, but the fruit itself never comes into contact with sources of radiation. Irradiation may be referred to as a "cold pasteurization" process, as it does not significantly raise the temperature of the treated fruits. Since fruits contain 80-95% water and their intercellular spaces (about 20% of total volume) contain oxygen, the high energy gamma rays generate copious amounts of free radicals from those of water and oxygen in the fruit immediately after radiation. The free radicals in turn bring about the breakage of the genetic material (DNA) of the insects and spoilage microorganisms, thus destroying them. But after a short period (2-3 days) the free radicals get scavenged off or converted into harmless molecules. Gamma radiation also delays ripening and senescence by decreasing the activity of the cell wall degrading enzyme pectin methyl esterase (PME) and the activity of ACC-oxidase involved in ethylene synthesis. Activities of some other enzymes like polygalactosidase, cellulase and 1-aminocyclopropane-1-carboxylate oxidase are also negatively affected. Ripening of bananas is inhibited at irradiation doses of 0.25-0.35 kGy, and the irradiated fruit later can be ripened by treatment with ethylene. Similar results have been reported for mango, papaya, guava and several other tropical and subtropical fruits. But ionizing radiation at doses above 1 kGy can induce various types of physiological disorders in fruits like internal browning in avocados; skin discoloration and stem darkening in grapes; skin damage in bananas; internal cavities in lemon and lime etc.

Table 4 : Doses of irradiation for different fruits for postharvest disease control

Fruit crop	Minimum dose required (Gy)	Maximum dose tolerated (Gy)
Apple	150	100-150
Apricot, peach, nectarine	200	50-100
Avocado	—	15
Lemon	150-200	25
Orange	200	200
Strawberry	200	200
Grapes	—	25-50

For the control of postharvest diseases generally, a minimum dose of 1.75 kGy is required for effective inhibition of postharvest fungi. So to avoid such a high dose, combination treatment such as heat + irradiation is sometime used. Such combination has been shown to be effective for control of brown rot on stone fruits and anthracnose on mango and papaya. Irradiation at doses below 1 kGy is an effective insect disinfestations treatment against various species of fruit flies, mango seed weevil, navel orange worm, codling moth, scale insects and other insect species of quarantine significance in marketing fresh fruits. Most insects are sterilized at doses of 0.05-0.75 kGy; some adult moths survive 1 kGy, but their progeny are sterile. For mango, irradiation at a dose between 0.25-0.75 kGy is used for quarantine treatment and shelf life extension.

25.2.17 Nitric oxide (NO)

Nitric oxide is a gaseous free radical with relatively long half life, amounting to 3-5 seconds in biological systems. It is very reactive species and forms other oxides (NO_2, N_2O_3 and N_2O_4) in presence of atmospheric oxygen. Post harvest treatment of fruits with a low concentration of NO gas can extend post harvest shelf life but the application of NO by release from a gas cylinder is practically difficult. To solve this problem, the solid NO-donor compound-diethylenetriamine nitric oxide may be used in sachets for liberating nitric oxide in storage chambers. The use of NO can be made for delaying in fruit ripening and improvement in retention of texture during storage (Zhu *et al.,* 2010). The better retention of cellular component such as pigments, titratable acidity and antioxidants are retained well due to reduction in the degree of disintegration of cellular membranes with lesser electrolyte leakage. The chilling injury is reduced in the fruits kept in cold storage and symptoms like fresh browning and translucency are reduced (Singh *et al.*,2009) It is highly potent molecule for induction of resistance in produce against systemic infection. It also protects from microbial infections such as *Penicillium italicum, Rhizopus nigricans, Aspergillus niger and Monilinia fructicola*. This resulted in post harvest shelf life extension of treated fruits. Due to maintenance of cellular compartmentation, the enzymes do not come into contact with the substrate, thus preventing the rate of reaction of enzymes such as polyphenole oxidase. This new direction emanated from research on NO would allow more recent formulations for the use in enhancing self life of fruits in eco-friendly manner.

25.3 CONCLUSION

Being the bio-material, the quality retention and loss prevention of fruits during handling storage and retailing always remained a critical issue for researchers, policy makers, processors and consumers. In past, there were several chemicals in use for post harvest treatment of fruits, but due to advancement in technology, health consciousness of consumers and environmental concern has phased out these chemical with the adoptions of improved ones. In last couple of years, some environmental and consumer friendly post harvest treatment materials like ozonized water, essential oils, edible waxes, 1-Methylcyclopropane, nitric oxide (NO), irradiation and polyamines are gaining popularity world over. Ozonized water is an ideal and cheap sanitizer which is being used for washing of produce. In developed countries, 1-Methylcyclopropane (1-MCP) is commercially used for fruit treatment in order to minimize ethylene evolution during transit and storage. Edible waxes are now being obtained from plant (carnauba wax, zein protein, alovera, guar gum) and sea fauna (chitosan). These are strong physiological activity retardant and antifungal substances. Polyamines like putrescine, spermidine and spermine are very useful molecules and now being gainfully utilized for minimizing mechanical damage, chilling injuries during storage and for overall of extension of fruits. Nitric oxide is a potential post harvest treatment agent now being utilized in post-harvest management of fresh horticultural produce. It acts in gaseous form, which can be achieved by the use of DETANO and SOP (available in solid form).

25.4 REFERENCES

Anon. (2010). UN food agency, updated pesticide standards. Press release from U.S.Embassy, Tokyo, Japan.

Banks, N.H. 2004. Some effects of TAL Pro-long coating on ripening bananas. *J. Expt. Bot.,* 35 (150): 127-137.

Barkai-Golan,R., Phillip,D.J.,2001. Postharvest fruit treatment of fresh fruits and vegetables for decay control. *Plant Diseases* 75.1085-1089

Castillo, A., Mckenzie, K.S., Lucia, L.M. and Acuff, G.R. (2003). Ozone treatment for reduction of *Escherichia coli* and *Salmonella* serotype *Typhimurium* on beef carcass surfaces. *J Food Prot.,* 66(5): 775-779.

Dou, H., Jones, S. and Ritenour, M. (2005). Influence of 1-MCP application and concentration on post-harvest peel disorders and incidence of decay in citrus fruit. *J. Hort Sci Biotech.*, 80(6): 786-792.

Fravel, D.R., 2005. Commercialization and implementation of biocontrol. *Annual Review of Plant Biology* 43, 337-359.

Gomez-Galindo, F., Herpppich, W., Gekas, V. and Sjoholm, I. (2004). Factors affecting quality and postharvest properties of vegetables: Integration of water relations and metabolism. *Crit. Rev.Fod Sci.Nutr.*, 44:139-154.

Gutierrez,S.M., Trinchero,D.G., Cerri,M.A., Vilella, F. and Sozzi, O.G. (2008). Different responses of goldenberry fruit treated at four maturity stages with the ethylene antagonist 1-methylcyclopropene. *Postharvest Biol. Tech.,* 48(2): 199-205.

Jiang, Y., and Li, Y. (2001). Effects of chitosan on postharvest life and quality of longan fruit. *Food Chem.,*73: 139-143.

Khan, A.S. and Singh, Z.(2007). 1-MCP regulates ethylene biosynthesis and fruit softening during ripening of 'Tegan Blue' plum. *Postharvest Biol.and Tech.,* 43(3): 298-306.

Khan, A.S., Singh, Z. and Abbasi, N.A. (2007). Pre-storage putrescine application suppresses ethylene biosynthesis and retards fruit softening during low temperature storage in 'Angelino' plum. *Postharvest Biol.&Tech.*, 46: 36-46.

Malik, A.U. and Singh, Z. (2005). Pre-storage application of polyamines improves shelf-life and fruit quality of mango. *Journal of Horticultural Science and Biotechnology.*, 80 (3): 363-369

Manenoi,A., Bayogan, E.R.V., Thumdee, S. and Paull, R.E.(2007). Utility of 1-methylcyclopropene as a papaya postharvest treatment. *Postharvest Biol.Tech.,*44(1):55-62.

Martinez-Romero, D., Serrano,M., Carbonell,L., Burgos,F. and Valero, D. (2002). Effects of postharvest putrescine treatment on extending shelf life and reducing mechanical damage in apricot. *J of Food Sci.*, 67 (5):1706-1711.

McHugh, T.H., and Senesi, E. (2000). Apple wraps: A novel method to improve the quality and extend the shelf life of fresh-cut apples. *J. Food Sci.*, 65(3): 480-485.

Mehta, K. (2003). Use of ozone for fresh and processed foods. *Processed food industry.*, 6(8):16-19.

Mirdhghan, S.H., Rahemi, M., Castillo, S., Martínez-Romero, D., Serrano, M. and Valero, D. (2007). Pre-storage application of polyamines by pressure or immersion improves shelf-life of pomegranate stored at chilling temperature by increasing endogenous polyamine levels. *Postharvest Biol. and Tech..*, 44(1):26-33.

Perez, A.G, Sanz, C., Rios J. and Olias, J.M. (1999). Effects of ozone treatment on postharvest strawberry quality. *J Agric Food Chem.*, 47: 1652-1656.

Plooy du,W., Combrinck, S., Botha, B.(2009). Essential oil amended coatings as alternatives to synthetic fungicides in citrus postharvest management *Postharvest Biol. and Technol.*, 53 : 117-122.

Regnier, T., du Plooy,W., Combrinck, S., Botha, B.(2008). Fungitoxicity of *Lippia scaberrima* essential oil and selected terpenoid components on two mango postharvest spoilage pathogens. *Postharvest Biol. and Technol.* 48, 254–258.

Singh, S.P. and Pal, R.K.(2008). Response of climacteric-type guava (*Psidium guajava* L.) to postharvest treatment with 1-MCP. *Postharvest Biol. and Tech.,*47(3):307-314.

Singh, S.P., Zora,S. and Swinny,E.E. (2009). Postharvest nitric oxide fumigation delays fruit ripening and alleviates chilling injury during cold storage of Japanese plum (*Prunus saliciana*) . *Postharvest Biol. and Tech.,*53 :101-108.

Sisler, E.C. and Serek, M. (1997). Inhibitors of ethylene responses in plants at the receptor level: recent developments, *Physiol. Plant.,* 100: 577–582. .

Smith, T. A. (1985). Polyamines. *Annu Rev. Plant Physiol,*. 36 :117–143.

Valero, D., Martinez, D., Riquelme, F. and Serrano, M. (1998). Polyamine response to external mechanical bruising in two mandarin cultivars. *Hort Science.*, 33 (7): 1220-1223.

Valero, D., Martínez-Romero, D. and Serrano, M. (2002). The role of polyamines in the improvement of the shelf life of fruit. *Trends in Food Sci. and Technl*., 13(6-7):228-234.

Zhao, Y., Shao, X.F., Tu, K., Chen, J.K., 2007. Inhibitory effect of *Bacillus subtilis* B10 on the diseases of postharvest strawberry. *Journal of Fruit Science* 24, 339-343.

Zhu, L., Zhou,J. And Zhu,S. (2010). Effects of a combination of nitric oxide treatment and intermitten warming on prevention of chilling injury of peach fruit during storage. *Postharvest Biol. Technol*., 121: 165-170.

Chapter – 26

Waste Management in Context to Precision Horticulture

B. N. Patel and G. D. Patel

26.1 INTRODUCTION

India is the second largest producer of fruits and vegetables and ranks next to Brazil and China, respectively. With the commensurate total production of 63.5 and 87.2 million tonnes of fruits and vegetables, the country shares about 11 per cent of total production under fruits and 13 per cent of total production under vegetables, respectively (Anon., 2008) in global scenario. Horticultural produce undergoes spoilage at the time of harvesting, handling, storage, marketing and processing resulting in huge wastage. Efficient management of this waste can help preserving essential nutrients of our food and feeds and bringing down the production cost of processed product, besides minimizing pollution hazards and purifying the environmental condition. Recycling of horticultural waste is one of the most important aspects of utilizing it in a number of new ways to yield new products and meeting the requirements of essential products required to mankind.

In advance countries like USA, Australia, Brazil and Israel, more than 50 per cent of fruits and vegetables productions are processed into different products, whereas, in India it is less than 2.0 per cent of total production. Generally, post harvest losses in fruits and vegetables range between 25-30 per cent of total production, valued at more than 57,000 crores annually due to lack of availability of appropriate post harvest infrastructure (Guleria and Verma, 2008). Horticultural crops are grown all over the country incurring wastes everywhere. There are different stages of waste from farm to fork.

- Waste at field level: dropped fruits, trimmed (pruned) twigs, bird and insect damage fruits as well as injured and bruised due to faulty packaging, excess leaves (cauliflower, cabbage, turmeric, banana etc.,) utilized and under utilized fruits culls and their part.

- Waste during transit, storage and marketing: damaged and spoiled fruits, unsold wrapping materials etc.
- Waste at processing industries and personal uses (kitchen): removed cores, peels, seeds, stones, pods, pomace, spoiled fruits and vegetable and their products.

With small holdings of farmers, the amount of waste in individual form is small except in large pocket growing individual type of fruits and vegetables. These waste have attracted the attention of scientist and a large volume of data have been generated on various fruits and vegetables constituents, mode of extraction of some of the products and their end uses.

Important wastes of horticultural crops and their by-products

Crops	Nature of the waste	By-products
Mango	Trimmed twigs	Fuel purpose, improving soil health.
	Immature fruits	Preparation of chutney, pickles and amchur.
	Peels, stones, fibre, Hull and kernel	Starch, fat, pectin, vinegar, syrup and alcoholic beverages.
Banana	Banana pseudostem	Fibre ,yarn, fabric, paper, handicraft
	Scutching Waste	Micro crystalline cellulose powder and Vermicompost.
	Central core of pseudostem	Candy, pickle, chips and RTS.
	Sap	As liquid fertilizer.
Pineapple	Peel core, trimmed shreds, leaves	Juice, wine, syrup, bromeline cattle feed, Biogas, fibre etc.
Citrus	Peels, pomace and seeds	Pectin, essential oil, seed oil, cattle feed, peel, candy etc.
Grape	Stem, seeds and seed hulls	Cream of tartar, seed oil, tannin from culls, vinegar, pectin, wines and stock feed
Guava	Pulp waste	jam mixes, cheese
Apple	Pomace, cores and peels	Juice, wine, vinegar, pectin, and animal feed, mixed in jelly preparation
Coconut	Leaves, inflorescence and nuts waste	Vermicopost, handicraft material, mulch.

Contd.

Tomato	Core, peel, seed and solid waste	Animal feed, seed oil, fats
Potato	Peel, coarse solids	Animal feed and single Cell Protein
Peas	Pea vine	Rich in carbohydrates
Onion, garlic	Outer leaves	vermicopost
Cassava	Stem with attached root	Animal feed
Okra	Stem	Mucilage powder
Beet root	Peel	Natural food colour
Watermelon	Rind	Mixed jam
Carrot	Pomace	Carrot based condensed milk product, natural colour.
Marigold	Wastage flowers	Poultry feed, organic manure , natural dyes

In India, about 4400 processing units are running at present, out of that, 90 per cent units are in small sector with annual production limit between 50-250 tonnes. These units do not have any organized method of waste disposal. The most of the solid wastes are dumped outside the unit, which are eaten and scattered by the stray cattle. An efficient disposal and recycling of these wastes could helps to combat pollution hazards, add vital nutritional components in food and feed as well as bring down the production cost of production of processed food. They could also provide valuable raw materials for gas reactors, organic manures and extraction of several types of useful by products.

For controlling waste disposal and strengthen the utilization of waste, there was a strict enforcement of the Environmental Protection Act 1986, discarding solid waste outside the factory premises as huge dumps and liquid wastes in the municipal sewage system may attract a fine of Rs. One Lac and/or imprisonment for 5 to 7 years.

Nutritional composition of different horticultural wastes has been quoted by Singh and Singh (2005). They reported that these waste are rich source of vital constituents like carbohydrates, fats, protein, minerals, and fibers etc, Mango seed kernel is rich in carbohydrates, fats, protein and minerals. Orange, pumpkin and melons seeds can provide fats and mineral matters. Possible by products from the waste of processing units have been described by Singh *et. al.* (2009) and could help to develop by-products on the basis of availabilities of waste. Anon, (2005) described the process of some potential value added product from different fruit processing waste. Different valued-added products like animal feed, vermicompost and vital nutrients were prepared from different horticultural wastes.

26.2 DIFFERENT USES OF PROCESSING WASTE

26.2.1 Pectin production

Pectin is an important constituent widely distributed in various plant parts. It is present in the middle lamella of cell walls in an insoluble form and act as nature's cement in the holding the adjacent cells together. Majority of the pectin produced in the world is extracted from the citrus peels and utilized for jam making and pharmaceutical industries. Maini (2001) reported that maximum pectin yield and jelly grade was obtained from peel of citrus *Galgal*. Maximum purest pectin was obtained from mangosteen rind and lime peel as compared to the other source of waste reported by Madhav and Pusphalatha (2002). Like that, good quality and higher quantity of pectin was available from mango ripe fruit peels (Tandon *et.al.*,1991).

26.2.2 Starch production

Banana pseudostem, having a yield of 20 to 25 tonnes from about 1,000 banana plants, when extracted, gives about 5% edible starch. Likewise, 1, 40,000 tonnes of the starch may also be available from mango seed kernels (Singh, *et al.,* 2009). Vegetable waste from roots of cassava and tuber crops viz. potato, sweet potato, yam and amorphophalous are rich source of starch.

26.2.3 Fats production

The industrial processing of mango end up with the considerable proportion of the stone and peels, which are discarded and end up in wastes. The mango seed kernel contain 8.85g/100g. of waste. Sweet orange seed contain higher amount of fat (36.9g/100g) followed by jackfruit (inner and outer portion) i.e 11.82g/100g Singh and Singh, (2005). Like wise in vegetable crops, the industrial processing of tomato generates solid wastes after extraction process, which consists of peels, seeds cores and culls which contain 10 per cent fats.

26.2.4 Essential oils production

India producs about 3 million tonnes of citrus fruits like mandarins, sweet oranges, limes, lemons etc. and generats large wastes around processing units. Citrus wastes are rich source of clouding agents, essential oils, pectin and a variety of the by-products. Sud and Badyal (1993) reported maximum yield in terms of peel oil (2.05 ml/100g peels) from orange peel. Ahmed *et al.*(2006) noted that malta peel had the highest oil yield (1.21 per cent) followed by lemon peel (1.12 per cent). Regarding processing units, mandarin essential oil is extracted in small quantity in different processing units at Banglore (Karnataka), Nagpur (MH), Abohar (Punjab), and Sikkim. Lime oil is extracted at Uttrayan, Jalgoan and Kodur.

Mango seed kernel also contains essential oil, which is used in manufacture of soaps, shaving cream and other cosmetic products (Amin *et.al.*,1984). Tomato seed, a major part of solid waste are an unexplored non-traditional source of oil having an oil content of up to 38 per cent (Sogi *et al.,* 1998). Bhullar and Sogi (2000) also studied the physico chemical properties of tomato seed oil.

26.2.5 Animal feed production

Orange wastes in the form of peels, rages and a seed is dried in California and Florida for the use of livestock feed. It contains lower amount of crude proteins, fibre and fat, but rich in carbohydrates, often, it is called as carbohydrate concentrate. The liquid- concentrate made from the crude citrus waste is called molasses. Its use is similar to cane molasses and it is widely used for feeding dairy and beef cattle. Growing yeast on it and using it for animal feed is further helpful. Fresh apple pomace used @ 10.2 kg per day per animal as a partial replacement of fodder without any detrimental effect on health and milk yield of the animals.(Shah and Masoodi,1994).

Ripe banana peels, which constitute about 20 to 30 per cent, are found littered all around the streets causing pollution and skiding risks. In Vietnam, these peels are sun dried and used as poultry feed. Pineapple bran has a great demand as a cattle feed in Hawaiin Island. Proximate composition of the silage made out of the pineapple tops and leaves with molasses and urea was almost comparable to any of the commonly used grass silage. This has been reported by Singh *et. al.* (2009).

Major waste like mango seed kernel is a good source of carbohydrate and it may replace around 20 per cent corn being grown for feeding the dairy and beef cattle. Drying of mango kernels and peel may make a high class energy food for animal feed. Sossidou *et.al.*(2008) concluded that composted vegetable waste can be transformed into value added soil media for free-range hens in accordance with on farm welfare and sustainability principles. Nobakht and Safamehr (2007) reported that dried tomato pomace(DTP) can be used as an alternative feed stuff in lying hens diets, at inclusion levels up to 10 per cent without negative effect on performance and egg quality.

26.2.6 Production of natural colours

Due to there carcinogenic effects, coal tar dyes are being fast replaced by natural colours or plant pigments for the use in food products. Natural colours of blue grapes skin, *kokum, phalsa*, *jamun*, beet root, carrot and tomato have thoroughly been investigated for there nature, concentration, extractability, stability and suitability as food colours.

Methods for making colour concentrates and powders from them have been developed. A new fat soluble yellow pigment known as “garcinol” was isolated from the rind of *kokum* (Singh *et al.* 2009). Central Committee on Food Standards is actively considering to permit the use of *kokum* and safflower pigments in food industry.

26.2.7 Biogas production

Biomass consists of agricultural, forest, crop residues, solid and liquid wastes from industries etc. Sewage and sludge can be utilized for production of biogas through microbial technology. Some of the fruit wastes (both solid and liquid) can be usefully diverted for biogas generation and making manures. Fruits are not suited for active fermentation in generators due to their acidic nature. Neutralizing agents like lime slurry has to be added along with fruit waste for the continuous active fermentation and to maintain the efficiency of gas generators (Singh *et al.* 2009). The methane yields from the mango peels, orange wastes, pomegranate rotten seeds and lemon pressings were significantly higher than the cellulose (Gunaseelan, 2004).

The ensiled mango peel was a useful substrate for anaerobic digestion for methane generation, and the biogas yield per kg volatile solids added was 0.68 m^3, with a methane content of 52 per cent (Madhukara *et al.*, 1993).The total bio-gas yield from the various fruit/vegetable wastes was 158kg dry matter which is three times more than using mixture of cow dung (Kumar *et al.*, 1993).

26.2.8 Waste for composting

Fruits waste like banana, mango, sapota, vegetables, dairy products, bread, eggshells, meats and products which are biodegradable can be composted. They are rich source of nutrients and this type of composting paves way for waste utilization as well as reuse of waste into wealth.

Vermicompost: It is excreta of earthworms which can be used as organic fertilizers. It contains macro nutrients, micro nutrients, enzymes and growth regulators. Banana waste like suckers and leaves along with dung in a ratio of 70:30(Banana waste : Dung), respectively can be used for vermicomposting.

26.2.9. Other uses of waste

- A number of beverages such as cider, beer, wine an brandy and vinegar can be obtained from the fermentation of fruit wastes.
- RTS beverage, vinegar and wine obtained from pineapple processing waste were having higher organoleptic and keeping quality (Omana *et al.*, 1993 at Vellanikkara).
- A good quality vinegar with 4.5 to 5.0 per cent acetic acid and characterized by mild mango flavour was obtained (Ethiraj and Suresh, 1992)
- Apple pomace has been utilized for production of cider.
- The fermented papaya yielded ethanol contents of 3.83 to 5.19 per cent (v/v).(Akin-Osanaiye *et.al.,* 2005).

26.3 SIGNIFICANT ACHIEVEMENT IN WASTE MANAGEMENT OF BANANA AT NAU, NAVSARI, GUJARAT

- 22 Raspador units at NAU and farmers field extracted 1000 and 3000kg fibre, respectively.
- About 64 kg yarn was supplied to MANTRA textile industry, Surat for different type of fabric manufacture.
- In 2008-09, 80 kg yarn was prepared at Kolkotta by CIRCOT and 2000 kg fibre sent for yarn making during 2009-10.
- Handmade paper and board unit is established at NAU and expected to produce about 200-300kg paper.

- Prepare Cellulose powder was developed from banana pseudostem fibre.
- Attempts have been made to prepared different recipes from central core viz,. candy, pickles, chips and RTS. Process for candy preparation has been standardized and analyzed at CFTRI, Mysore.
- For *Labeo rohita,* up to 30% is showing better substitution of fish feed with vermicompost than fish feed alone.
- Enriched SAP as liquid fertilizer(higher amount of potassium) and growth promoter (GA, Cytokinin), vermiwash and vermicompost on nursery seedling, banana and sugarcane, respectively.

26.4 CONSTRAINTS

- Majority of processing units are small and amount of waste produced is negligible.
- Lack of proper knowledge of utilization or recycling of processing waste.
- Lack of proper waste processing instruments.
- Poor coordination between scientists and processors.
- Lack of availability of technically experienced man power.
- Lack of proper infrastructure facility such as proper storage and packaging technology.
- The fruit processing is highly seasonable, so the supply of waste is irregular and non uniform.

26.5 CONCLUSION

From the foregone discussion it can be inferred that fruits and vegetables wastes can be definitely utilized as value added product viz., fat, starch, enzymes, clouding agents, oil, pectin, ethanol, vinegar, wine, juice, cosmetics, animal feed, compost, natural gases and edible products and minimize the pollution. Hence, it needs to be commercialized in sake of reduction the environmental problems.

24.6 REFERENCES

Ahmed, M. M:, Salim, U. R:, Faqir, M, A. and Bajwa, E. E. (2006). Comparative Physical Examination of Various Citrul Peel Essential Oils. *International. J. Agri. Biol., Vol.* 8, *No. 2, 20.*

Akin-Osanaity, B.C., Nzelibe,H.C. and Agbagi,A.S.I. (2005). Production of ethanol from Carica papaya (pawpaw) agro waste: effect of saccharification and different treatments on ethanol yield. *African J. Biotechno.* 4(7) : 657-659.

Amin. H. D.; Bhuva, H.P.; Katrodia, J.S. and Naik. A.G (1984). Seminar on "*Khet Aadharit Udhyog Na Prashno Ane Bhavi*" organized by Guj. Agri. University.

Anonymous (2008). *Indian Horticulture Database-2008,* NHB, Ministry of Agriculture, Government of India. Bangalore, pp. 372.

Bhullar, J.K. and Sogi, D.S.(2000). Shelf life Studies and Refining of Tomato Seed Oil. *J. Food Sci. Technol.* 37(5) : 542-544.

Ethiraj, S. and Suresh, E.R. (1992). Studies on the Utilization of Mango Processing Waste for Production of Vinegar. *J.Fd. Sci. Technol.*, 29 (1) : 48-50.

Guleria, S.P.S and Verma, A.K. (2008). Utilization of Byproducts Related to Fruits and Vegetable Processing Industry. *Indian Fd. Pack.*, Nov.-Dec : 128-135.

Gunaseelan, V.N. (2004). Biochemical methane potential of Fruit and Vegetable Soil Waste Feed Stocks. *Biomass and Bioenergy*, 26 (4) : 389-399.

Kumar, S.; Jain, M.C. and Mani, S.B. (1993). In: Golden Jubilee Symposium on "*Horticultural research-changing scenario*", Bangalore, pp. 372.

Madhav, A. and Pushpalatha P. B. (2002). Charecterization of Pectin Extracted from Different Fruit Wastes.. *J. Trop. Agri,* 40 : 53-55

Madhukara, K.; Krishna, N.; Raju, N.R. and Srilatha, H.R. (1993). Utilization of Mango Waste in Biogas generation. *Process Biochemistry*, 28 (2) : 119-123.

Maini, S. B. (2001). In: Proceedings National Seminar on "*New horizons in production and post-harvest management of tropical and subtropical fruits*", Bangalore, pp. 183-190.

Nobakht, A. and Safarmehr, A.R. (2007). The Effect of Inclusion of Different Levels of Dried Tomato Pomace in Laying Hens Diet on Performance and Plasma and Egg Yolk Cholesterol contents. *J. Anim. Vet. Advances*: 6 (9):1101-1106.

Omana, P.; Sudhadevi, P.K. and Radhakrishnan, T.C. (1993). In: Golden Jubilee Symposium on "*Horticultural research-changing scenario*", *Sci. Comm.* 1984-85. pp. 90-96.

Shah, G. H. and Mansoodi, F.A. (1994). Studies on the Utilization of Wastes from Apple Processing Plants. *Indian Fd.Pack,* 48 : 47-52.

Singh, D. K. and Singh, S. K. (2005). Physiology and Postharvest Management of Horticultural produce, pp. 33-38.

Singh, V.;Jinny.; Dashora, L. K. and Shukla, K.B. (2009). An outlook on Fruits and Vegetables Dispensation Waste and its Exploitation. *Agrovet Buzz,* 1 (5) : 14-17.

Sogi, D.S.; Kiran, J. and Bawa, A.S. (1998). Characterization and Utilization of Tomato Seed Oil From Tomato Processing Waste. *J. Food Sci. Technol.* 36(3) : 248-249.

Sossidou, E. N., Rose, S. P. and Tserveni-Goussi, A. S. (2008). The Use of Composted Vegetable Waste as a Soil Media for Free-Range Laying Hens. *Folia Veterinaria*, 52 (2): 69-70.

Sud, R.G and Badyal, J.M. (1993). Utilization of Citrus Rind for Peel Oil: Stage of Fruit and Evolution of Methods. *Indian Fd.Pack,* 47 (4) : 48-49.

Tondon, D.K.; Kalra, S.K.; Singh, B. P. and Neelima Garg. (1991). Characterization of Pectin from Mango Fruit Waste. *Indian Fd. Pack*, 45 (4) : 9-12.

Chapter – 27

Safety and Quality Issues in Horticulture

Sanjai K. Dwivedi and Deepa H. Dwivedi

27.1 INTRODUCTION

In India, the horticulture sector contributes around 29 per cent of the GDP and 37 per cent of the total exports of agricultural commodities from about 13.08% of area. India has been placed as the second largest producer of fruits and vegetables; largest producer and consumer of cashew nut, tea, spices; third largest producer of coconut; fourth largest producer and consumer of rubber and sixth largest producer of coffee in the world. Several new programmes area in operation and significant developments in area expansion, production, value-addition and exports have taken place. Today, as a result of better synergy between research agencies, technological and policy initiatives by the governmental agencies, proactive participation of private players and higher degree of diversification to horticulture, this sector has emerged as a sustainable and viable proposition for even the small and marginal farmers including different stakeholders in the commodity chain. In spite of the remarkable progress made in the horticulture Sector in India, there is lot of scope in improvement in safety and quality aspects of horticultural produce.

The concept of biosafety encompasses a range of measures, policies and procedures for minimizing potential risks likely to appear to the environment and human health. Such safeguards must be put in place now. Biosafety is currently being promoted in a variety of ways by industry, governments and civil society. Quality and safety become prime concern because most of the horticultural produce are consumed daily in fresh, cooked and processed form and contribute major share in our daily diet. To ensure biosafety and quality of horticultural produce we need to ensure: a)Supply of quality inputs, b)Adoption of good cultural practices, c) Handling and processing of horticultural produce in scientific /hygienic manner, d)Packing, storage, transportation and delivery etc. in such a way that it reaches at consumers' table without deteriorating the quality.

To ensure such aspects various preventive and certification measures have been initiated by government and certification agencies all over the globe including India , such as prevention of Food Adulteration Act 1954, Agricultural Produce Marketing Act, 1937, Food Safety Standards Association of India (FSSAI), Food Safety System Certification (FSSC:2200), International Food Certification (IFC), Food Chain Traceability, Hazard Analysis and Critical Control Point (HACCP), Food Product Orders (FPO), etc. The major emphasis under these certifications/regulations is to ensure quality and safety of the produce and minimize the risks by reducing the biological, chemical, physical hazards like microorganism contamination, level of heavy metal/chemicals/ drugs beyond the critical limits, residual effects of toxic substances etc. Brief of a few certification/regulations are listed below.

27.2 GOOD AGRICULTURAL PRACTICES FOR ENSURING QUALITY PRODUCTION

Good Agricultural Practices (GAPs) are a collection of principles to apply for on-farm production and post-production processes, resulting in safe and healthy food and non-food agricultural products, while taking into account economical, social and environmental sustainability. GAPs may be applied to a wide range of farming systems and at different scales. They are applied through sustainable agricultural methods, such as integrated pest management, integrated fertilizer management and conservation agriculture. They rely on four principles: a) Economically and efficiently produce sufficient (food security), safe (food safety) and nutritious food (food quality) b) Sustain and enhance natural resources; c) Maintain viable farming enterprises and contribute to sustainable livelihoods; d) Meet cultural and social demands of society.

The concept of GAPs has changed in recent years because of rapidly changing agriculture, globalization of world trade, food crisis, nitrate pollution of water, appearance of pesticide resistance, soil erosion etc. GAPs applications are being developed by governments, NGOs and private sector to meet farmers and transformers needs and specific requirements. However, many think these applications are only rarely made in holistic or coordinated way.

They provide the opportunity to assess and decide on which farming practices to follow at each step in the production process. For each agricultural production system, they aim at allowing a comprehensive management strategy, providing for the capacity for tactical adjustments in response to changes. The implementation of such a management strategy requires knowing, understanding, planning, measuring, monitoring, and record-keeping at each step of the production process. Adoption of GAPs may result in higher production, transformation and marketing costs, hence finally higher costs for the consumer. To minimize production costs and maintain the quality of agri-food, various agencies offer a series of advise publications to benefit the farmers. GAPs also require maintaining a common database on integrated production techniques for each of the major agro-ecological area, thus to collect, analyze and disseminate information of food practices in relevant geographical contexts (Anon, 2009).

27.3 ACCREDIATION OF NURSERIES FOR SUPPLY OF QUALITY PLANTING MATERIAL

Genuine planting material is the most important input in horticulture for obtaining quality produce. Non availability of quality planting material is the major challenge in the expansion of horticulture in the country. Requirements of planting material is estimated to growing at the rate 4 per cent per annum and there had been requirement of about 1414 million of fruit plants seedlings during the year 2011-2012. There are about 6300 fruit plant nurseries in the country to meet this requirement. This also necessitates having a network of fruit nurseries which confirm to the model nursery standards in terms of infrastructure, quality of seed and planting materials and adoption of nursery management practices. Keeping in view the importance of planting material for the success of any orchard, the Ministry of Agriculture, Govt. of India has identified National Horticulture Board as a Nodal Agency for the purpose of rating of horticulture nurseries and awarding accreditation certificate in the country. In order to supply required planting material and ensure quality of the supplied items, accrediation procedures of fruit nurseries have been initiated by National Horticulture Board (NHB). Crop specific nurseries are to be accredited to ensure adoption of Good Nurseries Production Practices. As such the Board has prepared detailed guidelines for implementing the system of rating of horticulture nurseries giving due weightage to capacity of production of pedigreed, disease free planting material under disease free production environment by trained and skilled personnel who follow standard nursery production manual. The rating is determined by the assessment committee headed by an eminent subject matter specialist who adjudges nursery from single to five star scales. The accrediation is primarily based on location, qualified skilled human resource, procedures adopted, mother plant block, layout, health and hygiene conditions, etc. Data recording/record keeping procedure is also one of the major inspection points. It is assumed that a model fruit nursery will yield approximately 50,000 plants per year from one hectors area. Nurseries are initially accredited for two years and renewed after every two years.

So far, National Horticulture Board has received 123 proposals for rating and recognition of fruit plant nurseries, out of which 66 are from private fruit plant nursery growers and 67 are from ICAR Institutes / Centers, SAU's and State Department. of Horticulture. In all, 50 nurseries have been inspected and visited by the Assessment Team (Mistry, 2010). These nurseries have been rated and accredited under the system by giving different rating according to the infrastructure available and good nursery management practices being followed.

27.4 QUALITY AND SAFETY ASPECTS OF FRESH FRUITS AND VEGETABLES

Diseases caused by food borne pathogens constitute a major public health concern. The emphasis world over is on safety aspects of fresh fruits and vegetables. In response to consumer concerns, independent third party inspections of farms to ensure that fruits and vegetables are being grown, harvested and packed using GAPs, GHPs, GMPs have become important. Compliance is being sought for EUREPGAP/GLOBALGAP, HACCP, IS/ISO-22000:2005-Food Safety Management Systems Standards.

Grade standards of fruits and vegetables have been notified in Fruits and Vegetables Grading & Marking Rules, 2004 under the provisions of the Agricultural Produce (Grading & Marking) Act, 1937. These standards are harmonized with international standards such as that of Codex Alimentations Commission, European Commission, etc.

The Organic Agricultural Produce Grading and Marking Rules, 2009 provide for grading and certification of all kinds of fruits and vegetables grown organically as per the standards/requirements prescribed in National Programme for Organic residues, heavy metals, microbial load, etc., Codex Alimentaruis Commission has adopted Code of Hygienic Practices for Fresh Fruits and Vegetables CAC/RCP 53-2003 which addresses GAPs, GHPs and GMPs that help control microbial, chemical and physical hazards. Looking into the sensitivity of fresh leafy vegetables, a separate Annexure on it has been added to the document. In recent years, viruses have been increasingly recognized as important causes of food borne diseases. The Codex Committee on Food Hygiene is working on framing Guidelines to minimize the risk of illness arising from the presence of human enteric viruses in food. The Committee is also updating the Codex document "Principles for the Establishment and Application of Microbiological Criteria for Foods (CAC/GL 21-1997)" by introducing new risk management matrices namely, Food safety Objectives, Performance Objective and Performance Criteria (Khurana, 2010). The new matrices will act as a means to establish and communicate target limits for food borne hazards at specific points in the food chain that will achieve a particular level of health protection.

27.5 STANDARDS IN COLD CHAIN MANAGEMENT

The first cold storage in India was constructed in the year 1940 in Uttar Pradesh for storage of potato. Now there are more than 6000 cold stores across the country, mostly for potato storage in bulk and the majority of them are based on old technology, except a few modern stores. In order to introduce modern energy efficiency technology in the cold chain sector, Government of India has been taken the first step in creating technical standards for ; a) cold storages for storage of fresh horticulture produce which do not require pre-cooling b) Multi-commodity Cold storages for short term and long term storage of fresh horticulture products which require pre-cooling and varying storage requirements and Control Atmosphere (CA) Cold Storage. These have been notified by Govt. of India for their implementation linking the subsidy to the level of technology. In the second phase, two additional standards drafted, have been framed after getting public comments. These Technical Standards are specific to a) Fruit ripening chambers/units, and b) special equipments for carriage of perishable food stuff including fresh horticulture produce. Other links of the cold chain have also been attempted to ensure system integrity. Standards aim at improving design, construction and operational practices as thrust (Sharma, 2010). The standards cover many important aspects involved in project designing. These standards are gaining wider public acceptability paving way towards fulfilment of the third generation requirements. It is expected that these steps will improve the quality of the cold storages.

27.6 STANDARDS IN FOOD CHAIN MANAGEMENT

With the improvement in the living standard and life style of people in emerging economies there has been increase in the number of consumers stimulating demand for higher value-added products. These are consumers that are fully aware of health, nutrition and quality issues and how they impact their daily lives. The planet's population continues to grow, with a billion people added every 12 years. This is accompanied by both an increased demand for food and consumption of the planet's resources. In Asia, China and India in particular, not only population growth but fast-paced, turbulent economic development is being witnessed. This is also the case in other countries with emerging economies.

As a result, the food and beverage industries, are at the forefront in dealing with challenges that have the potential to influence the future or the planet, the sustainability of economic growth and the availability of natural resources. The food and beverages industries are moving to satisfy demands from international consumers and to open new market opportunities by operating across borders and in multiple continents, Consumers consider food safety and quality fundamental and take them for granted. But managing food safety and quality risks is no longer sufficient. Sustainable performance, from a social, environmental and economic point of view, has become an indispensable prerequisite for all businesses. As consumers make their selections based on environmental, health and ethical parameters, it is of particular importance to companies in the food and beverage industries to provide a managed, clear and transparent response to the demand for sustainable performance expressed by consumers and society at large.

27.7 STANDARDS OF CODEX

Codex is the common name for the Codex Alimentarius Commission (CAC). The Codex Alimentarius Commission was cleared by FAO and WHO in the year 1963. Its mandate is to develop food standards, guidelines, recommendations and codes of practice with the objective of protecting the health of consumers and ensuring fair practices in food trade. Currently, 183 countries are members of Codex and one member organization, the European Union. The presentation of codex aims to explain the objectives of Codex and inform various stakeholders and the government officials about the important issues, currently, being debated and having an impact on food business within and outside India. It is, therefore, necessary to understand Codex so that stakeholders could effectively participate in the international food standard setting process. All the views are considered by the National Codex Contact Point of India (Food Safety and Standards Authority of India-FSSAI) and India's position is presented in the Codex meetings. It is important for the stakeholders to contribute in work of codex in India. It will also useful to the future leader and members of delegates from India in presenting India's position to the various meeting of Codex (Tanwar, 2010). The codex standard setting process along with initiatives undertaken by competet authorities of India to harmonize the domestic standards with international standards.

27.8 REFERENCES

Anonymous (2009). Good Agricultural Practices. http//en.wikipedia.Org.

Anonymous (2003). Report of the expert consultation on a Good Agricultural Practices (Approach) *FAO Agriculture Department Report.* Rome (Italy) 10-12 Nov.2003.

Deulgaonkar, C (2010). Emerging Standards in Food Chain. Book of Abstract, IHC – 2010 held in Delhi, 18-21 Nov.2010.p-119

Khurana SC (2010). Recent Developments in Quality and Safety aspects of fresh fruits and vegetables. *Book of Abstract*, IHC – 2010 held in Delhi, 18-21 Nov.2010. p. 117

Mistry NC (2010). Recognition and accrediattion of planting material. *Book of Abstract,* IHC – 2010 held in Delhi, 18-21 Nov.2010.p. 113.

Sharma, RK (2010). Emerging Standards in cold Chain management. *Book of Abstract*, IHC –2010 held in Delhi, 18-21 Nov.2010.p. 115.

Tanwar A (2010). Codex and its relevance to India. *Book of Abstract*, IHC – 2010 held in Delhi, 18-21 Nov.2010. p. 121

❑❑❑

Chapter – 28

Participatory Approaches in Precision Farming

Atul Chandra

28.1 INTRODUCTION

Precision Farming has assumed great significance in the country for making the use of available resources more judiciously and efficiently. It is farm management strategy which utilizes precise information to increase profit and to reduce enviornmental impact. It takes into account in field variability in soil fertility and crop conditions. In nutshell, it may be defined as management of input variables such as application rates, cultivars selections, tillage practices, irrigation scheduling, optimization of soil and plant nutrient through nutrient management practices and management of pest and diseases by reducing the quantities of pesticides. Development of new technologies has become integral part of precision farming.

Precision farming was widely accepted firstly as a management practice in corn belt of mid west United States of America which helped in increasing the yield and significant reduction in production costs. Besides, crops like wheat and soybean have also been included in its purview. In south east parts of U.S.A., research is currently on in the crops of citrus, cotton, groundnut, vegetables and even livestock. Precision farming management practices have also integrated controlling and monitoring of field information using geographical information system (GIS), Global Positioning System (GPS), sensors, Variable Rate Technology (VRT) and yield monitoring. These components are highly sophisticated. GIS software stores data pertaining to soil type, nutrient levels etc in layers and assign that information to the particular field location. It can also be used to analyze characteristics between layers to develop application maps or other management options. Global Positioning System is a referencing device capable of identifying sites within a field. Global positioning system (GPS) helps in storing information regarding field location in terms of latitude and longitude. Maps showing variability of nutrient levels soil types, topography, pest incidence and yield can be created by virtue of it. It is a set of 24 satellites in high altitude orbit above the earth for pin pointing objects on the surface of the earth. These satellites continuously transmit radio signals which are picked up and deciphered by special receivers. A GPS receiver requires at least four satellites to determine its position on earth. Receivers are mounted on tractors/top of cab or on some high point to get a clear view of the spy. Crop stress, soil properties, pest incidence etc. are determined by the help of sensors through traveling over the field by tractors or combine. These can be used for measurement of soil conductivity, fertility status, weeds etc. Another type of sensor viz Remote sensors (aerial/ satellite) can provide instead maps of field conditions. Computer control device and associated hardware are part of Variable Rate Technology (VRT) for application of fertilizers, pesticides etc on the basis of field characteristics of the specific location. In precision farming, information is needed about mean characteristics of small, relatively homogenous management zones. It is data acquisition of the farms to find the soil, vegetation and other parameters which are amenable to remote sensing. Remote sensing techniques are helpful in providing continuous acquired data of agricultural crops. Remote sensors image vegetation growing on different soil types with different water availability. The following applications can be made by relief sensors:

1. Soil Properties sensing : Soil texture, soil structure, soil moisture, soil nutrients.
2. Crop sensing: Plant population, crop stress and nutrient stress.
3. Yield monitoring system: Crop yield, crop moisture.
4. Variable rate technology system: Fertilizers flow, weed detection.

Besides, crop condition assessment, agricultural drought assessment, pest and diseases, land capability etc studies can also be made. These studies are aided by sensors such as IRS WiFS/LISS-111/PAN and Radar sat Scan SAR.

For software solutions for precision farming include Linux which is a good platform for performing research oriented work. Besides, GRASS, Xldleas and Santis are three packages which are helpful in precision farming (Kamble, 2006).

28.2 PRECISION FARMING PROSPECTS IN INDIA

Presently in India, use of GPS in agriculture is limited. Farmers in the country are poor and land holding size is small. However, it has large potential for improving the agricultural production in the country. Its use may help Indian farmers to harvest the fruits of frontier technologies without compromising the quality of land and produce (Kamble, 2006). It is pertinent here to note that precision farming is a system rather than a single piece of equipments or technology. GPS by itself may have little value to the farmer unless it is combined with a yield monitor or VRT. Further more, it may not be of profitable proposition for low value commodities and contrary to it, it is highly remunerative for high valued horticultural crops.

In reality, precision farming concept in toto can not be replicated in our country as it is possible in developed countries like USA and European countries but the concept and its implementation can be greatly beneficial. The end user of precision farming is a farmer. He/she can not afford to follow it by own resources. Several government, non-government organisations, public and private sectors are involved in making it a success. So there is participatory approach in research and development of precision farming. Among centre of ISRO, M.S.Swaminathan Research Foundation, Chennai, Indian Agricultural Reasearch Institute, New Delhi and Project Directorate of Farming System Research, Modipuram have started actively working in this direction and are expected to provide the frontier technologies to the farmers in near future. ISRO has initiated *Gramsat* Project in Orissa. This project is aimed to empower the people especially the poor and marginalized by awareness building and access to information and services. To achieve this, a net work of one way video and two way audio forecasting the yield of mono and multiple crops is being done. Area estimates and crop inventory is being done for rice, banana, chillies, cotton, maize, sugarcane and tobacco. Satellite data can delineate different crops grown in the same area and an inventory of each of the crop can be developed (Kamble, 2006).

28.3 PARTICIPATORY APPROACH FOR PRECISION FARMING IN THE COUNTRY

Tata *Kisan Kendras* are using remote sensing technology to analyse soil, information on crop health conditions, pest attack and prediction of crop yield. It results into better crop condition, more production and better economic returns to the farmers. Efforts of ISRO, MS Swaminathan Research Foundation, Chennai and IARI, New Delhi are likely to turn green revolution into evergreen revolution by virtue of precision agriculture.

28.4 ROLE OF DEPARTMENT OF HORTICULTURE, GOVT. OF RAJASTHAN

Government of Rajasthan through National Horticulture Mission and RKVY schemes is contributing significantly to promote Precision Farming in the state (Anon, 2009). For that, following schemes are being implemented on participatory basis:

1. Development of Fruit orchards.
2. Cultivation of Roses, Zerbera, Gladiolus, Tuberose etc.
3. Improved cultivation of Cumin, Chilli, Garlic, Coriander through Technology Demonstrations at farmer's fields.
4. Cultivation of aromatic and medicinal plants such as Ashwagandha, Aloe vera, Mehandi *(Lawsonia alba),* Isabgol etc.
5. Financial assistance to the extent of 100% unto Rs 10.0 lacs for development of water storage structure and its plastic lining under development of community water reservoirs.

6. Promotion of construction of protected structures such as green house, shade net house, poly house, plastic tunnels and mulching for soil moisture conservation.
7. Integrated pest management in horticultural crops.
8. Development of model nurseries and small nurseries
9. Subsidy on plant protection equipments.
10. Subsidy for creation of seed processing infrastructure.
11. Subsidies of cold storage, value addition units of fruits, vegetables, seed spices.
12. Subsidies for drip irrigation, sprinkler set etc.

28.5 PARTICIPATORY APPROACH FOR PLANTATION OF DATE PALM TISSUE CULTURE PLANTS

Commercial plantation of date palm tissue culture plants of elite varieties such as Barhee, Khuneizi, Khalas, Medjool, Khadrawy etc and male is being done on participatory basis at farmers fields in Rajasthan by Atul Ltd, a member of *Lal Bhai* group of companies by the financial assistance of Government of Rajasthan to the tune of 202 crores under RKVY. Scientists from Swami Keshwanand Rajasthan Agricultural University, Bikaner, ICAR Institutes viz. CAZRI, Jodhpur and Jaisalmer and Central Institute for Arid Horticulture (CIAH) Bikaner are technically supporting the programme. Under the programme, more than 1.0 lacs of date palm tissue cultured plants have been planted at farmers fields in the state of Rajasthan. Two demonstration centres viz. one at *Bhojka* (Jaisalmer) and another at *Khara* (Bikaner) are being developed and plantation of 14125 and 5270 date plants, respectively of the above mentioned varieties has been done. In the state of Gujarat, plantation of date palm tissue culture plants is being done at farmers fields by Kutch Crop Servies Ltd C/o Excel Crop Care Ltd, Gajod and more than 550 acres plantation has been done in Kutch district under Government financially assisted programme (Chandra, 2010). National Horticulture Board and S.D. Agricultural University, Sardar Krushinagar, Gujarat are supporting it. For populariztion of datepalm tissue cultured plantation, National Horticultural Board has supported organisation of seminar cum field day in Kutch, Gujarat and national workshop on datepalm tissue cultured plants at CIAH, Bikaner in the year 2010. A technical seminar on 'Advances in datepalm' was organised at Jaipur on April 15, 2009 for enriching the knowledge on datepalm tissue culture to the scientists, extension workers, farmers and govt. officials.

28.6 PRECISION FARMING DEVELOPMENT BY PRECISION FARMING DEVELOPMENT CENTRE

A total of 17 precision farming development centres (PFDCs) are working all over the country under Ministry of Agriculture, Govt. of India. These centres are developing technologies, transferring technologies to farmers through training and awareness programme, demonstrations, preparation of articles, development of package of practices, creating awareness about micro irrigation and protected cultivation, conducting surveys for collection of data and feedback about implementation. In Rajasthan, it was functioning since 1987. Till 1995, it functioned from CTAE, Udaipur and thereafter it is functioning at ARS, Bikaner. Scientific and systematic research on aspects such as micro-irrigation, fertigation and plastic mulch on horticultural crops viz. *ber*, pomegranate, *bael*, *aonla*, kinnow, sweet orange, tomato, brinjal, bottle gourd and seed spices viz. fennel, cumin, fenugreek etc. is in progress for developing package and practices under arid regions of Rajasthan. Precision Farming Development Centre has developed technologies related to efficient utilization of precious inputs like water, fertilizers and chemicals etc for on horticultural crops through drip irrigation, mulching and high density plantation (Anon, 2010). Rajasthan spreads over 342 lakh ha geographical area which is 10.4% of total land mass and 5.5 per cent population of the country. The state has hardly 1% of total national water resources of the state 57ha area comes under desert which is 61% of total area under desert of the country. The rains varies from 141 mm to 999 mm with average annual rainfall of 575 mm with mean daily maximum temperature of 42 ^{0} C during summers and 23.6 ^{0} C in winters. Arid and semi-arid areas are sandy having low water holding capacity, high infiltration rate, shallow depth and very low fertility. Considerable area has hard pan at varying depths and some soils are gypsiferous. Under these adverse climatic and soil conditions, precision farming holds great promise. The state has 10.39 lakh ha area under horticultural crops. Out of that, 0.50 lakh ha has been reported under seed spices. Nearly 8.43 lakh ha area has been covered in the state under micro-irrigation. PFDC, Bikaner is coordinating with state micro-irrigation committee (SMIC) by providing technical support for implementing MI scheme in the state of Rajasthan. Considerable work on precision farming is also being done by MPUAT, Udaipur particularly on protected vegetables cultivation and CIAH, Bikaner on drip irrigation and efficient utilization of fertilizers and chemicals. SKRAU, Bikaner has recently signed MOU with M/S Jain irrigation system Ltd, Jalgaon MS, mainly on collaborative research on precision farming.

28.7 ROLE OF ICAR INSTITUTES IN PRECISION FARMING

CIAH, Bikaner has developed varieties of fruits and vegetables suitable under arid and semi arid conditions. Prominent among these are Goma Kirti, Thar sevika, Thar Bhubh Raj of ber, Goma Aihwarya of aonla, Thar Shoba of khejri, AHW-19 AHW-65, Thar Manak of Mateera, Thar Samridhi of bottle gourd, AHK-119, AHK-200 of kachri and Goma Manjri of cluster bean. Integrated nutrient and water management recommendations have been developed by the institute. *Ber* and *aonla* based cropping systems and multistorey cropping system has been developed. Crop geometry in high density plantation of *ber* and *aonla* has been standardized. Institute has developed composite farming system (Anon, 2007a).

28.8 ROLE OF SAU'S

College of Agriculture at Bikaner, Jobner, Udaipur besides Horticulture and Forestry, Jhalawar are contributing in development of Precision Farming through research projects and post graduate thesis research work. Considerable work has been done by M.Sc and Ph.D. students on precision farming aspects such as varietal evaluation, drip irrigation, fertigation, nutritional survey of orchards, integrated nutrient management, protected cultivation etc. Department of Horticulture, College of Agriculture, Bikaner successfully completed research project on "Efficient water management in vegetables through pressurized irrigation system in canal command areas of north Rajasthan. Through application of 0.70 volume of water by drip, there was 46 per cent increase in yield and 30 per cent saving of water in bottle gourd hybrid. In tomato, application of 0.55 volume of water through drip had 68 per cent higher fruit yield over surface irrigation(Anon, 2007b). Research conducted by M.Sc. students at Bikaner revealed best vegetative growth of bearing ber trees by soil application of 500g nitrogen in tree basin and in kagzi lime tree by application of 300g nitrogen as basal dose/tree and foliar spray of urea (1.5 per cent) and zinc sulphate (0.5 per cent). In brinjal crop, application of 40 kg ferrous sulphate per hectare in soil before transplanting resulted maximum plant growth and yield. In cabbage, 45cm × 45cm spacing with 180 kg nitrogen/ha in soil application gave best vegetative growth and head yield. At Department of Horticulture, SKN College of Agriculture, Jobner, IPM on *ber* and *aonla* was standardized.

28.9 FARMERS PARTICIPATION

Innovative farmers contribution in development and popularizing precision farming technologies have been appreciable. Progressive farmer has developed technology for establishing plants with less quantity of water by use of sub soil moisture. Progressive farmer have also demonstrated efficient utilization of limited water resources through drip irrigation system for cultivation of kinnow in canal command area. The list of such farmers whose contribution is worth very long in the state.

From the above discussion, it can be concluded that precision farming in farm management strategy utilizing precision information gathered through GIS, GPS, sensors, VRT and yield monitoring gadgets. In Indian conditions, presently it is not being practised into the land holding being smallar and economy of farmers. However, efficient utilization of irrigation water through micro irrigation (drip/ sprinkler etc.), fertigation, integrated nutrient management, integrated pest management, development of improved varieties etc. are being much emphasized by Government policies and schemes and programmes have been initiated. Precision farming approach is participatory involving different Government Departments (Ministry of Agriculture, Deptt. of Agriculture and cooperation GOI, ICAR, New Delhi, State Deptt of Agriculture and Horticulture, Project Directorate on farming of Cropping System Research, Modipuiram) NGOs, such as TATA Kissan Kendra etc. M.S.Swaminathan Research Foundation, Chennai and SAUs etc. Precision Farming Development Centres spread over the country are giving valuable inputs in growth and development of precision farming. Role of SAUs and ICAR institutes developing different technologies and crop varieties have been sufficient. Post graduate research work has contributed a lot in precision farming. Innovative farmers and farmers led extension has developed confidence in general farmers to adopt it in the fields. Thus precision farming can prosper the country by participatory approach.

28.10 REFERENCES

Anonymous (2007a).*CIAH profile* CIAH /TECH/PUB./No.31, compiled and edited by Dr. T.A.More, Dr.R. Bhargava, Dr.R.S.Singh, Dr.D.K.Samadhia, CIAH, Bikaner.

Anonymous (2007b). *Vegetable cultivation under drip irrigation*. Technical Bulletin no 1, Compiled by Dr .P. K.Yadav, Er.J.K.Gaur, Dr. Jitendra Singh and Atul Chandra, Department of Horticulture, College of Agriculture, Rajasthan Agriculture University, Bikaner.

Anonymous (2009).National Horticulture Mission. *Gist of Annual Action Plan 2009-2010*. Rajasthan Horticulture Development Society, Directorate of Horticulture, Pant Krishi Bhawan, Jaipur, Rajasthan.

Anonymous (2010). *PFDC*, Bikaner. Agriculture Research Station, Beechwal, SKRAU, Bikaner.

Chandra, Atul (2010). Public – Private partnership for plant material production for commercial plantation by the farmers of the country. National workshop on Date palm held at Central Institute of Horticulture, Bikaner on 4th October, 2010.

Kamble, Bahurao Dashrath(2006). Precision Farming in Indian Agriculture Scenario. URN. Rains, Glen.C and Thomas, Danial L. (2009). Precision Farming: An Introduction. URN www.caes.uga edu/ publications/ pub detail. Cfm7pk id=6285 23 March, 2009. (Accessed on 24 November, 2009)

Chapter – 29

Strengthening Women Participation in Precision Farming in Horticulture

Kanak Lata and Sanjay Singh

29.1 INTRODUCTION

Some historians believe that it was women who first domesticated crop plants and thereby initiated the art and science of farming. While men went out hunting in search of food, women started collecting seeds from the native flora and began cultivating those of interest from the point of view of food, feed, fodder, fibre and fuel. This view is strengthened by the fact that women have been traditionally seed selectors. Even today this tradition has continued in many parts of the developing world. Women have played and continue to play a pivotal role in the conservation of basic life support system such as land, water, flora and fauna. They have protected the health of the soil through organic recycling and promoted crop security through the maintenance of varietal diversity and genetic resistance. In many hill and remote areas of developing countries such as *Himalyan* region, agriculture is largely in hands of women. Men tend to go to towns and cities in search of salaried jobs for fulfilment of family income. Therefore, without total intellectual and physical participation of women, it will not be possible to popularize alternative systems of land management for shifting cultivation, arrest gene and soil erosion, and promote the care of the soil and the health of economic plants and farm animals. Women also tend to look at problems in their totality from the sowing of the crop to its ultimate utilization either as food, feed or raw material for industry. Since they are usually involved in of marketing, they generally possess a clear understanding of market preferences and prejudices. Therefore, for an economically and ecologically sustainable agriculture the involvement of women farmers and farm women (agriculture farmers with and without assets) in the process of modernization of farming practices and village industries is absolutely essential. Horticulture is a major part of Indian agriculture. Although belated, their importance has now been recognized in India not only to provide the much needed balanced nutrition and gainful occupation to our populace but also to develop viable internal and export trade. Fruits like banana and vegetables like potato being energy rich are considered prominent food crops e.g. banana can produce as high as 37 million calories per hectare. Unfortunately the component of fruits and vegetables in Indian diet is hardly six per cent which should be raised at least five times so that the per capita per day consumption mathches the recommended dose of 85g. fruits, 75 – 125g. leafy vegetables 85g. other vegetable and 85g. roots and tubers. To achieve this present production of fruits and vegetables would need to be pushed up considerably by increasing the production levels and the area both.

29.2 THE PRECISION FARMING

With precision efforts, there is great scope to improve the yield levels in fruits since the present productivity is far lower than the potentials. This is also evident from the fact that yield levels in some fruits and vegetables in India are one third to one fifth to those obtained in the developed countries. Therefore, the technologies developed as a result of research efforts need to be quickly adopted in our production system. Since women play a prominent role in national horticultural productivity, it is necessary to identify the appropriate technologies for them, the constraints in their adoption and the need of new technology development. Ignorance about the new technologies, lack of skill for their adoption and in case of some operations, need of more physical power, element of drudgery and attention at odd periods come in the way of their adoption by farm women.

The earning capacity of females in agricultural employment is substantially low all over the world. The female earning varies from 64.9 per cent in UK, 64.9 per cent in India and 110.8 per cent in Denmark.

Pandit Jawaharlal Nehru at the centenary Celebrations of M. Karve on 18th April, 1955 said, "One of the trust measures of a nation's advancement is the state of its women. For out of the women comes the new generation, and it is from their lips and laps that begins to learn. Social revolution forms the basis of economic revolution which leads to progress. Women play the most important part in social revolution".

The important trades for horticultural productivity consist of nursery technology related to ornamental plants, fruits and vegetables, flower- gardens, homestead gardens fruit and vegetable processing industry and marketing of horticultural produce. At present, in North India, women mostly participate in fruit and vegetable processing, flower gardening, ornamental nursery, homestead gardening, kitchen gardening and to some extent in vegetable growing. In south India, besides these operations, women also participate in vegetable and flower marketing and nursery technology (Table1)

Table 1 : Women participation in horticulture trade

Trade	North India	South India
Ornamental nursery	50	60
Fruits and vegetable nursery	40	50
Fruit growing	30	40
Vegetable growing	50	50
Flower gardening	60	80
Homestead gardening	75	80
Fruits and vegetable processing	80	90
Fruit marketing	20	40
Vegetable and fruit marketing	40	60

The extent of participation of women in different operations in horticultural production systems depicts interesting regional differences in India. Goyal *et. al* (2003) and Waris (2008) reported the rate of participation of women in weeding, drying and storage in the field of agriculture. In northern states like Haryana, women are hardy and perform work related to heavy implements like spade, machines and those requiring greater skill, e.g. pruning and training, raising nursery plants, lifting and planting of plants, spraying, grading, packing, marketing etc. In the southern states although women do not take part in operations requiring heavy manual work, they participate to some extent in nursery technology planting, irrigation, grading, packing etc. In general, major operations looked after by them are sowing and planting, irrigation, lifting of small vegetables and ornamental seedlings in nursery, application of manures and fertilizers, weeding and hoeing harvesting of fruits, vegetables and flowers, transport of produce to short distances and processing of horticultural produce.

Training for farm women on institutional basis was started in India under the centrally sponsored schemes of farmers training and educational programmes in 1966-67 which now cover all the districts operated by state governments, in these programmes. However, training component for farm women remain restricted to only home science aspects like nutrition, health care and some new recipes. For this bilateral assistance is being obtained from the government of Denmark, DANIDS is assisting government of India to develop leadership qualities amongst farm women so that they can guide other farm women and to provide follow-up through extension support which will be integrated in the existing T and V system. The government of Netherlands is to provide assistance in many states of India. All these centers are to be provided lady training officer and female demonstrator. Major emphasis in these projects is being given on food crops, oilseed and pulses and that on horticultural component is negligible. The present state of horticultural education and training in women is rather dismal particularly in the northern states of India. The situations are somewhat better in the southern states. In the northern states, there are hardly few women taking part in the horticultural education institutions. This is of a great significance since women would pick up the improved technologies faster if imparted by women trainers.

There are many *Mahila Mandals* and SHG with a membership of several lacs rural women. A study of the working of SHG members 300 in 2009 has shown that 20 out of 30 SHG were involved in economic activities like kitchen gardening, community gardening, fruit orchards, compost making, nursery growing, but most of the members reported lack of guidance and proper training facilities. All over the India, 578 KVKs are working hard to provide the guidance of new horticultural technologies along with agriculture, animal husbandry and allied subjects.

29.3 PRODUCTION TECHNOLOGIES

The traditional area in which women already participate to great extent include fruit and vegetable processing, ornamental nursery, besides light operations in fruit and vegetable growing. Women can effectively participate in several activities and can take part in horticultural production system. As a result of research efforts, many new technologies have been developed and are considered appropriate for farm women. These are given below:

29.3.1 Nursery technologies

29.3.1.1 Raising plants in polythene containers

Traditionally plants were raised in nursery beds which had high requirement of space and care for longer period. Modern techniques of raising grafted fruit plants in polythene containers using detached vegetative propagation methods are more elegant and conducive for adoption by farm women.

29.3.1.2 Raising plants in polytubes

Raising plants in poly tubes is a very simple and drudgery effective method. Women can take part in all operations right from filling of tubes, sowing, raising seedlings, and budding operation until disposal of budding. In case of citrus plants, budded plants are raised in polythene tubes which could easily be adopted. The tubes' raised plants can be easily lifted, packed and transplanted easily.

29.3.1.3 Propagation benches with sub-surface irrigation units

One of the most prominent operations done by farmwomen is irrigation of nursery plants by a watering can. The amount of drudgery involved in conventional method is obvious. This can be easily avoided by provision of propagation benches on which the plant containers having holes at their bottoms can be placed for irrigation by capillarity. The water level in the soil on the benches is maintained by simple water level cabinets as used in every modern household, connected to a water source.

29.3.1.4 Standard growing medium

Potting medium consists of mixtures of sand: clay: FYM or other similar meterial in suitable proportion ratio or loom: Very little care is at present given to sterilization of the medium and addition of the component which increases its water holding capacity. Now a days, potting mixture is available which is ready to use. There is need to make the farm women aware to use this media so as to reduce the drudgery in respective operations to prepare such mixtures, avoid difficulty in obtaining the different components and to avoid risk of infection of the propagules by diseases.

29.3.1.5 Mist chamber

Traditionally, frequent sprinkling of water has been in use to prolong the life of the planted propagules so that they regenerate into complete plants. Now several designs of mist chambers or humid chambers are available to suit different conditions of climate. A simple underground trench with polythene covers or above ground polythene chambers are cheaper structures. In these cases humidity is kept high by sprinkling water and the excessive heat is avoided either by partial shading or by provision of exhaust fans. The other designs with automatic timer devices to put the mist nozzles on and off at calibrated and pre-decided intervals are expensive but more efficient.

29.3.1.6 Tissue culture

Tissue culture techniques have now been developed in several horticultural crops like banana, grapes, papaya, date palm, fig, orchid and in many ornamental plants which can be commercially exploited. The various operations in a tissue culture unit relate to preparation of growing media, aseptic transfer into the media tubes maintenance, cutting and transfer of differentiated tissue until final separation and transfer of the plantlets into the soil medium. Thus all the operations are conducive for work by women but require a high degree of skill. Now a days, tissue culture technique has been commercially exploited. Women have successfully adapted the technique and have been considered better technicians than their male counterparts.

29.3.1.7 Use of bottom heat

In temperate regions and during the severe winters in the sub-tropical region of the country, these elegant techniques can now be used to increase rooting. Women can be involved in this work after skill training.

29.4 FRUIT PRODUCTION TECHNOLOGIES

29.4.1 Knowledge of quality cultivar

Although the country already possesses a rich germplasm resource of high yielding commercial varieties of fruits, several new varieties have now been developed having good productivity and desirable quality characters. Women should be made aware about these improved cultivars.

29.4.2 Plantation of orchards

Plantation of new orchard involves operations like layout, digging of pits, filling of pits and planting and care of young plants. Although layout and planting requires some skill but women can easily be trained in these operations. However, digging of pits requires more physical efforts and may not suit women. This operation can be easily mechanized. A post whole digger attached to tractor can dig 1m deep and 1m wide pit in 1 – 2m depending on soil type.

Of the after care operations, frequent watering of the young plantation involves great amount of drudgery and is often performed by women. The modern techniques like pitcher irrigation, micro-tricklets, double wall pots etc. can greatly reduce the drudgery as well as the irrigation cost considerably. These techniques not only eliminate the drudgery for the farm women but are highly economic owing to saving on water and the reduction in mortality of plants.

29.4.3 High density plantation

Orchards operations can now be very easily carried out by women in orchards owing to the small stature of trees of newly bred varieties. High density plantation is amenable to higher and quicker returns. Thus 1600 plants of mango and 500 grape vines can be planted per hectare. Similarly 5000 plants of banana and 53000 to 63500 plants of pineapple can be planted per hectare by paired row planting. In such intensive planting system women can remain engaged profitably even on small holdings.

29.4.4 Training and pruning technique

Although all fruit plants require some amount of training but the fruits like grape can be trained to suit different cultivars and situation. Bower system is the most expensive but head system is much cheaper. The reduction in yields in the later system can be compensated to some extent by the high density plantation. The farm women can be easily trained in the skill of training. Similarly pruning also requires good amount of skill in fruits like grape, ber, peach, fhalsa etc.

29.4.5 Use of growth regulators

Gibbrellic acid for cluster thinning and berry enlargement in grape, ethephon and NAA to induce uniform growing and harvesting in pineapple NAA for control of mal formation in mango, NAA, 2, 4-D and 2, 4, 5-T for reducing fruit drop in mango, citrus and apple have given commercial success and can be widely adopted. Besides this, growth regulators like IBA and NAA are effective in inducing rooting in cuttings for propagating new plants

29.4.6 Orchard management

The different involved are application of manures and fertilizers, foliar feeding, irrigation, weeding, intercropping etc. Farm women are traditionally very intimately involved in all these operations which incidentally also have a great amount of drudgery. Women often don't prefer operations requiring a lot of heavy work like mixing of manure; diverting water for irrigation etc. Irrigation is often given at odd times or during night time which does not suit women. Fortunately modern techniques are now available which can greatly help. Part of the nutritional need can be met by foliar feeding particularly for micro nutrients.

29.4.7 Trickle irrigation

Although expensive but it is very economic in term of water use efficiently. Moreover, once installaed, it can be easily operated by skilled farmwomen. The cost of installation can be greatly reduced by use of gravity flow in place of a pump. Simple trickles in place of costly nozzles, etc.

29.5 VEGETABLE GROWING AND ORNAMENTAL GARDENING

29.5.1 Seed production

Seed production, particularly hybrid seeds, already become a profitable business both in vegetable as well as ornamental gardening.

In recent years, hybrid seed production in vegetable is gaining importance. The operations like sowing rouging controlled pollination, harvesting and seed processing are done by women. In the process of controlled pollination emasculation is very tedious as well as dedicate operation. However the technique has now been improved upon development of male sterile lines in tomatoes and onion has avoided the need of emasculation and caging. Block planting of the two parents can be done in case of cucurbitaceous crops and male flowers from female plant can be removed produce hybrid seeds. In cruciferous crops like cauliflower having protogyny, bud pollinations can be used for this purpose for controlled pollinations. After emasculation, bagging or caging is another important operation which can be done efficiently by women. Certain specialized operations like selection and preparation of planting materials in biannual crops can also be done by women. The seed processing's operations like harvesting , threshing , cleaning , dragging , sieving and packing , sampling , labeling are best done by women .

29.5.2 Vegetable and ornamental plants nursery

Traditionally women perform most of the operations related to these flourishing horticultural trades. The large number demand of plants of annual and perennial flowers, foliage plants besides several vegetables crops has increased over the years particularly near the cities. There is a great demand of seeds, bulbs and corms both in internal as well as export market. The bulbs of amaryllis, tuberose, gerbera, lilies, gladiolus, cannas etc. can be easily transported. Modernization in operations like pot culture, green house/ glass house management, packing etc, will significantly add to the efficiency in operation of these nurseries and would remove a lot of respective difficult operation for women.

29.5.3 Vegetable growing technology

Some of the significant achievement which is likely to provide more income to women growers is as follows:-

Early varieties all the year round bearing varieties processing suitable varieties. Attractive color, aroma and pungency rich varities are sure to add to the income of vegetable growers. Several new vegetable production technologies like offseason production, glass house / green house protected cultivation enable transplanting to catch early and production of vegetable in adverse climate is useful in providing sustainable and assured income to the women growers. Multiple cropping sequences for rainfed and irrigated conditions have also been worked out. Drip irrigation is highly profitable in vegetable crops since they are sown in lines in an intensive cropping system. Foliar feeding of nutrients has also given significant response in almost all vegetable crops and thus the technique could be utilized to ensure ease in fertilizer application commonly done by women. Interculture in vegetable crops is an important operation which require great amount of labour and it is commonly performed by women. New implements have been developed which can decrease the necessity of the great physical effort. Similarly harvesting by clippers in some vegetable and flowers commonly accomplished by women would be more trouble free and efficient.

29.5.4 Homestead garden

Since nutrition and health care of family has been the domain of women in both rural and urban areas, raising and maintenance of home garden can be undertaken by women which can become important basis for the removal of rampant malnutrition. Moreover the production of the nutritive horticultural crops which are often perishable can be produced at the consumption site itself so as to eliminate the great amount of spoilage losses. Some very good models of nutrition gardens have been developed by different horticultural institution in the country. At the Indian Institute of Horticultural Research efficient model has been developed for adoption by population group of different sizes considering that 200m plot would be enough for an average family of 5-6 members. Fruit trees like aonla, guava, acid lime, oranges, papaya banana, pomegranate, jamun, sapota, mango etc being rich in vitamins and minerals, could be important component of such gardens. Efficient management of a homestead garden consists of adoption of crop rotation and continuous and efficient use of all the available space including fencing and ridge space etc.

29.5.5 Technologies for growing of ornamental and medicinal plants

The important commercial flowers of India are rose, jasmine, tuberose, gladiolus, bougainvillea among the perennial flowers and marigold, chrysanthemum, nasturtium, petunia etc in annual flowers. Among the other ornamental plants the important ones are hibiscus, crotons, orchids, anthuorium, rubber plant, aurocaria, palms, acalypha and several others shrubs and trees. Several new varities of these plants has been evolved to suit various uses like garland making, bouquets, wreaths, flower vases, for presentations and decorations etc and the agro-techniques for their cultivation have also been standardized which can be easily adopted. One of the important industries based on flower gardening is perfumery like rose oil, rose water essence and several other perfumery product from other flowers. Women play an important role not only in the growing of these crops but also in the related trades and industries. The medicinal plants of commercial importance are vinca, khus, and safed musali. Ocimum, mint etc. These have gained importance in recent years and can provide a great amount of occupation and employment to women.

29.6 MUSHROOM PRODUCTION

In India , the present internal consumption of mushroom is estimated to be about 250 tonnes per annum which is expected to rise with increasing awareness about its high nutritive value. Commercial mushroom cultivation in our country is very less. So far marchella is collected from coniferous forest of Himachal Pradesh, Jammu Kashmir and Uttarakhand by women workers. Now standard techniques of mushroom cultivation are available. Mushroom cultivation is mostly home bound activity and requires very little space and thus can be easily done by women. However its cultivation should be handled only after proper training. Training facilities are available at Krishi Vigyan Kendra, Solon; Himachal Pradesh which is very famous for mushroom research activities. Of the several verieties of mushroom, the paddy straw mushroom (*Pleurotus.*spp.) is very easy to cultivate. It can be grown in polythene bags and complete its life cycle from spawning to harvest in just 30-40 days. A net profit of about Rs. 20,000 can be obtained from medium size mushroom farm taking 20 crops per annum using 800 kg (600 bags) straw for each crop. The button mushroom requires more careful attention but fetches higher income. For this various steps like preparation of compost, pasteurizations, cooling, spawning, casing, picking etc have been standardized. A net income of Rs. 71,000 can be obtained from a medium size farm having 600 trays taking 4-5 crops per annum.

29.7 POST-HARVEST TECHNOLOGIES

From time immemorial, women have engaged themselves in post harvest aspects like preparation of dehydrated products, preserves, pickles, chutneys, preserve, beverages etc. from fruits and vegetables. Post harvest operations were the exclusive domain of farm women irrespective of the farm size (Manju, 2002). Improved technologies for the preparation of horticulture products from different horticultural crops are now available which can be adopted by women. At present, most of processing units are of cottage scale. Now large industrial units are also developing particularly for the preparation of concentrated juices and beverages. The other important post-harvest technology operation like grading, packing, marketing, storage of produce, so far did not receive much attention. However great advancement have been made on these aspects also and thus women can adopt these technologies with respect to grading of produce. Pre-packing treatments like curing, hot water and fungicidal dips, pre cooling, waxing can be attempted on priority basis by women. Similarly techniques of packaging in corrugated fibre board boxes and storage in zero energy cooling chambers or refrigerated stores can also be adopted by women to minimize post-harvest losses. At present, there is scarcity of both zero energy chambers as well as cold storage units in the country which need to be improved.

29.8 APICULTURE

Only a few women take part in bee keeping perhaps owing to fear complex and ignorance. Since bee keeping requires tender handling and can increase productivity if orchards communities, it can be easily adopted by farm women. The exotic *Apis mellifera* is much more productive than the indigenous *Acerana indica.* The operation like hiving of bees and bee swarms, occasional feeding, division of colonies by mass rearing of queens, uniting the colonies, queen introduction, prevention of absconding , swarm control, honey extraction etc can be easily learnt by women.

29.9 SERICULTURE

This is an important trade in which women already participate. Sericulture is largely based on mulberry and castor plantation and involves operations like collection and storage of eggs at low temperature and their hatching to coincide with appearance of foliage in the plantation and feeding tender leaves to the worms. Clearing and changing of the feeding trays, bailing of cocoons, reeling of silk, etc can be easily done by women

29.10 DISEASE AND PEST CONTROL

Chemical control of diseases and pests is usually practised particularly in well managed horticultural gardens. Control measures are available for most of the diseases. Similarly, integrated pest management practices involving insecticides, light traps, trap crops, pheromones, and sterility hormones are also available. Women do not usually participate in spraying operations although with some training they can easily perform these operations. To avoid pollution by chemical residues in the produce and health hazard by toxic chemical to the operators, modern techniques of biological control have been developed. The operations like rearing of parasites / predators their release in the field are simple and can be performed by women. Use of proper root stocks can also eliminate the risk of disease and nematodes.

29.11 DEVELOPMENTAL NEEDS

For the strengthening of women participation in precision farming the following activities are essential :

Training: Exact training needs of different categories of women should be identified. Training programmes should incorporate the following aspects. Training in different horticultural trade like nursery management, tissue culture, apiculture, sericulture, mushroom growing, processing of produce besides several other skilled jobs for production of fruits, vegetables and flowers should be organized. The programme of *Krishi Vigyan Kendras* should be expanded and the training program for women should be intensified to enhance their productivity, liberate them from drudgeries of farm and home activities and also to improve their skills leading to employability. The training content for farm women in the existing programme needs to break new grounds in conformity with the changing times and advancement of technologies. The agricultural extraction organization should try to eliminate the sex bias in offering the training content and skills to farm women. Technologies cannot remain in isolation nor create cultural shock; it needs to be considered in the techno- economic perspective along with socio culture traditional and environmental factors. The exciting adult literacy/education programmes must therefore focus on liberating farm women for these sociological impediments. The training programme to farm women would be more acceptable and useful if these are imparted by women trainers to enable free participation and discussion.

To ensure group support and mutual consent for adoption of the new technologies training programmes should be organized in groups of farm women from homogeneous social strata. Education tour of such exclusive women groups would be useful to expose them to the development taking place in different aspects of horticulture in different part of country.

29.12 CREDIT AND INPUT

The role of women in horticulture can be enhanced if the access of women to input agencies like back, cooperative societies, self help group, and market yards are increased. This can be done by:

- Suitably, amending the existing rules for credit by bank specially NABARD, the compliance of which seems difficult because of condition of collateral security of immovable properties for credit. Since in rural area the immovable property is commonly held by male head it becomes further difficult to offer it. This inhibits credit facilities to women. The cooperation for development of women being set up in different state should take up this matter at appropriate level with the Reserve Bank of India, NABARD and State Cooperative Development.
- SHGs are very good source for farm women to financial support. Organizing SHGs and women cooperatives on functional basis is important so they can arrange credits, technical inputs and marketing facilities etc. The credit lending instruction should support horticultural projects floated by these groups.

29.13 MARKETING

Efficiency in marketing of perishable horticultural products largely decides the economics of production and avoids gluts and spoilage at production sites. Therefore marketing system incorporating grading, packaging, transport and disposal of produce to market centres should be properly organized. For these women cooperative should be revitalized with a focus on horticulture related activities. At present, there are very few women cooperative in the country and none is functioning to the desired level owing to poor membership and lack of leadership. This needs improvement.

Facilities should be created for short term and long term storage of perishable horticultural produce at appropriate locations.

29.14 PUBLICITY

Good publicity is required to make the farm women and their organization aware of the facilities available to them for increasing horticultural productivity. For this:

- Video cassettes on the cultivation of all horticultural crops should be prepared and made available to all the training organizations in different institutions and departments in the country
- Film division/Doordarshan should prepare documentaries of different aspects where women can participate in horticultural activities and this should be relayed throughout the country.
- There is dearth of adequate, scientific national data base with regard to farm women role expectations, role performance and implications of transfer of technology. Research programme of 'home science' institution all over the country need to be geared in this direction.

29.15 DRUDGERY, TIME AND LABOUR MANAGEMENT

- For relieving women from the drudgeries of the household work related to child care and enchasing the productivity of farm women, the existing child care facilities through crèches and *balwadis* in the rural area need to be expanded in terms of coverage and these must be made available to women in conformity with their requirement at different farming season.

Although farm women have been participating in horticultural production in the country but that has so far remained poor particularly because of ignorance, lack of skills and owing to different work condition. The available modern technology can eliminate many of these constraints to successful horticultural trades and operate them successfully with great ease. It is, however, necessary to modernize some of the farm technology. Above all, training in various skills to farm women would go a long way in increasing horticultural activities in different parts of the country.

29.16 REFERENCES

Goyal, G; Randhawa, V.; Pannu, K. and Kaur, R (2003). Empowerment of women in agriculture. *Rajasthan. J. Ext. Edu.*, 11 : 5 - 6.

Manju Suman (2002). Involvement of women in agricultural activities. *Maharashtra J. Ext. Edu.* 21(1) : 105-106.

Waris, Amtul (2008), Farm Women's participation in agricultural activities in semi-arid region and Andhra Pradesh. *Green Farming*, 2(2) : 137-138.

FSC
www.fsc.org
MIX
Papier | Fördert gute Waldnutzung
FSC® C083411

Zeitfracht Medien GmbH
Ferdinand-Jühlke-Straße 7
99095 Erfurt, Deutschland
produktsicherheit@kolibri360.de